# The Bark Canoes and Skin Boats of North America

*Edwin Tappan Adney*
*and*
*Howard I. Chapelle*

SMITHSONIAN INSTITUTION PRESS
WASHINGTON, D.C.
1983

Reprinted, 1983.

*Library of Congress Cataloging-in-Publication Data*

Adney, Edwin Tappan, 1868–1950.
The bark canoes and skin boats of North America (by) Edwin Tappan Adney and Howard I. Chapelle. Washington, Smithsonian Institution, 1964.

xiv, 242 p. illus. 29 cm. (U.S. National Museum. Bulletin 230)

At head of title: Museum of History and Technology.
Bibliography: p. 231–234.
ISBN 0-87474-204-8 (cloth); 1-56098-296-9 (paper)

1. Indians of North America—Boats. 2. Canoes and canoeing—North America. I. Chapelle, Howard Irving, joint author. II. U.S. Museum of History and Technology. III. Title. (Series)

Q11.U6 no. 230 970.6623829 64–62636
E98.B6A3

Manufactured in the United States of America

97 96 95 5 4

*Special acknowledgment*
*Is here gratefully made to The Mariners' Museum, Newport News, Virginia, under whose auspices was prepared and with whose cooperation is here published the part of this work based on the Adney papers; also to the late Vilhjalmur Stefansson, for whose* ENCYCLOPEDIA ARCTICA *was written the chapter on Arctic skin boats*

# *Contents*

# *Illustrations*

| Figure | | Page |
| --- | --- | --- |

# *The Bark Canoes and Skin Boats of North America*

# INTRODUCTION

Figure 1

FUR-TRADE CANOE ON THE MISSINAIBI RIVER, 1901. (*Canadian Geological Survey photo.*)

THE BARK CANOES of the North American Indians, particularly those of birch bark, were among the most highly developed of manually propelled primitive watercraft. Built with Stone Age tools from materials available in the areas of their use, their design, size, and appearance were varied so as to create boats suitable to the many and different requirements of their users. The great skill exhibited in their design and construction shows that a long period of development must have taken place before they became known to white men.

The Indian bark canoes were most efficient watercraft for use in forest travel; they were capable of being propelled easily with a single-bladed paddle. This allowed the paddler, unlike the oarsman, to face the direction of travel, a necessity in obstructed or shoal waters and in fast-moving streams. The canoes, being light, could be carried overland for long distances, even where trails were rough or nonexistent. Yet they could carry heavy loads in shallow water and could be repaired in the forest without special tools.

Bark canoes were designed for various conditions: some for use in rapid streams, some for quiet waters, some for the open waters of lakes, some for use along the coast. Most were intended for portage in overland transportation as well. They were built in a variety of sizes, from small one-man hunting and fishing canoes to canoes large enough to carry a ton of cargo and a crew, or a war-party, or one or more families moving to new habitations. Some canoes were designed so that they could be used, turned bottom up, for shelter ashore.

The superior qualities of the bark canoes of North America are indicated by the white man's unqualified adoption of the craft. Almost as soon as he arrived in North America, the white man learned to use the canoe, without alteration, for wilderness travel. Much later, when the original materials used in building were no longer readily available, canvas was substituted for bark, and nails for the lashings and sewing; but as long as manual propulsion was used, the basic models of the bark canoes were retained. Indeed, the models and the proportions used in many of these old bark canoes are retained in the canoes used today in the wildernesses of northern Canada and Alaska, and the same styles may be seen in the canoes used for pleasure in the summer resorts of Europe and America. The bark canoe of North America shares with the Eskimo kayak the distinction of being one of the few primitive craft of which the basic models are retained in the boats of civilized man.

It may seem strange, then, that the literature on American bark canoes is so limited. Many possible explanations for this might be offered. One is that the art of bark canoe building died early, as the Indians came into contact with the whites, before there was any attempt fully to record Indian culture. The bark canoe is fragile compared to the dugout. The latter might last hundreds of years submerged in a bog, but the bark canoe will not last more than a few decades. It is difficult, in fact, to preserve bark canoes in museums, for as they age and the bark becomes brittle, they are easily damaged in moving and handling.

Some small models made by Indians are preserved, but, like most models made by primitive men, these are not to any scale and do not show with equal accuracy all parts of the canoes they represent. They are, therefore, of value only when full-sized canoes of the same type are available for comparison, but this is too rarely the case with the American Indian bark canoes. Today the builders who might have added to our knowledge are long dead.

It might be said fairly that those who had the best opportunities to observe, including many whose profession it was to record the culture of primitive man, showed little interest in watercraft and have left us only the most meager descriptions. Even when the watercraft of the primitive man had obviously played a large part in his culture, we rarely find a record complete enough to allow the same accuracy of reproduction that obtains, say, for his art, his dress, or his pottery. Once lost, the information on primitive watercraft cannot, as a rule, be recovered.

However, as far as the bark canoes of North America are concerned, there was another factor. The student who became sufficiently interested to begin research soon discovered that one man was devoting his lifetime to the study of these craft; that, in a field with few documentary records and fewer artifacts, he had had opportunities for detailed examination not open to younger men; and that it was widely expected that this man would eventually publish his findings. Hence many, who might otherwise have carried on some research and writing, turned to other subjects. Practically, then, the whole field had been left to Edwin Tappan Adney.

Born at Athens, Ohio, in 1868, Edwin Tappan Adney was the son of Professor H. H. Adney, formerly a colonel in a volunteer regiment in the Civil War but then on the faculty of Ohio University. His mother was Ruth Shaw Adney. Edwin Tappan Adney did not receive a college education, but he managed to pursue three years' study of art with The Art Students' League of New York. Apparently he was interested in ornithology as well as in art, and spent much time in New York museums, where he met Ernest Thompson Seton and other naturalists. Being unable to afford more study in art school, he went on what was intended to be a short vacation, in 1887, to Woodstock, New Brunswick. There he became interested in the woods-life of Peter Joe, a Malecite Indian who lived in a temporary camp nearby. This life so interested the 19-year-old Ohioan that he turned toward the career of an artist-craftsman, recording outdoor scenes of the wilderness in pictures.

He undertook to learn the handicrafts of the Indian, in order to picture him and his works correctly, and lengthened his stay. In 1889, Adney and Peter Joe each built a birch-bark canoe, Adney following and recording every step the Indian made during construction. The result Adney published, with sketches, in *Harper's Young People* magazine, July 29, 1890, and, in a later version, in *Outing*, May 1900. These, so far as is known, are the earliest detailed descriptions of a birch-bark canoe, with instructions for building one. Daniel Beard considered them the best, and with Adney's permission used the material in his *Boating Book for Boys*.

In 1897, Adney went to the Klondike as an artist and special correspondent for *Harper's Weekly* and *The London Chronicle*, to report on the gold-rush. He also wrote a book on his experience, *Klondike Stampede*, published in 1900. In 1899 he married Minnie Bell Sharp, of Woodstock, but by 1900 Adney was again in the Northwest, this time as special correspondent for *Colliers* magazine at Nome, Alaska, during the gold-rush of that year. On his return to New York, Adney engaged in illustrating outdoor scenes and also lectured for the Society for the Prevention of Cruelty to Animals. In 1908 he contributed to a Harper's Outdoor Book for Boys. From New York he removed to Montreal and became a citizen of Canada, entering the Canadian Army as a Lieutenant of Engineers in 1916. He was assigned to the construction of training models and was on the staff of the Military College, mustering out in 1919. He then made his home in Montreal, engaging in painting and illustrating. From his early years in Woodstock he had made a hobby of the study of birch-bark canoes, and while in Montreal he became honorary consultant to the Museum of McGill University, dealing with Indian lore. By 1925 Adney had assembled a great deal of material and, to clarify his ideas, he began construction of scale models of each type of canoe, carrying on a very extensive correspondence with Indians, factors and other employees (retired and active) of the Hudson's Bay Company, and with government agents on the Indian Reservations. He also made a number of expeditions to interview Indians. Possessing linguistic ability in Malecite, he was much interested in all the Indian languages; this helped him in his canoe studies.

Owing to personal and financial misfortunes, he and his wife (then blind) returned in the early 1930's to her family homestead in Woodstock, where Mrs. Adney died in 1937. Adney continued his work under the greatest difficulties, including ill-health, until his death, October 10, 1950. He did not succeed in completing his research and had not organized his collection of papers and notes for publication when he died.

Through the farsightedness of Frederick Hill, then director of The Mariners' Museum, Newport News, Virginia, Adney had, ten years before his death, deposited in the museum over a hundred of his models and a portion of his papers. After his death his son Glenn Adney cooperated in placing in The Mariners' Museum the remaining papers dealing with bark canoes, thus completing the "Adney Collection."

Frederick Hill's appreciation of the scope and value of the collection prompted him to seek my assistance in organizing this material with a view to publication. Though the Adney papers were apparently complete and were found, upon careful examination, to contain an immense amount of valuable information, they were in a highly chaotic state. At the request of The Mariners' Museum, I have assembled the pertinent papers and have compiled from Adney's research notes as complete a description as I could of bark canoes, their history, construction, decoration and use. I had long been interested in the primitive watercraft of the Americas, but I was one of those who had discontinued research on bark canoes upon learning of Adney's work. The little I had accomplished dealt almost entirely with the canoes of Alaska and British Columbia; from these I had turned to dugouts and to the skin boats of the Eskimo. Therefore I have faced with much diffidence the task of assembling and preparing the Adney papers for publication, particularly since it was not always clear what Adney had finally decided about certain matters pertaining to canoes. His notes were seldom arranged in a sequence that would enable the reader to decide which, of a number of solutions or opinions given, were Adney's final ones.

Adney's interest in canoes, as canoes, was very great, but his interest in anthropology led him to form many opinions about pre-Columbian migrations of Indian tribes and about the significance of the decorations used in some canoes. His papers contain considerable discussion of these matters, but they are in such state that only an ethnologist could edit and evaluate them. In addition, my own studies lead me to conclude that the mere examination of watercraft alone is insufficient evidence upon which to base opinions as far-reaching as those of Adney. Therefore I have not attempted to present in this work any of Adney's theories regarding the origin or ethnological significance of the canoes discussed. I have followed the same practice with those Adney papers which concern Indian language, some of which relate to individual tribal canoe types and are contained in the canoe material. (Most of his papers on linguistics are now in The Peabody Museum, Salem, Massachusetts.)

The strength and weaknesses of Adney's work, as shown in his papers, drawings, and models, seem to me to be fully apparent. That part dealing with the eastern Indians, with whom he had long personal contact, is by far the most voluminous and, perhaps, the most accurate. The canoes used by Indians west of the St. Lawrence as far as the western end of the Great Lakes and northward to the west side of Hudsons Bay are, with a few exceptions, covered in somewhat less detail, but the material nonetheless appears ample for our purpose. The canoes used in the Canadian Northwest, except those from the vicinity of Great Slave Lake, and in Alaska were less well described. It appears that Adney had relatively little opportunity to examine closely the canoes used in Alaska, during his visit there in 1900, and that he later was unable to visit those American museums having collections that would have helped him with regard to these areas. As a result, I have found it desirable to add my own material on these areas, drawn largely from the collections of American museums and from my notes on construction details.

An important part of Adney's work deals with the large canoes used in the fur trade. Very little beyond the barest of descriptions has been published and, with but few exceptions, contemporary paintings and drawings of these canoes are obviously faulty. Adney was fortunate enough to have been able to begin his research on these canoes while there were men alive who had built and used them. As a result he obtained information that would have been lost within, at most, the span of a decade. His interest was doubly keen, fortunately, for Adney not only was interested in the canoes as such, he also valued the information for its aid in painting historical scenes. As a result, there is hardly a question concerning fur trade canoes, whether of model, construction, decoration, or use, that is not answered in his material.

I have made every effort to preserve the results of Adney's investigations of the individual types in accurate drawings or in the descriptions in the text. It was necessary to redraw and complete most of Adney's scale drawings of canoes, for they were prepared for model-building rather than for publication. Where his drawings were incomplete, they could be filled in from his scale models and notes. It must be kept in mind that in drawing plans of primitive craft the draftsman must inevitably "idealize" the subject somewhat, since a drawing shows fair curves and straight lines which the primitive craft do not have in all cases. Also, the inboard profiles are diagrammatic rather than precise, because, in the necessary reduction of the full-size canoe to a drawing, this is the only way to show its "form" in a manner that can be interpreted accurately and that can be reproduced in a model or full size, as desired. It is necessary to add that, though most of the Adney plans were measured from full-size canoes, some were reconstructed from Indian models, builders' information, or other sources. Thanks to Adney's thorough knowledge of bark construction, the plans are highly accurate, but there are still chances for error, and these are discussed where they occur.

Although reconstruction of extinct canoe types is difficult, for the strange canoes of the Beothuk Indians of Newfoundland Adney appears to have solved some of the riddles posed by contemporary descriptions and the few grave models extant (the latter may have been children's toys). Whether or not his reconstructed canoe is completely accurate cannot be determined; at least it conforms reasonably well to the descriptions and models, and Adney's thorough knowledge of Indian craftsmanship gives weight to his opinions and conclusions. This much can be said: the resulting canoe would be a practical one and it fulfills very nearly all descriptions of the type known today.

Adney's papers and drawings dealing with the construction of bark canoes are most complete and valuable. So complete as to be almost a set of "how-to-do-it" instructions, they cover everything from the selection of materials and use of tools to the art of shaping and building the canoe. An understanding of these building instructions is essential to any sound examination of the bark canoes of North America, for they show the limitations of the medium and indicate what was and what was not reasonable to expect from the finished product.

In working on Adney's papers, it became obvious that this publication could not be limited to birch-bark canoes, since canoes built of other barks and even some covered with skins appear in the birch bark areas. Because of this, and to explain the technical differences between these and the birch canoes, skin-covered canoes have been included. I have also appended a chapter on Eskimo skin boats and kayaks. This material I had originally prepared for inclusion in the *Encyclopedia Arctica*, publication of which was cancelled after one volume had appeared. As a result, the present work now covers the native craft, exclusive of dugouts, of all North America north of Mexico.

In my opinion the value of the information gathered by Edwin Tappan Adney is well worth the effort that has been expended to bring it to its present form, and any merit that attaches to it belongs largely to Adney himself, whose long and painstaking research, carried on under severe personal difficulties, is the foundation of this study.

Howard Irving Chapelle
*Curator of Transportation,*
*Museum of History and Technology*

## *Chapter One*

# EARLY HISTORY

THE DEVELOPMENT of bark canoes in North America before the arrival of the white men cannot satisfactorily be traced. Unlike the dugout, the bark canoe is too perishable to survive in recognizable form buried in a bog or submerged in water, so we have little or no visual evidence of very great age upon which to base sound assumptions.

Records of bark canoes, contained in the reports of the early white explorers of North America, are woefully lacking in detail, but they at least give grounds for believing that the bark canoes even then were highly developed, and were the product of a very long period of existence and improvement prior to the first appearance of Europeans.

The Europeans were most impressed by the fact that the canoes were built of bark reinforced by a light wooden frame. The speed with which they could be propelled by the Indians also caused amazement, as did their light weight and marked strength, combined with a great load-carrying capacity in shallow water. It is remarkable, however, that although bark canoes apparently aroused so much admiration among Europeans, so little of accurate and complete information appears in their writings.

With two notable exceptions, to be discussed later, early explorers, churchmen, travellers, and writers were generally content merely to mention the number of persons in a canoe. The first published account of variations in existing forms of the American bark canoe does not occur until 1724, and the first known illustration of a bark canoe accurate enough to indicate its tribal designation appeared only two years earlier. This fact makes any detailed examination of the early books dealing with North America quite unprofitable as far as precise information on bark canoes is concerned.

The first known reference by a Frenchman to the bark canoe is that of Jacques Cartier, who reported that he saw two bark canoes in 1535; he said the two carried a total of 17 men. Champlain was the first to record any definite dimensions of the bark canoes; he wrote that in 1603 he saw, near what is now Quebec, bark canoes 8 to 9 paces long and 1½ paces wide, and he added that they might transport as much as a pipe of wine yet were light enough to be carried easily by one man. If a pace is taken as about 30 inches, then the canoes would have been between 20 and 23 feet long, between 40 and 50 inches beam and capable of carrying about half a ton, English measurements. These were apparently Algonkin canoes. Champlain was impressed by the speed of the bark canoes; he reported that his fully manned longboat was passed by two canoes, each with two paddlers. As will be seen, he was perhaps primarily responsible for the rapid adoption of bark canoes by the early French in Canada.

The first English reference that has been found is in the records of Captain George Weymouth's voyage. He and his crew in 1603 saw bark canoes to the westward of Penobscot Bay, on what is now the coast of Maine. The English were impressed, just as Champlain had been, by the speed with which canoes having but three or four paddlers could pass his ship's boat manned with four oarsmen. Weymouth also speaks admiringly of the fine workmanship shown in the structure of the canoes.

When Champlain attacked the Iroquois, on what is now Lake Champlain, he found that these Indians had "oak" bark (more probably elm) canoes capable of carrying 10, 15, and 18 men. This would indicate that the maximum size of the Iroquois canoes was about 30 to 33 feet long. The illustrations in his published account indicate canoes about 30 feet long; but early illustrations of this kind were too often the product of the artist's imagination, just as were the delineations of the animals and plants of North America.

As an example of what may be deduced from other early French accounts, Champlain in 1615, with a companion and 12 Indians, embarked at La Chine in two bark canoes for a trip to the Great Lakes. He stated that the two canoes, with men and baggage aboard, were over-crowded. Taking one of these canoes as having 7 men and baggage aboard, it seems apparent that it was not much larger than the largest of the canoes Champlain had seen in 1603 on the St. Lawrence. But in 1672, Louis Joliet and Father Jacques Marquette traveled in two canoes, carrying a total of 5 French and 25 Indians—say 14 in one canoe and 16 in the other. These canoes, then, must have been at least 28 feet long over the gunwales, exclusive of the round of the ends, or about 30 feet overall. The Chevalier Henri de Tonti, one of La Salle's officers, mentions a canoe carrying 30 men—probably 14 paddlers on each side, a steersman, and a passenger or officer. Such a capacity might indicate a canoe about 40 feet over the gunwales, though this seems very long indeed; it is more probable that the canoe would be about 36 feet long.

Another of La Salle's officers, Baron de LaHontan, gave the first reasonably complete account that has been found of the size and character of a birch-bark canoe. This was written at Montreal June 29, 1684. After stating that he had seen at least a hundred bark canoes in his journeys, he said that birch-bark canoes ranged in length from 10 to 28 *pieds* and were capable of carrying from 2 to 14 persons. The largest, when carrying cargo, might be handled by three men and could carry 2,000 pounds of freight (20 quintals). These large canoes were safe and never upset. They were built of bark peeled in the winter; hot water was thrown on the bark to make it pliable, so that it could be rolled up after it was removed from the tree. The canoes were built of more than one piece of bark as a rule.

The large canoes, he reports, were 28 *pieds* long, 4½ *pieds* wide and 20 *pouces* deep, top of gunwale to top of frames on bottom. The last indicates "inside" measurement; in this the length would be over the gunwales, not overall, and the beam inside the gunwales, not extreme. He also says the canoes had a lining or sheathing of cedar "splints" or plank and, inside this, cedar ribs or frames. The bark was the thickness of an *écu* (this coin, a crown, was a little less than ⅛ inch thick), the sheathing the thickness of two *écus*, and the ribs of three. The ends of the ribs were pointed and these were seated in holes in the underside of the gunwales. There were 8 crosspieces (thwarts) between the gunwales (note: such a canoe would commonly have 9 thwarts; LaHontan may have erred here).

The canoes were convenient, he says, because of their great lightness and shallow draft, but they were easily damaged. Hence they had to be loaded and unloaded afloat and usually required repairs to the bark covers at the end of each day. They had to be staked down at night, so that a strong wind might not damage or blow them away; but this light weight permitted them to be carried with ease by two men, one at each end, and this suited them for use on the rivers of Canada, where rapids and falls made carrying frequently necessary. These canoes were of no value on the Lakes, LaHontan states, as they could not be used in windy weather; though in good weather they might cross lakes and might go four or five leagues on open water. The canoes carried small sails, but these could be used only with fair winds of moderate force. The paddlers might kneel, sit, or stand to paddle and pole the canoes. The paddle blade was 20 *pouces* long, 6 wide, and 4 *lignes* thick; the handle was of the diameter of a pigeon's egg and three *pieds* long. The paddlers also had a "setting pole," to pole the canoes in shoal water. The canoes were alike at both ends and cost 80 *écus* (La Hontan's cost 90), and would last not more than five or six years. The foregoing is but a condensed extract of LaHontan's lively account.

In translating LaHontan's measurements a *pied* is taken as 12.79 inches, a *pouce* as about 1⅛ inches. The French fathom, or *brasse*, as used in colonial Canada, was the length from finger-tip to finger-tip of the arms outstretched and so varied, but may be roughly estimated as about 64 inches; this was the "fathom" used later in classing fur-trade canoes for length. In English measurements his large canoe would have been about 30 feet long over the gunwales and, perhaps, almost 33 feet overall, 57½ inches beam inside the gunwales, or about 60 inches extreme beam. The depth inside would be 21 or 21¾ inches bottom to top of gunwale amidships. LaHontan also described the elm-bark canoes of the Iroquois as being large and wide enough to carry 30 paddlers, 15 on a side, sitting or standing. Here again a canoe about 40 feet long is indicated. He said that these elm-bark canoes were crude, heavy and slow, with low sides, so that once he and his men reached an open lake, he no longer feared pursuit by the Iroquois in these craft.

From the slight evidence offered in such records as these, it appears that the Indians may have had, when the Europeans first reached Canada, canoes at least as long as the 5-fathom or 5½-fathom canoe of later

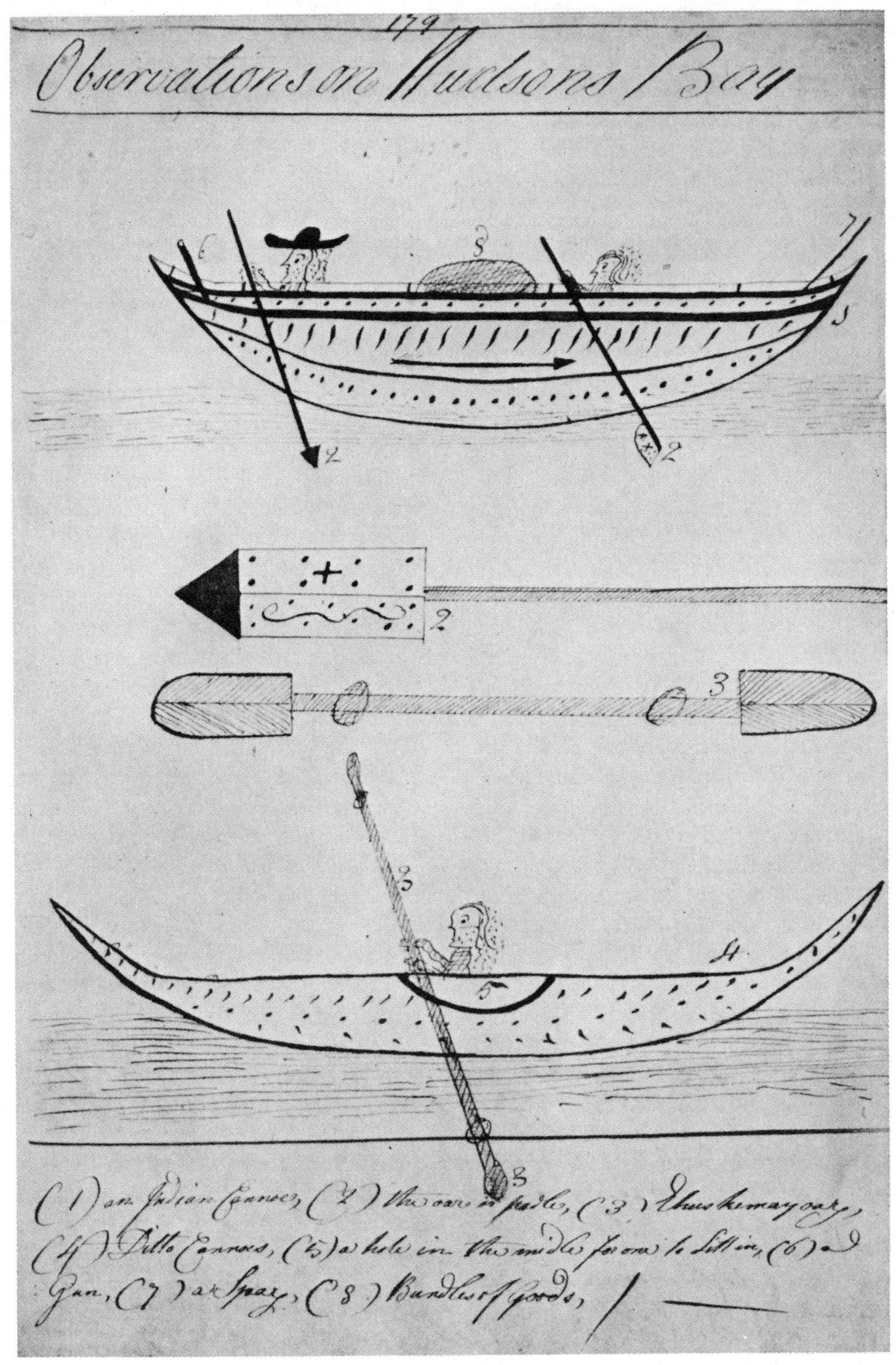


Figure 2

Page From a Manuscript of 1771, "Observations on Hudsons Bay," by Alexander Graham, Factor, now in the archives of the Hudson's Bay Company in London. The birch-bark canoe at the top, the kayak below, and the paddles are obviously drawn by one not trained to observe as an artist.

times. It appears also that these dimensions applied to the canoes of the Great Lakes area and perhaps to the elm-bark canoes of the Iroquois as well. Probably there were canoes as short as 10 feet, used as one-man hunting and fishing boats, and it is plainly evident that canoes between this length and about 24 feet were very common. The evidence in La Salle's time, in the last half of the seventeenth century, must be taken with some caution, as French influence on the size of large canoes may have by then come into play. The comparison between the maximum length of the Iroquois canoes, inferred from the report of Champlain, and that suggested by LaHontan, might indicate this growth.

Beginning as early as 1660, the colonial government of Canada issued *congés* or trading licenses. These were first granted to the military officers or their families; later the *congés* were issued to all approved traders, and the fees were used for pensions of the military personnel. Records of these licenses, preserved from about 1700, show that three men commonly made up the crew of a trading canoe in the earliest years, but that by 1725 five men were employed, by 1737 seven men, and by 1747 seven or eight men. However, as LaHontan has stated that in his time three men were sufficient to man a large canoe with cargo, it is evident that the *congés* offer unreliable data and do not necessarily prove that the size of canoes had increased during this period. The increase in the crews may have been brought about by the greater distances travelled, with an increased number of portages or, perhaps, by heavier items of cargo.

The war canoe does not appear in these early accounts as a special type. According to the traditions of the eastern Micmac and Malecite Indians, their war canoes were only large enough to carry three or four warriors and so must not have exceeded 18 feet in length. These were built for speed, narrow and with very sharp ends; the bottom was made as smooth as was possible. Each canoe carried the insignia of each of its warriors, that is, his personal mark or sign. A canoe carrying a war leader had only his personal mark, none for the rest of the crew. It is possible to regard the large canoes of the Iroquois as "war canoes" since they were used in the pursuit of French raiders in LaHontan's time. However, the Iroquois did not build the canoes primarily for war; in early times these fierce tribesmen preferred to take to the warpath in the dead of winter and to raid overland on snowshoes. In open weather, they used the rough, short-lived and quickly built elm-bark canoes to cross streams and lakes or to follow waterways, discarding them when the immediate purpose was accomplished. Probably it was the French who really produced the bark "war canoes," for they appear to have placed great emphasis on large canoes for use of the military, as indicated by LaHontan's concern with the largest canoes of his time. Perhaps large bark canoes were once used on the Great Lakes for war parties, but, if so, no mention of a special type has been found in the early French accounts. The sparse references suggest that both large and small canoes were used by the war parties but that no special type paralleling the characteristics of the Micmac and Malecite war canoes existed in the West. The huge dugout war canoe of the Indians of the Northwest Coast appears to have had no counterpart in size among the birch or elm bark canoes.

Except for LaHontan, the early French writers who refer to the use of sail agree that the canoes were quite unfitted for sailing. It is extremely doubtful that the prehistoric Indians using bark canoes were acquainted with sails, though it is possible that the coastal Indians might have set up a bush in the bow to utilize a following wind and thus lighten the labor of paddling. However, once the Indian saw the usefulness of a sail demonstrated by white men, he was quick to adopt it; judging from the LaHontan reference, and the use of sails in canoes must have become well established in some areas by 1685.

One of the most important elements in the history of the canoe is its early adoption by the French. Champlain was the first to recommend its use by white men. He stated that the bark canoe would be very necessary in trade and exploration, pointing out that in order to penetrate the back country above the rapids at Montreal, during the short summer season, and to come back in time to return to France for the winter (unless the winter was to be spent in Canada) the canoe would have to be used. With it the small and large streams could be navigated safely and the numerous overland carries could be quickly made. Also, of course, Indians could be employed as crews without the need of training them to row. This general argument in favor of the bark canoe remained sound after the desirability of going home to France for the winter had ceased to influence French ideas. The quick expansion of the French fur trade in the early seventeenth century opened up the western country into the Great Lakes area and to the northward. It was soon discovered that by using canoes on

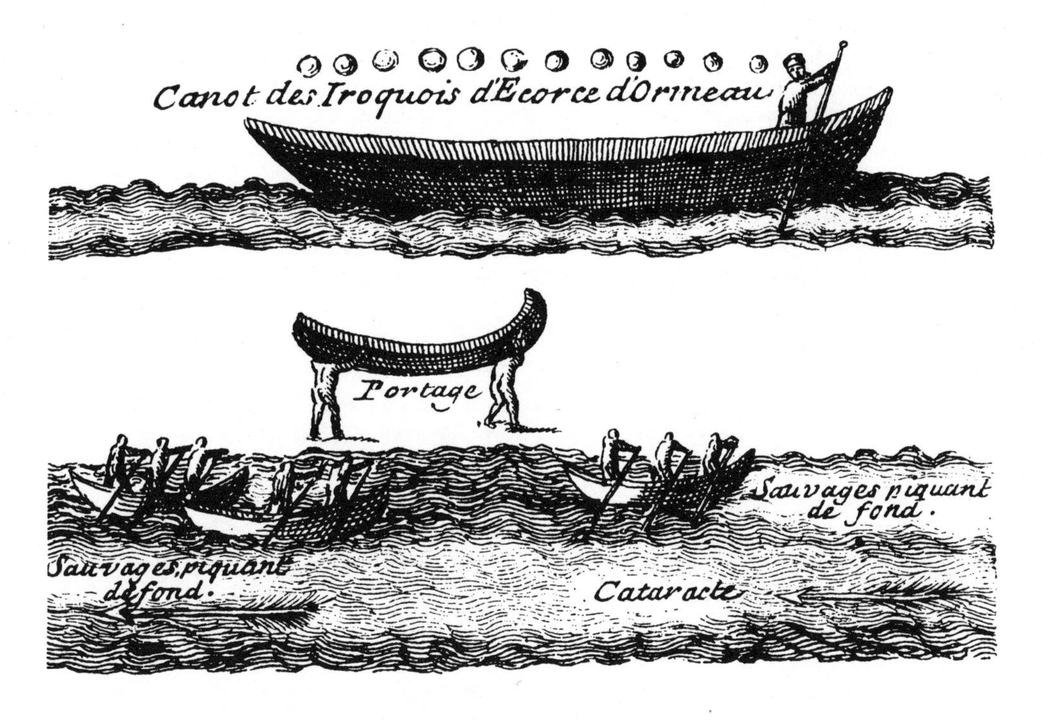

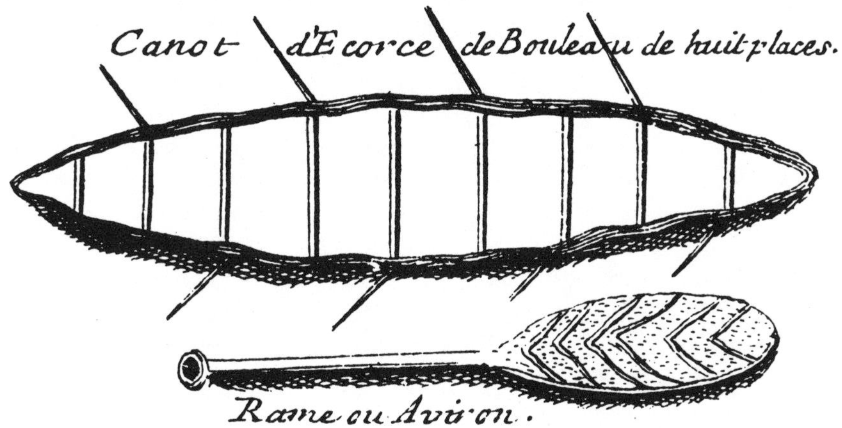


Figure 3

CANOES FROM LAHONTAN'S *Nouveaux Voyages . . . dans l'Amerique Septentrionale*, showing crude representations typical of early writers.

the ancient canoe route along the Ottawa River goods could reach the western posts on the Lakes and be transported north early enough to reach the northernmost posts before the first freeze-up occurred. The use of sailing vessels on the Lakes did not enable this to be accomplished, so that until the railroads were built in western Canada, the canoe remained the mode of transport for the fur trade in this area. Even after the railways were built, canoe traffic remained important, until well into the first half of the twentieth century as part of the local system of transportation in the northwestern country of Canada.

The unsatisfactory illustrations accompanying early published accounts have been mentioned. The earliest recognizable canoe to be shown in an illustration is the reasonably accurate drawing of a Micmac canoe that appears in Bacqueville de la Poterie's book, published in 1722. LaFiteau, another Frenchman, in 1724 published a book that not only contains recognizable drawings but points out reasons for the variation in the appearance of bark canoes:

> The Abenacquis, for example, are less high in the sides, less large, and more flat at the two ends; in a way they are almost level for their whole extent; because those who travel on their small rivers are sure to be troubled and struck by the branches of trees that border and extend over the water. On the other hand, the Outaouacs [Ottawas] and the nations of the upper country having to do their navigation on the St. Lawrence River where there are many falls and rapids, or especially on the Lakes where there is always a very considerable swell, must have high ends.

His illustrations show that his low-ended canoes were of Micmac type but that his high-ended canoes were not of the Ottawa River or Great Lakes types but rather of the eastern Malecite of the lower St. Lawrence valley. This Jesuit missionary also noted that the canoes were alike at the ends and that the paddles were of maple and about 5 feet long, with blades 18 inches long and 6 wide. He observed that bark canoes were unfitted for sailing.

The early English settlers of New England and New York were acquainted with the canoe forms of eastern Indians such as the Micmac, Malecite, Abnaki, and the Iroquois. Surviving records, however, show no detailed description of these canoes by an English writer and no illustration until about 1750. At this time a bark canoe, apparently Micmac, was brought from Portsmouth, New Hampshire, to England and delivered to Lord Anson who had it placed in the Boat House of the Chatham Dockyard. There it was measured and a scale drawing was made by Admiralty draftsmen; the drawing is now in the Admiralty Collection of Draughts, in the National Maritime Museum at Greenwich. A redrawing of this plan appears opposite. It probably represents a war canoe, since a narrow, sharp-ended canoe is shown. The bottom, neither flat nor fully round, is a rounded V-shape; this may indicate a canoe intended for coastal waters. Other drawings, of a later date, showing crude plans of canoes, exist in Europe but none yet found appear as carefully drawn as the Admiralty plan, a scale drawing, which seems to be both the earliest and the most accurate 18th-century representation of a tribal type of American Indian bark canoe.

Due to the rapid development of the French fur trade, and the attendant exploration, a great variety of canoe types must have become known to the French by 1750, yet little in the way of drawings and no early scale plans have been found. This is rather surprising, not only because the opportunity for

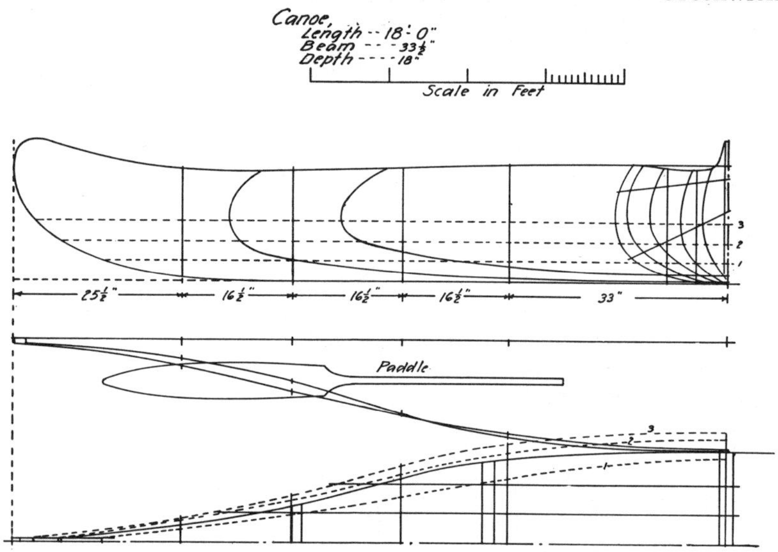

Figure 4
LINES OF AN OLD BIRCH-BARK CANOE, probably Micmac, brought to England in 1749 from New England. This canoe was not alike at both ends, although apparently intended to be so by the builder. (*From Admiralty Collection of Draughts, National Maritime Museum, Greenwich.*)

observation existed but also because a canoe factory was actually operated by the French. The memoirs of Colonel Franquet, Military Engineer-in-Chief for New France, contain extensive references to this factory as it existed in 1751.

The canoe factory was located at Trois Rivières, just below Montreal, on the St. Lawrence. A standard large canoe was built, and the rate of production was then 20 a year. Franquet gives as the dimensions of the canoes the following (converted to English measurement): length 36 feet, beam about 5½ feet, and depth about 33 inches. Much of his description is not clear, but it seems evident that the canoe described was very much like the later *grand canot*, or large canoe, of the fur trade. The date at which this factory was established is unknown; it may have existed as early as 1700, as might have been required by the rapid expansion of the French trade and other activities in the last half of the previous century. It is apparent from early comments that the French found the Indian canoe-builders unreliable, not to say most uncertain, as a source of supply. The need for large canoes for military and trade operations had forced the establishment of such a factory as soon as Europeans could learn how to build the canoes. This would, in fact, have been the only possible solution.

Of course, it must not be assumed that the bark canoes were the only watercraft used by the early French traders. They used plank boats as well, ranging from scows to flat-bottomed bateaux and ship's boats, and they also had some early sailing craft built on the Great Lakes and on the lower St. Lawrence. The bateau, shaped much like a modern dory but with a sharp stern, was adopted by the English settlers as well as the French. In early colonial times this form of boat was called by the English a "battoe," or "Schenectady Boat," and later, an "Albany Boat." It was sharp at both ends, it usually had straight flaring sides with a flat bottom, and was commonly built of white pine plank. Some, however, had rounded sides and lapstrake planking, as shown by a plan of a bateau of 1776 in the Admiralty Collection of Draughts. Early bateaux had about the same range of size as the bark canoes but later ones were larger.

After the English gained control of Canada, the records of the Hudson's Bay Company, and of individual traders and travellers such as Alexander Henry, Jr., and Alexander MacKenzie, at the end of the eighteenth century, give much material on the fur-trade canoes but little on the small Indian canoes. In general, these records show that the fur-trade canoe of the West was commonly 24 feet long inside the gunwales, exclusive of the curves of bow and stern; 4 feet 9 inches beam; 26 inches deep; and light enough to be carried by two men, as MacKenzie recorded, "three or four miles without resting on a good road." But the development of the fur-trade canoes is best left for a later chapter.

The use of the name "canoe" for bark watercraft does not appear to been taken from a North American Indian usage. The early French explorers and travellers called these craft *canau* (pl. *canaux*). As this also meant "canal," the name *canot* (pl. *canots*) was soon substituted. But some early writers preferred to call the canoe *ecorse de bouleau*, or birchbark, and sometimes the name used was merely the generic *petit embarcation*, or small boat. The early English term was "canoa," later "canoe." The popular uses of canoe, canoa, *canau*, and *canot* are thought to have begun early in the sixteenth century as the adaptation of a Carib Indian word for a dugout canoe.

## Summary

It will be seen that the early descriptions of the North American bark canoes are generally lacking in exact detail. Yet this scanty information strongly supports the claim that bark canoes were highly developed and that the only influence white men exercised upon their design was related to an increase in size of the large canoes that may have taken place in the late seventeenth and early eighteenth centuries. The very early recognition of the speed, fine construction, and general adaptability of the bark canoes to wilderness travel sustain this view. The two known instances mentioned of early accurate illustration emphasize that distinct variations in tribal forms of canoes existed, and that these were little changed between early colonial times and a relatively recent period, despite steadily increasing influence of the European.

# *Chapter Two*

# MATERIALS and TOOLS

Bark of the paper birch was the material preferred by the North American Indians for the construction of their canoes, although other barks were used where birch was not available. This tree (*Betula papyrifera* Marsh.), also known as the canoe birch, is found in good soil, often near streams, and where growing conditions are favorable it becomes large, reaching a height of a hundred feet, with a butt diameter of thirty inches or more. Its range forms a wide belt across the continent, with the northern limits in Canada along a line extending westward from Newfoundland to the southern shores of Hudson Bay and thence generally northwestward to Great Bear Lake, the Yukon River, and the Alaskan coast. The southern limits extend roughly westward from Long Island to the southern shores of Lake Erie and through central Michigan to Lake Superior, thence through Wisconsin, northern Nebraska, and northwesterly through the Dakotas, northern Montana, and northern Washington to the Pacific Coast. The trees are both abundant and large in the eastern portion of the belt, particularly in Newfoundland, Quebec, the Maritime Provinces, Ontario, Maine, and New Hampshire, in contrast to the western areas. Near the limits of growth to the north and south the trees are usually small and scattered.

The leaves are rather small, deep green, and pointed-oval, and are often heart-shaped at the base. The edges of the leaves are rather coarsely toothed along the margin, which is slightly six-notched. The small limbs are black, sometimes spotted with white, and the large are white.

The bark of the tree has an aromatic odor when freshly peeled, and is chalky white marked with black splotches on either side of limbs or where branches have grown at one time. Elsewhere on the bark, dark, or black, horizontal lines of varying lengths also appear. The lower part of the tree, to about the height of winter snows, has bark that is usually rough, blemished and thin; above this level, to the height of the lowest large limbs, the bark is often only slightly blemished and is thick and well formed. The bark is made up of paper-like layers, their color deepens with each layer from the chalky white of the exterior through creamy buff to a light tan on the inner layer. A gelatinous greenish to yellow rind, or cambium layer, lies between the bark and the wood of the trunk; its characteristics are different from those of the rest of the bark. The horizontal lines that appear on each successive paper-like layer do not appear on the rind.

The thickness of the bark cannot be judged from the size of a tree and may vary markedly among trees of the same approximate size in a single grove. The thickness varies from a little less than one-eighth to over three-sixteenths inch; bark with a thickness of one-quarter inch or more is rarely found. For canoe construction, bark must be over one-eighth inch thick, tough, and from a naturally straight trunk of sufficient diameter and length to give reasonably large pieces. The "eyes" must be small and not so closely spaced as to allow the bark to split easily in their vicinity.

The bark can be peeled readily when the sap is flowing. In winter, when the exterior of the tree is frozen, the bark can be removed only when heat is applied. During a prolonged thaw, however, this may be accomplished without the application of heat. Bark peeled from the tree during a winter thaw, and early in the spring or late in the fall, usually adheres strongly to the inner rind, which comes away from the tree with the bark. The act of peeling, however, puts a strain on the bark, so that only tough, well-made bark can be removed under these conditions. This particular characteristic caused Indians in the east to call bark with the rind adhering "winter bark," even though it might have been peeled from a tree during the warm weather of early summer.

Since in large trees the flow of sap usually starts later than in small ones, the period in which good bark is obtainable may extend into late June in some localities. Upon exposure to air and moisture, the inner rind first turns orange-red and gradually darkens with age until in a few years it becomes dark brown, or sepia. If it is first moistened, the rind can be scraped off, and this allowed it to be employed in decoration, enough being left to form designs. Hence winter bark was prized.

To the eastern Indians "summer bark" was a poor grade that readily separated into its paper-like layers, a characteristic of bark peeled in hot weather, or of poorly made bark in any season. In the west, however, high-quality bark was often scarce and, therefore, the distinction between winter and summer bark does not seem to have been made. Newfoundland once had excellent canoe bark, as did the Maritime Provinces, Maine, New Hampshire, and Quebec, but the best bark was found back from the seacoast. Ontario and the country to the immediate north of Lake Superior are also said to have produced bark of high quality for canoe building.

The bark of the paper birch was preferred for canoe building because it could be obtained in quite large sheets clear of serious blemishes; because its grain ran around the tree rather than along the line of vertical tree growth, so that sheets could be "sewn" together to obtain length in a canoe; and because the bark was resinous and not only did not stretch and shrink as did other barks, but also had some elasticity when green, or when kept damp. This elasticity, of course, was lost once the bark was allowed to become dry through exposure to air and sunshine, a factor which controlled to some extent the technique of its employment.

Many other barks were employed in bark canoe construction, but in most instances the craft were for temporary or emergency use and were discarded after a short time. Such barks as spruce (*Picea*), elm (*Ulmus*), chestnut (*Castenea dentata L.*), hickory (*Carya* spp.), basswood (*Tilia* spp.), and cottonwood (*Populus* spp.) are said to have been used in bark canoe construction in some parts of North America. Birches other than the paper birch could be used, but most of them produced bark that was thin and otherwise poor, and was considered unsuitable for the better types of canoes. Barks other than birch usually had rough surfaces that had to be scraped away, in order to make the material flexible enough for canoe construction. Spruce bark had some of the good qualities of the paper birch bark, but to a far less degree, and was considered at best a mere substitute. The resinous barks, because of their structure could not be joined together to gain length, and their characteristic shrinkage and swelling made it virtually impossible to keep them attached to a solid framework for any great length of time.

Figure 5

OJIBWAY INDIAN carrying spruce roots, Lac Seul, Ont., 1919. (*Canadian Geological Survey photo.*)

The material used for "sewing" together pieces of birch bark was most commonly the root of the black spruce (*Picea mariana* (Mill.) B.S.P.), which grows in much of the area where the paper birch exists. The root of this particular spruce is long but of small diameter; it is tough, durable, and flexible enough for the purpose. The tree usually grows in soft, moist ground, so that the long roots are commonly very close to the surface, where they could easily be

Figure 6

Roll of Bark for a Hunting Canoe. Holding the bark is the intended builder, Vincent Mikans, then (in 1927), at age 100, the oldest Indian on the Algonkin Reserve at Golden Lake, Ont.

dug up with a sharp stick or with the hands. In some areas of favorable growing conditions, the roots of the black spruce could be obtained in lengths up to 20 feet, yet with a maximum diameter no larger than that of a lead pencil.

Other roots could be used in an emergency, such as those of the other spruces, as well as of the northern white-cedar (*Thuja occidentalis* L.), tamarack (hackmatack or eastern larch (*Laris laricina* (Du Roi) K. Koch) and jack pine (*Pinus banksiana* Lamb.), the last named being used extensively by some of the western tribes. Although inferior to the black spruce for sewing, these and other materials were used for sewing bark; even rawhide was employed for some purposes in canoe construction by certain tribes.

Canoes built of nonresinous barks were usually lashed, instead of sewn, by thongs of such material as the inner bark of the northern white cedar, basswood, elm, or hickory, for the reason stated earlier. Spruce root was also used for lashings, if readily available. Since sheets of birch bark were joined without employing a needle, the sewing actually could more correctly be termed lacing, rather than stitching. But for the nonresinous barks, which could stand little sewing or lacing, perhaps lashing is the better term.

Before steel tools became available to the Indians, the woodwork required in constructing a birch-bark canoe represented great labor, since stone tools having poor cutting characteristics were used. Selection

Figure

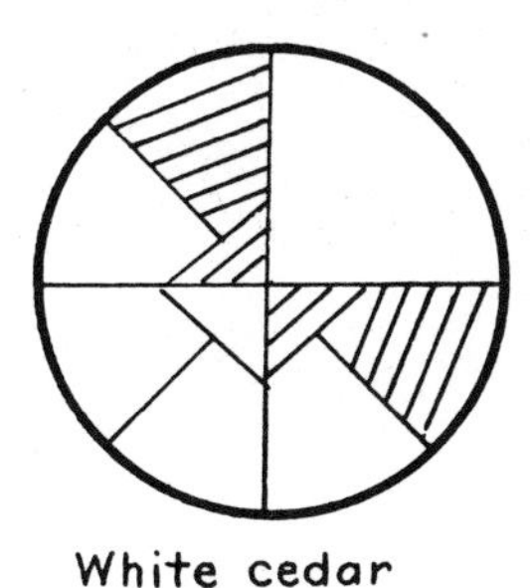

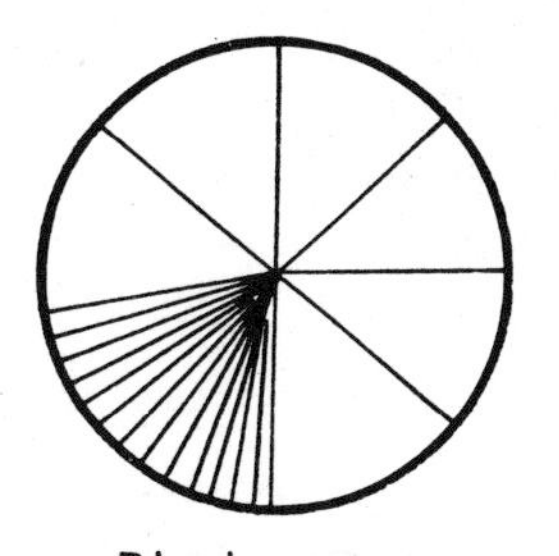

White cedar    Black spruce

Wood-splitting techniques

of the proper wood was therefore a vital consideration. In most sections of the bark canoe area, the northern white cedar was the most sought-for wood for canoe construction. This timber had the excellent characteristic of splitting cleanly and readily when dry and well-seasoned. As a result, the Indian could either utilize fallen timber of this species, windblown or torn up in spring floods; with the crude means available he chould fell a suitable tree well in advance of his needs; or he could girdle the tree so that it would die and season on the stump and then fell it at his convenience. If split properly, ribs of white cedar could be bent and set in shape by the use of hot water. In many areas the ribs, sheathing, and the gunwale members of bark canoes were made of this wood, as were also the headboards and stem pieces.

Black spruce was also employed, as it too would split well, although only when green. This wood also required a different direction in splitting than the white cedar. Ribs of black spruce could be bent and set in shape when this was done while the wood was green. In some areas black spruce was used in place of white cedar for all parts of a bark canoe structure.

Hard maple (usually either *Acer saccharum* Marsh. or *A. nigrum* Michx.), can be split rather easily while green; this wood was used for the crosspieces or thwarts that hold the gunwales apart and for paddles. Larch, particularly western larch (*Larix occidentalis* Nutt.), was used in some areas for canoe members. White and black ash (*Fraxinus americana* L. and *F. nigra* Marsh.), were also used where suitable wood of these species was available. In the northwest, spruce and various pines were employed, as was also willow (*Salix*). It should be noted that the use of many woods in bark canoe construction can be identified only in the period after steel tools became available; it must be assumed that the range of selection was much narrower in prehistoric times.

To make a bark cover watertight, it is necessary to coat all seams and to cover all "sewing" with a waterproof material, of which the most favored by the Indians was "spruce gum," the resin obtained from black or white spruce (*Picea mariana* or *P. glauca* (Moench) Voss). The resin of the red spruce (*Picea rubens* Sarg.) was not used, so far as has been discovered. The soft resin was scraped from a fallen tree or from one damaged in summer. Spruce gum could be accumulated by stripping a narrow length of bark from trees early in the spring and then, during warm weather, gathering the resin that appeared at the bottoms of the scars thus made. It was melted or heated in various ways to make it workable and certain materials were usually added to make it durable in use.

The most important aids to the Indian in canoe construction were his patience, knowledge of the working qualities of materials, his manual skill with the crude cutting, scraping, and boring instruments

Figure 8

Stone axe

known to him, and of course fire; time was, perforce, of less importance. The canoe builder had to learn by experience and close observation how to work the material available. The wood-working tools of the stone age were relatively inefficient, but with care and skill could be used with remarkable precision and neatness.

Felling of trees was accomplished by use of a stone axe, hatchet, or adze, combined with the use of fire. The method almost universally employed by primitive people was followed. The tree was first girdled by striking it with the stone tool to loosen and raise the wood fibers and remove the soft green bark. Above this girdle the trunk was daubed all around with wet earth, or preferably clay. A large, hot fire was then built around the base of the tree and, after the loose fibers were burned away and the wood well charred, the char was removed by blows from the stone tool. The process was repeated until the trunk was cut through enough for the tree to fall. The fallen trunk could be cut into sections by employing the same methods, mud being laid on each side of the "cut" to prevent the fire from spreading along the trunk. Fire could also be used to cut down poles and small trees, to cut them into sections, and to sharpen the ends into points to form crude wedges or stakes.

Stone tools were formed by chipping flint, jasper, or other forms of quartz, such as chalcedony, into flakes with sharp edges. This was done by striking the nodule of stone a sharp blow with another stone held in the hand or mounted in a handle of hide or wood to form a stone hammer. The flakes were then shaped by pressing the edges with a horn point—say, part of a deer antler—to force a chip from the flake. The chipping tool was sometimes fitted with a hide or wood handle set at right angles to the tool, so that its head could be hit with a stone or horn hammer. The flake being worked upon, if small, was often held in the hand, which was protected from the slipping of a chipping tool by a pad of rawhide. Heat was not used in chipping, and some Indians took care to keep the flake damp while working it, occasionally burying the flake for a while in moist soil. The cutting edge of a stone tool could be ground by abrasion on a hard piece of granite or on sandstone, but the final degree of sharpness depended upon the qualities of the stone being used as a tool. Slate could be used in tools in spite of its brittleness. In general, stone tools were unsuitable for chopping or whittling wood.

Figure 9

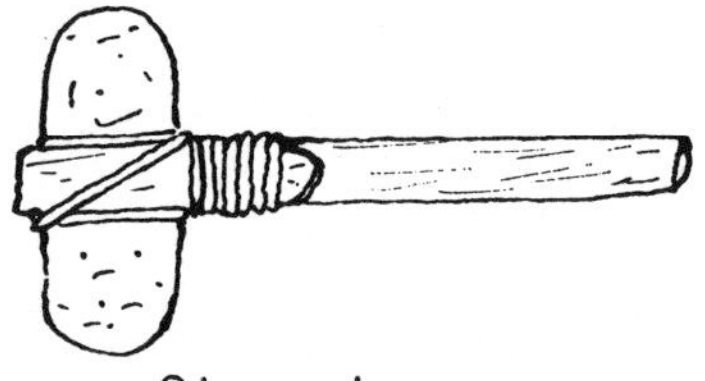

Stone hammer

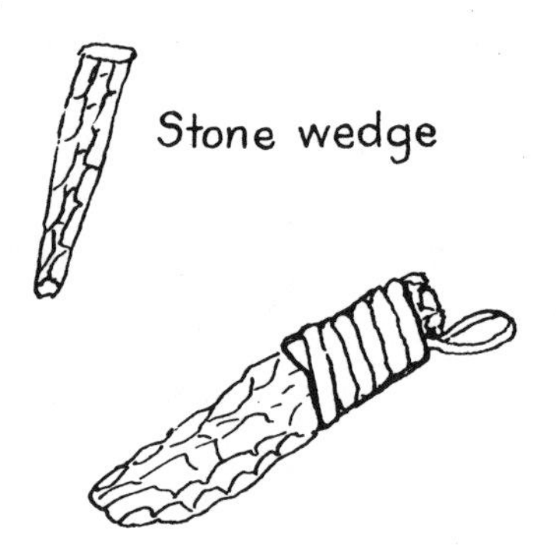


Stone knife with rawhide thong handle

Splitting was done by starting the split at the upper, or small end, of a balk of timber with a maul and a stone wedge or the blade of a stone axe, hatchet, or knife. The stone knives used for this work were not finished tools with wood handles, but rather, as the blade was often damaged in use, selected flakes fitted with hide pads that served as a handle. The tool was usually driven into the wood with blows from a wooden club or maul, the brittle stone tool being protected from damage by a pad of rawhide secured to the top, or head, of the tool. Once the split was started, it could be continued by driving more wedges, or pointed sticks, into the split; this process was continued until the whole balk was divided. White cedar was split into quarters by this method and then the heartwood was split away, the latter being used for canoe structural members. From short balks of the length of the longest rib or perhaps a little more, were split battens equal in thickness to two ribs and in width also equal to two, so that by splitting one batten two ways four finished ribs were produced. The broad faces of the ribs were as nearly parallel to the bark side of the wood as possible, as the ribs would bend satisfactorily toward or away from the bark side only. Black spruce, however, was split in line with the wood rays, from the heart outward toward the

Figure 10

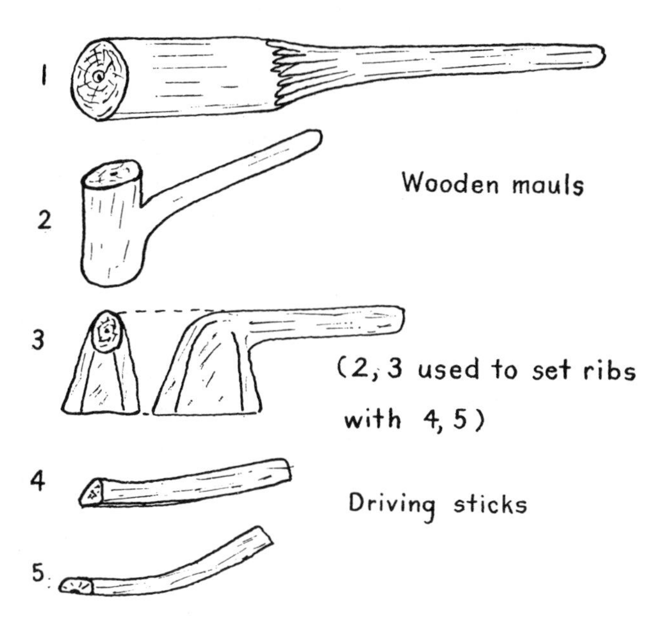


bark, so that one of the rib's narrow edges faced the bark side; only in this direction would the wood split readily and only when made this way would the ribs bend without great breakage.

Long pieces for sheathing and for the gunwale members were split from white cedar or black spruce. The splitting of such long pieces as these required not only proper selection of clear wood, but also careful manipulation of wood and tools in the operation. Splitting of this kind—say, for ribs in the finish cut—was usually done by first splitting out a batten large enough to form two members. To split it again, a stone knife was tapped into the end grain to start the split at the desired point, which, as has been noted, was always at the upper end of the stick, not at the root end. Once the split was opened, it was continued by use of a sharp-pointed stick and the stone knife; if the split showed a tendency to run off the grain as it opened, it could be controlled by bending the batten, or one of the halves, away from the direction the split was taking. The first rough split usually served to show the worker the splitting characteristics of a piece of wood. This method of finishing frame members in bark canoes accounts for the uneven surfaces that often mark some parts, a wavy grain producing a wave in the surface of the wood when it was finished.

If it were desired to produce a partially split piece of wood, such as some tribal groups used for the stems, or in order to allow greater curvature at the ends of the gunwale, the splitting was stopped at the desired point and a tight lashing of rawhide or bark was placed there to form a stop.

The tapering of frames, gunwales, and thwarts and the shaping of paddles were accomplished by splitting away surplus wood along the thin edges and by abrasion and scraping on all edges. Stone scrapers were widely employed; shell could be employed in some areas. Rubbing with an abrasive such as soft sandstone was used when the wood became thoroughly dry; hardwood could often be polished by rubbing it with a large piece of wood, or by use of fine sand held in a rawhide pad. By these means the sharp edges could be rounded off and the final shaping accomplished. Some stone knives could be used to cut wood slowly, saw fashion, and this process appears to to have been used to form the thwart ends that in many canoes were tenoned into the gunwales. A stone knife used saw fashion would also cut a bent sapling easily, though slowly. To cut and trim bark a stone knife was employed; to peel bark from a tree, a hatchet, axe, or chisel could be used.

Figure 11

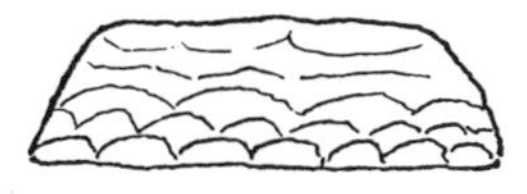

Stone scraper

Drilling was done by means of a bone awl made from a splinter of the shank-bone of a deer; the blade of this awl had a roughly triangular cross-section. The splinter was held in a wooden handle or in a rawhide grip. The awl was used not only to make holes in wood, but also as the punch to make holes for "sewing" in bark. Large holes were drilled by means of the bow-drill, in which a stone drill-point was rotated back and forth by the bow-string. Some Indians rotated the drill between the palms of their hands, or by a string with hand-grips at each end. The top of the drill was steadied by a block held in the worker's mouth, the top rotating in a hole in the underside of the block. With the bow-drill, however, the block was held in one hand.

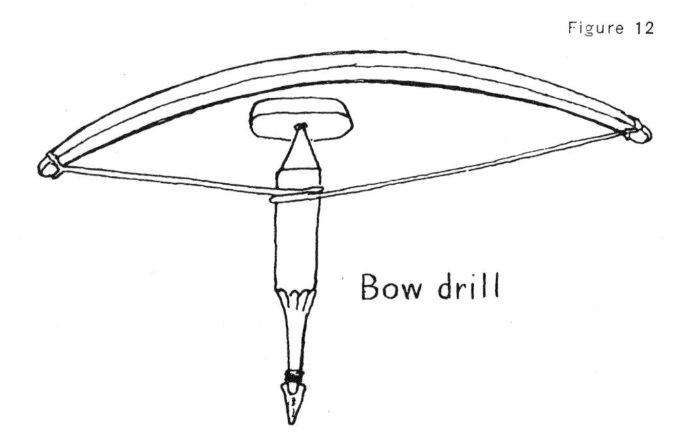

Figure 12

Peeling the bark from roots and splitting them was done by use of the thumbnail, a stone knife, or a clamshell. Biting was also resorted to. The end of a root could also be split by first pounding it with a stone, using a log or another stone as an anvil, to open the fibers at one end. Splitting a root was usually done by biting to start the split. Once this was done, half was held in the mouth and the other half between the thumb and forefinger of the right hand. Then the two parts were gradually pulled apart with the right hand, while the thumbnail of the left was used to guide the split. If the split showed a tendency to "run off," bending the root away from the direction of the run while continuing the splitting usually served to change the course of the split. If a root was hard to split, the stone knife came into play instead of the thumbnail. When the split reached arm's length, the ends were shifted in hand and mouth and the operation continued.

The use of hot water as an aid in bending wood was well known to some tribal groups before the white man came. Water was placed in a wooden trough, or in a bark basin, and heated to boiling by dropping hot stones into it. Some Indians boiled water in bark utensils by placing them over a fire of hot coals surrounded by stones and earth so that the flame could not reach the highly inflammable bark above the water-level in the dish. Stones were lifted from the fire with wooden tongs made of green saplings bent into a U-shape or made into a spoon-like outline. A straight stick and a forked one, used together, formed another type of tongs. The straight stick was placed in and under the fork; then, by forcing the latter under the stone and bringing the end of the straight stick hard against its top, the stone was held firmly, pincer-fashion.

The wood to be bent was first soaked in the boiling water, or the water was poured over it by means of a birch-bark or other dipper. When the wood was thoroughly soaked with boiling water, bending began, and as it progressed boiling water was almost continuously poured on the wood. When the wood had been bent to a desired form, it was secured in shape by thongs and allowed to cool and dry out, during which it would take a permanent set. Hard bends, as in gunwale ends and stem-pieces, were made by this means, usually after the wood had been split into a number of laminations in the area of the greatest bend. When the piece had been boiled and bent to its required form, the laminations were secured by wrapping them spirally with a thong of inner bark (such as basswood), of roots, or of rawhide.

Flat stones were used to weigh down bark in order to flatten it and prevent curling. Picked up about the canoe-building site, they had one smooth and fairly flat surface so that no harm came to the bark, and were of such size and weight as could be handled easily by the builder. Smooth stones from a stream appear to have been preferred. In preparation for building a canoe, the pins, stakes, and poles which were of only temporary use were cut or burned down in the manner mentioned and stored ready for use. Bark containers were made and filled with spruce gum, and the materials used in making it hard and durable were gathered. The building site was selected in the shade, to prevent the bark from becoming hard and brittle, and on ground that was smooth, clear of outcroppings of stone, and roots, or other obstructions, and firm enough to hold the stakes driven into it. The location was, of course, usually near the water where the canoe was to be launched.

When steel tools became available, the work of the Indian in cutting and shaping wood became much easier but it is doubtful that better workmanship resulted. The steel axe and hatchet made more rapid and far easier than before the felling and cutting up of trees, poles, and sticks; they could also be used in peeling bark. The favored style of axe among Cana-

Figure 13

Figure 14

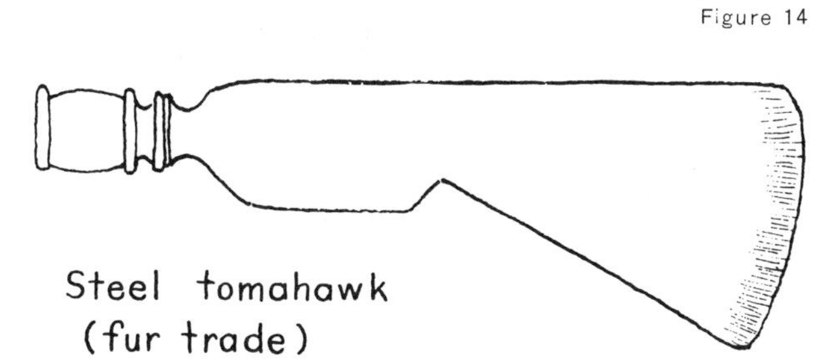

Steel tomahawk (fur trade)

dian Indians was what is known as the "Hudson Bay axe"; it is made as a fairly large or "full-axe," as a lighter "half-axe," and as a large hatchet, or hand-axe. The head of the blade is very narrow, the front of the blade vertical, while the back widens toward the cutting edge and the latter stands at a slightly acute angle to the front of the blade. This style of axe seems to follow the traditional form of the tomahawk and is popular because it cuts well, yet is lighter to carry than the other forms of axe. It is also called a "cedar axe" in some localities. In modern times, Indian hatchets are of the commercial variety, the "lathing" form being preferred because it holds somewhat to the old trade tomahawk in form of blade and weight. The traditional steel tomahawk, incidentally was an adaptation of one of the European forms of hatchet, sold in the early days of the fur trade.

Figure 15

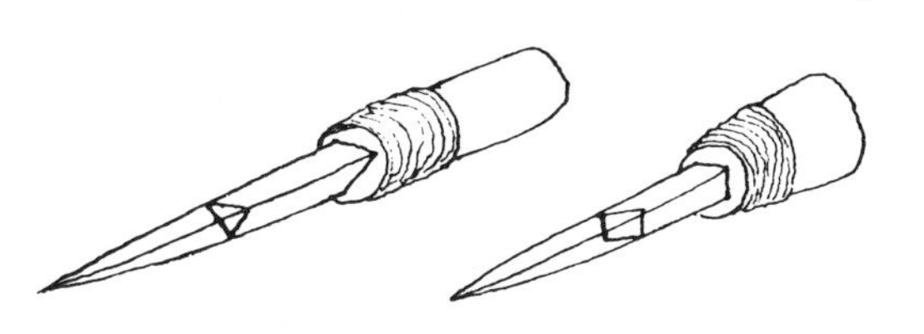

Steel canoe awls

The "canoe awl" of the fur trade was a steel awl with a blade triangular or square in cross-section, and was sometimes made of an old triangular file of small size. Its blade was locked into a hardwood handle, and it was a modern version of the old bone awl of the bark canoe builders, hence its name.

The plane was also used by modern Indians, but not in white man's fashion, in which the wood is held in a vise and smoothed by sliding the tool forward over the work. The Indian usually fixed the plane upside down on a bench or timber and slid the work over the sole, much as would be done with a power-driven joiner. However, the plane was not very popular among any of the canoe-building Indians.

Figure 16

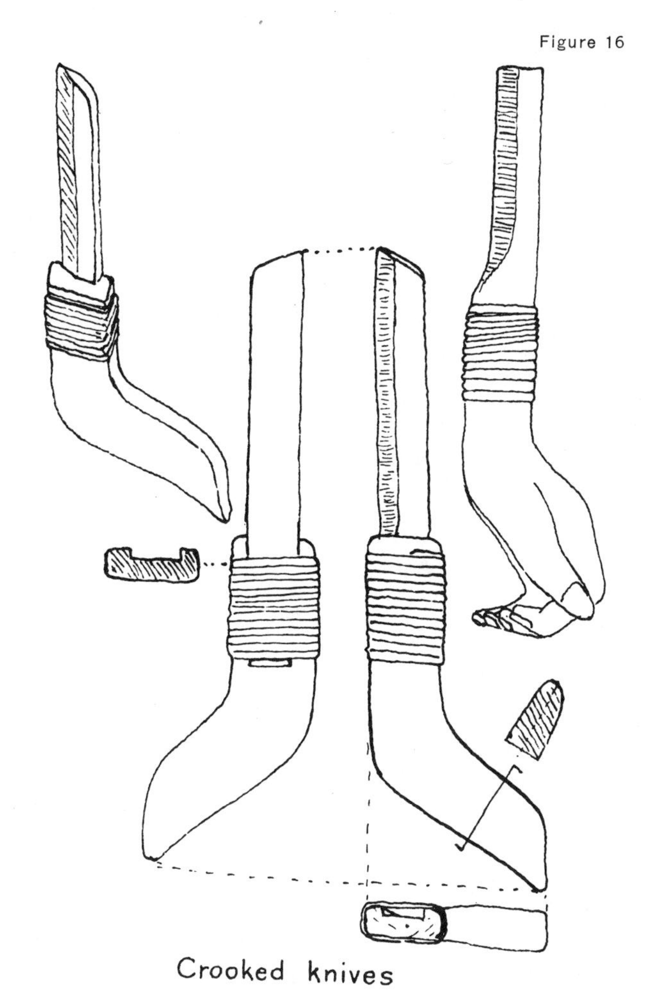

Crooked knives

The boring tool most favored by the Indians was the common steel gimlet; if a larger boring tool was desired, an auger of the required diameter was bought and fitted with a removable cross-handle rather than a brace.

One steel tool having much popularity among canoe-building Indians was the pioneer's splitting tool known as the "froe." This was a heavy steel blade, fifteen to twenty inches long, about two inches wide, and nearly a quarter inch thick along its back. One end of the blade ended in a tight loop into which a heavy hardwood handle, about a foot long, was set at right angles to the back edge of the blade, so that, when held in the hand, the blade was cutting edge down, with the handle upright. The froe was driven into the end of a balk of timber to be split by blows

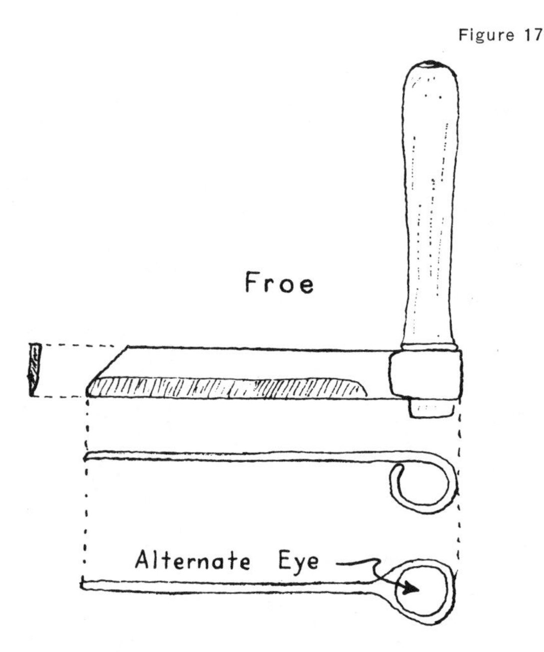


from a wooden maul on the back of its blade. Once the split was started, the maul was dropped and the hand that had held it was placed at the end of the blade away from the handle. By twisting the blade with the two hands the split could be forced open. The froe was a most powerful and efficient splitting tool when narrow, short plank, or battens, were required. The balk to be split was usually placed more or less end-up, as its length permitted, in the crotch of a felled tree, so as to hold it steady during the splitting. The pioneer used this tool to make clapboards and riven shingles; the Indian canoe builder found it handy for all splitting.

Another pioneer tool that became useful to the Indian canoe builder was the "shaving horse." A sort of bench and vise, it was used by Indians in a variety of forms, all based on the same principle of construction. Usually a seven-foot-long bench made of a large log flattened on top was supported by two or four legs, one pair being high enough to raise that end of the bench several feet off the ground to provide a seat for the operator. To the top of the bench was secured a shorter, wedge-shaped piece flattened top and bottom, with one end beveled and fastened to the bench and the other held about 12 inches above it by a support tenoned into the bench about thirty inches from the high end. Through the bench and the shorter piece were cut slots, about four feet from the high end of the bench and alined to receive an arm pivoted on the bench and extending from the ground to above the upper slot. The arm was shaped to overhang the slot on the front, toward the operator's end of the bench, and on each side. The lower portion of the arm was squared to fit the slot, and a crosspiece was secured to, or through, its lower end.

The worker sat astraddle the high end of the bench, facing the low end, with his feet on the crosspiece of the pivoted arm. Placing a piece of wood on top of the wedge-shaped piece, close to the head of the pivoted arm, he pushed forward on the crosspiece with his feet, thus forcing the head down hard upon the wood, so that it was held as in a vise. The wood could then be shaved down to a required shape with

Shaving Horse.

a drawknife or crooked knife without the necessity of holding the work. A long piece was canted on top of the bench so that the finished part would pass by the body of the worker, and, if it were necessary to shape the full length, it could be reversed.

Nails and tacks eventually came into use, though they were never used in all phases of the construction of a particular canoe. In the last days of bark canoe construction, the bark was tacked to the gunwales and, in areas where a gunwale cap was customarily employed, the cap was often nailed to the top of the gunwales.

The "bucksaw" also came into the hands of the Indians, but the frame of this saw was too awkward to carry, so the Indian usually bought only the blade. With a couple of nails and a bent sapling he could make a very good frame in the woods, when the saw was required. The ends of the sapling were slotted to take the ends of the blade and then drilled crosswise to the slot, so a nail could be inserted to hold the ends of blade and sapling together. With the end of the nail bent over, the frame was locked together and the tension was given to the blade by the bent sapling handle.

The "crooked knife" was the most important and popular steel tool found among the Indians building bark canoes. It was made from a flat steel file with one side worked down to a cutting-edge. The back of the blade thus formed was usually a little less than an eighth of an inch thick. The cutting edge was bevel-form, like that of a drawknife or chisel, with the back face quite flat. The tang of the file was fitted into a handle made of a crotched stick, to one arm of which the tang was attached, while the other projected at a slightly obtuse angle away from the back of the blade. The tang was usually held in place by being bent at its end into a slight hook and let into the handle, where it was secured with sinew lashing; wire later came into use for this lashing. The knife, held with the cutting edge toward the user, was grasped fingers-up with the thumb of the holding hand laid along the part of the handle projecting away from the user. This steadied the knife in cutting. Unlike a jackknife, the crooked knife was not used to whittle but to cut toward the user, and was, in effect, a one-hand drawknife. This form of knife is so satisfactory that it is to this day employed instead of a drawknife by many boat-builders in New Brunswick and Quebec. A variation in the crooked knife has the tip of the blade turned upward, on the flat, so that it can be used in hollowing

Figure 19

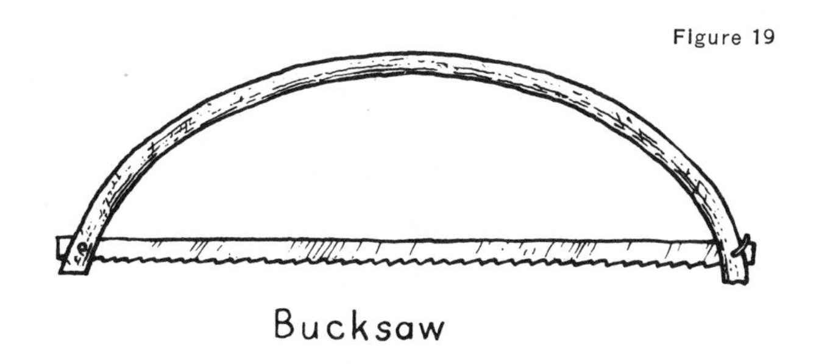

Bucksaw

out a wooden bowl or dish. The blades of crooked knives seen are usually about five-eighths inch wide and perhaps five or six inches long. Some are only slightly beveled along the cutting edge; others show this feature very markedly.

Awls, as well as chisels and other stone or bone blades, often had handles on their sides to allow them to be held safely when hit with a hammer. Some of the stone blades and chisels thus took the form of adzes and could be used like them, but only, of course, to cut charred or very soft wood. The sharpening of stone tools followed the same methods used in their original manufacture and was a slow undertaking.

To some Indians an efficient wood-cutting chisel was available in the teeth of the beaver. Each tooth was nearly a quarter inch wide, so two teeth would give a cut of nearly half an inch. The usual practice appears to have been to employ the skull as a handle, though some beaver tooth chisels had wooden handles. As used in making tenons in the gunwales, two holes, of a diameter equal to the desired width, were first drilled close enough together to make the length of the desired tenon, after which the intervening wood, especially if it was white cedar or black spruce, could be readily split out by means of either a beaver tooth or narrow stone chisel.

The maul was merely some form of wooden club; the most common type was made by cutting away part of the length of a small balk to form a handle, the remainder being left to form the head. The swelling of the trunk of a small tree at the ground, where the roots form, was also utilized to give weight and bulk to the head of a maul. It could be hardened by scorching the head in a fire. Another method of pounding and driving was to employ a stone held in one hand or both. Stone hammers were rarely employed, since the maul or a stone held in the hand would serve the purpose.

The birch tree that was to supply the bark was usually selected far in advance of the time of construction. By exploring the birch groves, the builder located a number of trees from which a suitable quantity of bark of the desired quality could be obtained. Samples of the bark of each tree were stripped from the trunk and carefully inspected and tested. If they separated into layers when bent back and forth, the bark was poor. If the "eyes" inside the bark were lumpy, the bark in their vicinity would split too easily; this was also true if they were too close together, but if the eyes on the inside of the bark appeared hollow there was no objection. Bark that was dead white, or the outer surface of which was marked by small strips partly peeled away from the layer below, would be rejected as poor in quality.

Preferably, bark was stripped from the selected trees during a prolonged thaw in winter, particularly one accompanied by rain, or as soon as the sap in the trees had begun to flow in early spring. If this was not possible, "winter" bark, as described on page 14, was used as long as it was obtainable. Only dire necessity forced the Indian to use bark of a poor quality. Fall peeling, after the first frosts, was also practiced in some areas. The work on the tree was done from stages made of small trees whose branches could be used in climbing, or from rough ladders constructed of short rungs lashed to two poles. When steel axes and hatchets were available the tree could be felled, provided care was taken to have it fall on poles laid on the ground to prevent damage to the bark in the fall and to keep the trunk high enough to allow it to be peeled. Felling permitted use of hot water to heat the bark, and thus made peeling possible in colder weather than would permit stripping a standing tree. Felling by burning, however, sometimes resulted in an uncontrolled fall in which the bark could be damaged.

Whether stone or steel knives were used, the bark was cut in the same manner, with the blade held at an angle to make a slashing cut; holding a sharp knife upright, so as to cut square to the surface of the bark, makes the tool stick and jump, and a ragged cut results. A stone or steel axe blade could also very readily be used in cutting bark; with such tools, it was customary to tap the head with a maul to make the cut. It was necessary to make only the longitudinal cut on the trunk of the birch tree, as the bark would split around the tree with the grain at the ends of this cut. Spruce and other barks, however, required both vertical and horizontal cuts.

Once the vertical cut was made to the desired length, one edge of the bark was carefully pried away from the wood with the blade of a knife. Then the removal of the bark could proceed more rapidly. Instead of starting the bark with a knife blade, some Indians used a small stick, one end of which was slightly bent and made into a chisel shape about three-quarters of an inch wide. This was used to pry the bark away, not only along the edge of the vertical cut, but throughout the operation of peeling. Another tool, useful in obtaining "winter" bark, which was difficult to strip from the tree, was a piece of dry, thick birch bark, about a foot square, with one edge cut in a slight round and beveled to a sharp edge. The beveled side was inserted beneath the bark and rocked on its curved cutting edge, thus separating the bark from the wood with less danger of splitting the bark. Spruce and other barks were removed from the tree with the same tools.

After the bark had been removed from the tree, it was handled with great care to avoid splitting it along the grain. Even in quite warm weather, the bark was usually heated slightly with a bark torch to make it flexible; sometimes hot water was applied if the inner rind was not to be used for decoration. Then the sheets were rolled up tightly in the direction of growth of the tree. This made a roll convenient for transporting and also helped to prevent the bark from curling. If the bark was not to be used immediately, it was carefully submerged in water so that it would not dry out before it was fitted to the canoe. Spruce and other resinous barks, which could not be stored, were used as soon as possible after they were stripped from the tree, the rough exterior surface being removed by scraping.

Roots for "sewing" were also gathered, split, and rolled up, then placed in water so they would remain flexible. Sometimes they were boiled as well, just before being used.

The spruce gum was gathered and tempered. Before metal kettles and frying pans became available to the Indians, it was heated in a number of ways. One method was to heat it in a wooden trough with hot stones. As the spruce gum melted easily, great temperature was not required. Stone and pottery containers were also used. Another method was to boil water in a bark container and drop in the spruce gum, which melted and floated on top of the water in such a consistency that it could be skimmed off with a bark spoon or dipper. Chips and dirt were skimmed off the hot gum with a strip of bark or a flat stick.

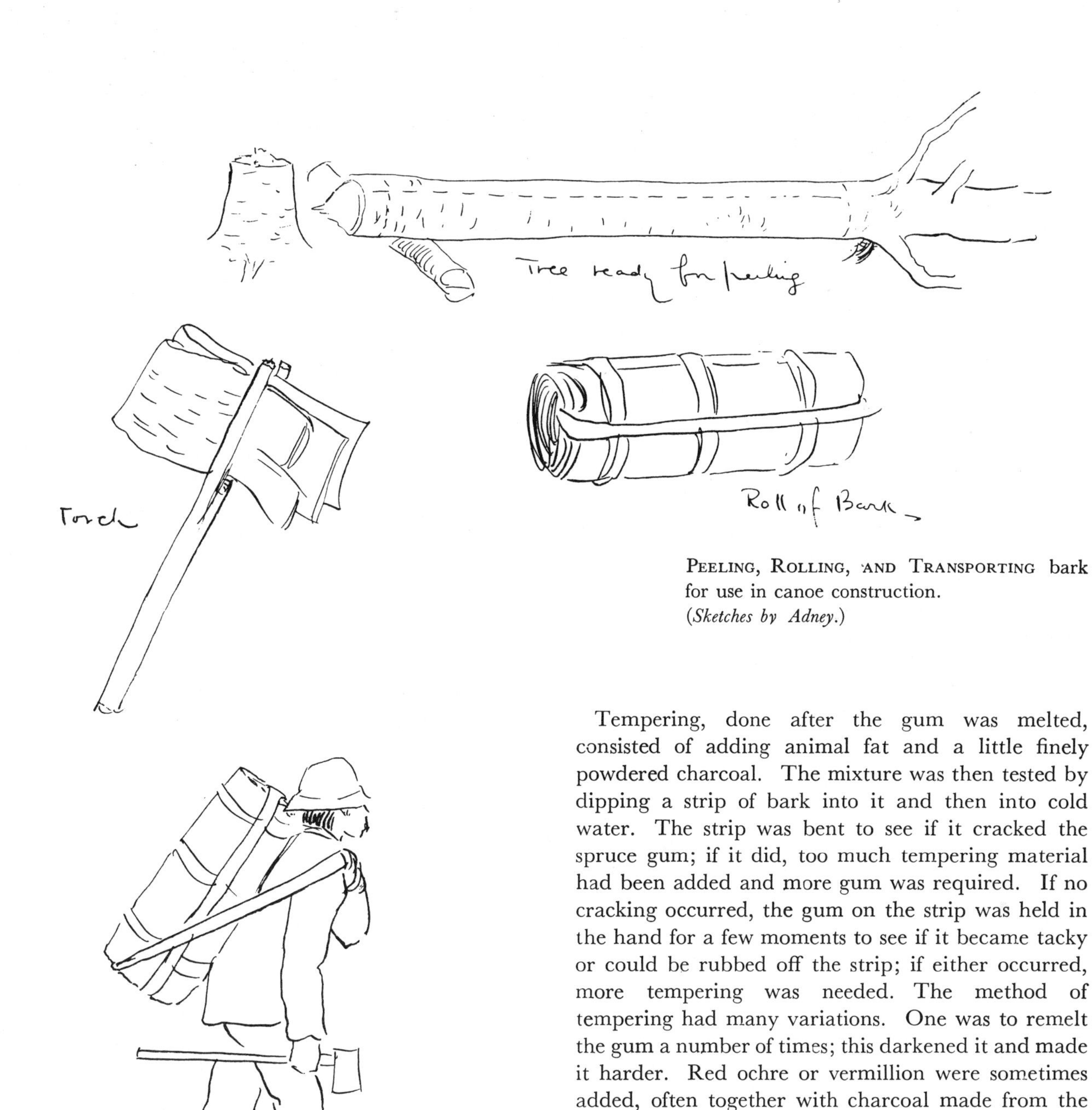


PEELING, ROLLING, AND TRANSPORTING bark for use in canoe construction.
(*Sketches by Adney.*)

Figure 20

Tempering, done after the gum was melted, consisted of adding animal fat and a little finely powdered charcoal. The mixture was then tested by dipping a strip of bark into it and then into cold water. The strip was bent to see if it cracked the spruce gum; if it did, too much tempering material had been added and more gum was required. If no cracking occurred, the gum on the strip was held in the hand for a few moments to see if it became tacky or could be rubbed off the strip; if either occurred, more tempering was needed. The method of tempering had many variations. One was to remelt the gum a number of times; this darkened it and made it harder. Red ochre or vermillion were sometimes added, often together with charcoal made from the willow. Instead of spruce gum, in some areas, pine resin was used, tempered with tallow and sometimes charcoal. The Indians in the East sometimes used remelted spruce gum to which a little tallow had been added, making a light brown or almost transparent mixture. Most tribal groups used gum that was black, or nearly so.

For repair work, when melted spruce gum could not be procured in the usual manner, hard globules and flakes of gum scraped from a fallen spruce tree were used. These could not be easily melted, so they were first chewed thoroughly until soft; then the gum was spread over a seam. This type of gum would not stick well unless it were smoothed with a glowing stick, and hence was used only in emergencies.

It is believed that before steel tools were available birch-bark canoes were commonly built of a number of sheets of bark rather than, as quite often occurred in later times, of only one or two sheets. The greater number of sheets in the early canoes resulted from the difficulty in obtaining large sheets from a standing tree. Comparison of surviving birch-bark canoes suggests that those built of a number of sheets would have contained the better bark, as large sheets often included bark taken from low on the trunk, and this, as has been mentioned, is usually of poorer quality than that higher on the trunk.

It is known that the early Indians carried on some trade in bark canoe building materials, as they did in stone for weapons and tools. Areas in which some materials were scarce or of poor quality might thus obtain replacements from more fortunate areas. Fine quality bark, "sewing" roots, and good spruce gum had trade value, and these items were sold by some of the early fur traders. Paint does not appear to have been used on early canoes, except, in some instances, on the woodwork. This use occurred mostly in the East, particularly among the Beothuks in Newfoundland. Paint was apparently not used on birch bark until it was introduced by white men in the fur trade.

## Summary

It will be seen that the Indian gathered all materials and prepared them for use with only a few simple tools, most of which could be manufactured at the building site and discarded after the work was completed. The only other tools he usually brought to the scene were those he normally required in his everyday existence in the forest. Some instruments used in canoe building, however, might be preserved; these were the measuring sticks on which were marked, by notches, certain measurements to be used in shaping a canoe. Also, some Indians used a building frame that shaped the bottom in plan view. These are best described when the actual building methods are examined.

BUILDING FRAME FOR A LARGE CANOE. Dotted lines show change in shape is caused by omitting cross-bars or by using short bars in ends. Note lashing at ends and method of fastening thwart with a thong.

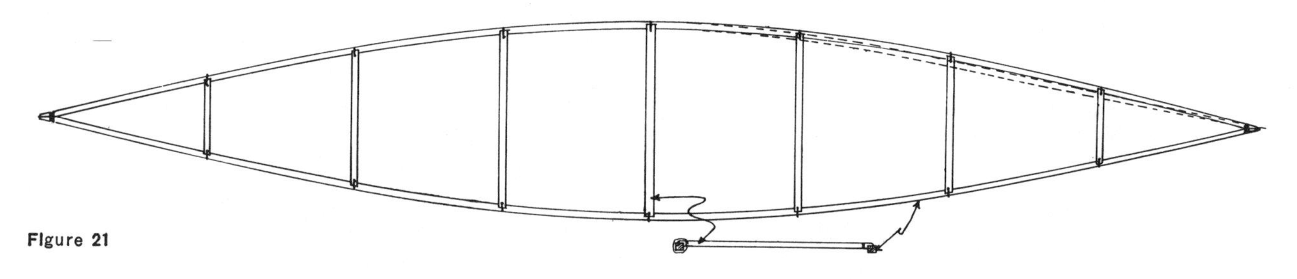

Figure 21

# *Chapter Three*

# FORM AND CONSTRUCTION

CLASSIFICATION OF THE TYPES of bark canoes built by the Indians is not a simple matter. Perhaps the most practical way is to employ the tribal designation, such as Cree canoe, Micmac canoe, accepting as a criterion the distinctive general appearance of the canoes used by each tribe. It must be emphasized, however, that this method of classification does not indicate the model, or "lines," employed. Both the model and the size of bark canoes were extensively affected by the requirements of use: lake, coastal, or river navigation; smooth, rough, or fast-running water; transportation of a hunter, a family, or cargo; the conditions and length of portages; and the permanence of construction desired. Canoes of various models, sizes, methods of construction, or decoration might be found within the limits of a single tribal classification. Also, within a given area, there might be apparent similarity in model among the canoes of two or three tribal groups. However, a classification based on geographical areas has been found to be impractical, because the movements of tribal groups in search of new hunting grounds tend to make tribal boundaries difficult to define.

## *Form*

The canoes of some tribal groups appear to be hybrids, representing an intermingling of types as a result of some past contact between tribes. Those of other groups are of like model, form, and even appearance, possibly owing to like conditions of employment. The effects of a similarity in use requirements upon inventiveness is seen in the applications for modern patent rights, where two or more applications can cover almost exactly the same device without the slightest evidence of contact between the applicants; there is no logical reason to suppose the same condition cannot apply to primitive peoples, even though their processes of invention might be very slow or relatively rare in occurrence.

The effects of migration of tribes upon their canoe forms can only be studied with respect to those comparatively recent times for which records and observations are available. From the limited information at hand it appears that the Indian, when he moved to an area where use requirements and materials available for building differed from those to which he had been accustomed, was often forced to modify the model, form, size, and construction of his canoe. In some instances this seems to have resulted in the adoption of another tribal form.

The distinctive feature that usually identifies the tribal classification of a bark canoe is the profile of the ends, although sometimes the profile of the gunwale, or sheer, and even of the bottom, is also involved. The bow and stern of many bark canoes were as near alike in profile as the method of construction would permit; nevertheless some types had distinct bow and and stern forms. Among tribes the form of the ends of the canoes varied considerably; some were low and unimpressive, others were high and often graceful.

Obviously practical reasons can be found for certain tribal variations. In some areas, the low ends appear to ensue from the use of the canoe in open water, where the wind resistance of a high end would make paddling laborious. In others the low ends appear to result from the canoe being commonly employed in small streams where overhanging branches would obstruct passage. Portage conditions may likewise have been a factor; low ends would pass through brush more easily than high. Types used where rapids were to be run often had ends higher than the gunwales to prevent the canoe from shipping water over the bow. The high, distinctive ends of the canoes most used in the fur trade, on the other hand, were said to have resulted from the necessity of employing the canoe as a shelter. When the canoe was turned upside down on the ground, with one gunwale and the tops of the high ends supporting it, there was enough headroom under the canoe to permit its use as a shelter without the addition of any temporary structure. The desirability of this characteristic in the fur-trade canoe can be explained by the fact that the crew travelled as many hours as possible each day, and rested for only a very short period, so that rapid erection of shelter lenghtened both the periods of travel and of rest.

Yet these practical considerations do not always explain the end-forms found in bark canoes. Canoes with relatively high ends were used in open waters, and similar canoes were portaged extensively. Possibly the Indian's consciousness of tribal distinctions led him to retain some feature, such as height of the end-forms, as a means of tribal recognition, even though practical considerations required its suppression to some degree.

The profile of the gunwales also varied a good deal among tribal types. Most bark canoes, because of the raised end-forms, showed a short, sharp upsweep of the sheer close to the bow and stern. Some showed a marked hump, or upward sweep, amidships which made the sheer profile follow somewhat the form of a cupid's bow. Many types had a straight, or nearly straight, sheer; others had an orthodox sheer, with the lowest part nearly amidships.

The bottom profiles of bark canoes showed varying degrees of curvature. In some the bottom was straight for most of its length, with a slight rise toward the ends. In others the bottom showed a marked curvature over its full length, and in a few the bottom was practically straight between the points at which the stems were formed. Some northwestern types had a slightly hogged bottom, but in these the wooden framework was unusually flexible, so that the bottom became straight, or even a little rockered when the canoe was afloat and manned.

The practical reasons for these bottom forms are not clear. For canoes used in rapid streams or in exposed waters where high winds were to be met many Indians preferred bottoms that were straight. Others in these same conditions preferred them rockered to varying degrees. It is possible that rocker may be desirable in canoes that must be run ashore end-on in surf. Of course, a strongly rockered bottom permits quick turning; this may have been appreciated by some tribal groups. Still other Indians appear to have believed that a canoe with a slightly rockered bottom could be paddled more easily than one having a perfectly straight bottom.

The midsections of bark canoes varied somewhat in form within a single tribal type, because the method of construction did not give absolute control of the sectional shape during the building, but, on the whole, the shape followed tribal custom, being modified only to meet use requirements. Perhaps the two most common midsection shapes were the U-form, with the bottom somewhat flattened, and the dish-shape, having rather straight, flaring sides combined with a narrow, flat, or nearly flat bottom. Some eastern canoes showed marked tumble-home in the topside above the bilge; often they had a wide and rather flat rounded bottom, with a short, hard turn in the bilge. A few eastern canoes, used mainly in open waters along the coast, had bottoms with deadrise—that is, a shallow V-form, the apex of the V being much rounded; the V-bottom, of course, would have aided in steering the canoe in strong winds. One type of canoe with this rising bottom had tumble-home topsides, but another, used under severe conditions, had a midsection that was an almost perfect V, the apex being rounded but with so little curvature in the arms that no bilge could be seen.

Generally speaking, the eastern canoes had a rather well rounded bottom with a high turn of the bilge and some tumble-home above, though they might have a flatter form when built for shallow-water use or for increased carrying capacity. A canoe built for speed, however, might be very round on the bottom, and it might or might not have some tumble-home in the topside. In the West, a flat bottom with flaring topsides predominated; fast canoes there had a very narrow, flat bottom with some flare, the width of the bottom and the amount of flare being increased to give greater capacity on a shallow draft. Some canoes in the Northwest had a skiff-form flat bottom and flaring sides, with the chine rounded off sharply.

The form of the sections near the ends of a canoe are controlled to a great extent by the form of midsection. In canoes having flat bottoms combined with flaring sides this form was usually carried to the ends, where it became a rather sharp V, giving fine lines for speed when the canoe was light, and only moderately increased resistance when it was loaded. Among eastern canoes having tumble-home topsides, the midsection form could be carried to the ends, gradually becoming sharper in canoes having "chin" in the profiles of the ends; in canoes having no chin, the sections necessarily took a pointed oval form close to the ends. A few canoes having flaring sides and chin ends showed a similar change in form. In all, however, the bow and stern showed a tendency toward fullness near the waterline.

Canoes with a strongly U-shaped midsection commonly carried this form to the ends, with increasing sharpness in the round of the U. The U-form predominated in the end-sections of eastern canoes, of course, though a few showed a V-form, as must be expected. The fairing of the end sections into the end profiles appears to have controlled this matter. The outline of the gunwales, in plan view, also influenced the form of the end-sections and of the level lines there. Some canoes, when viewed from above, showed a pinched-in form at the ends, this was caused by the construction of the gunwales or by the projection of the end-profile forms beyond the ends of the true structural gunwale members. Such canoes would have a very strong hollow in the level lines projected through their hull-form below the gunwales, and this could have been accentuated by any strong chin in the bow and stern shapes. On the other hand, many canoes showed no hollow, and the level lines were straight for some distance inboard of the ends, or were slightly convex. Full, convex level lines will appear below the waterline in canoes having a strongly rockered bottom.

It should be noted that the Indians were aware that very sharp-ended canoes usually were fast under paddle; hence they employed this characteristic in any canoe where high speed was desired. However, the degree of sharpness in the gunwales and at the level lines is not always the same at both ends, though the variation is sometimes too slight to be detected without careful measurement; it may at times have been accidental, but in many cases it appears to have been intentional.

Some eastern canoes having their greatest width, or beam, on the gunwales at midlength had finer level lines aft than forward, apparently to produce trim by the stern when afloat and manned. This made them steer well in rough water. Some northwestern canoes had their greatest beam abaft the midlength, giving them a long, sharp bow; the run was sometimes formed by sweeping up the bottom aft to a shallow stern, as well as by the double-ended form of the canoe. Despite a general similarity in the form of the ends, in some canoes the bow was marked by its greater height, in others, by the manner in which the bark was lapped at the seams, or by the manner of decoration. In a few with ends exactly alike the bow was indicated by the fitting of the thwarts such as, for example, by placing at the forward end a particular style of thwart, intended to hold the torch used in spearing fish at night, or to support a mast and sail.

In examining the lines, or model, of a bark canoe, the limitations imposed upon the builder by the characteristics of bark must be considered. The degree of flexibility, the run of the grain, and the toughness and elasticity of the bark used all influenced the form of canoes. The marked chin in the ends of some canoes, for example, resulted from an effort to offset the tendency of birch bark to split when a row of stitches lay in the same line of grain. The curved chin profile allowed the stitching to cross a number of lines of grain. Sometimes this tendency was avoided by incorporating battens into the coarse stitching; this style of sewing was particularly useful in piecing out birch bark for width in a canoe, where the sewing had to be in line with the grain. The Indians also employed alternating short and long stitching in some form for the same purpose. Spruce bark, as used in canoes in the extreme North and Northwest, could be sewn in much the same manner as birch bark, but with due regard for the longitudinal grain of the spruce bark.

The joining of two pieces of bark by root sewing or lacing, combined with the use of spruce gum to obtain watertightness, formed a seam that could be readily damaged by abrasion from launching the canoe, from pulling it ashore, or from grounding it accidentally. For this reason, seams below the waterline were kept at a minimum and were never placed along the longitudinal centerline of the bottom, where they would have formed a sharp apex to both the V-shaped midsection and to the dead-rise bottom form. Likewise, a seam was not used in forming the rocker of the bottom. Though seams had to be used to join the bark at bow and stern, the form of the canoe allowed the seams to be greatly strengthened and protected there.

The restrictions on form imposed by barks such as elm, chestnut, and hickory were very great. These barks, which are not as elastic as birch bark, were sometimes employed in a single large sheet. The sheets were not joined for length; canoes of this material were often formed by crimping, or lap folding, rather than by cutting out gores and then sewing the edges

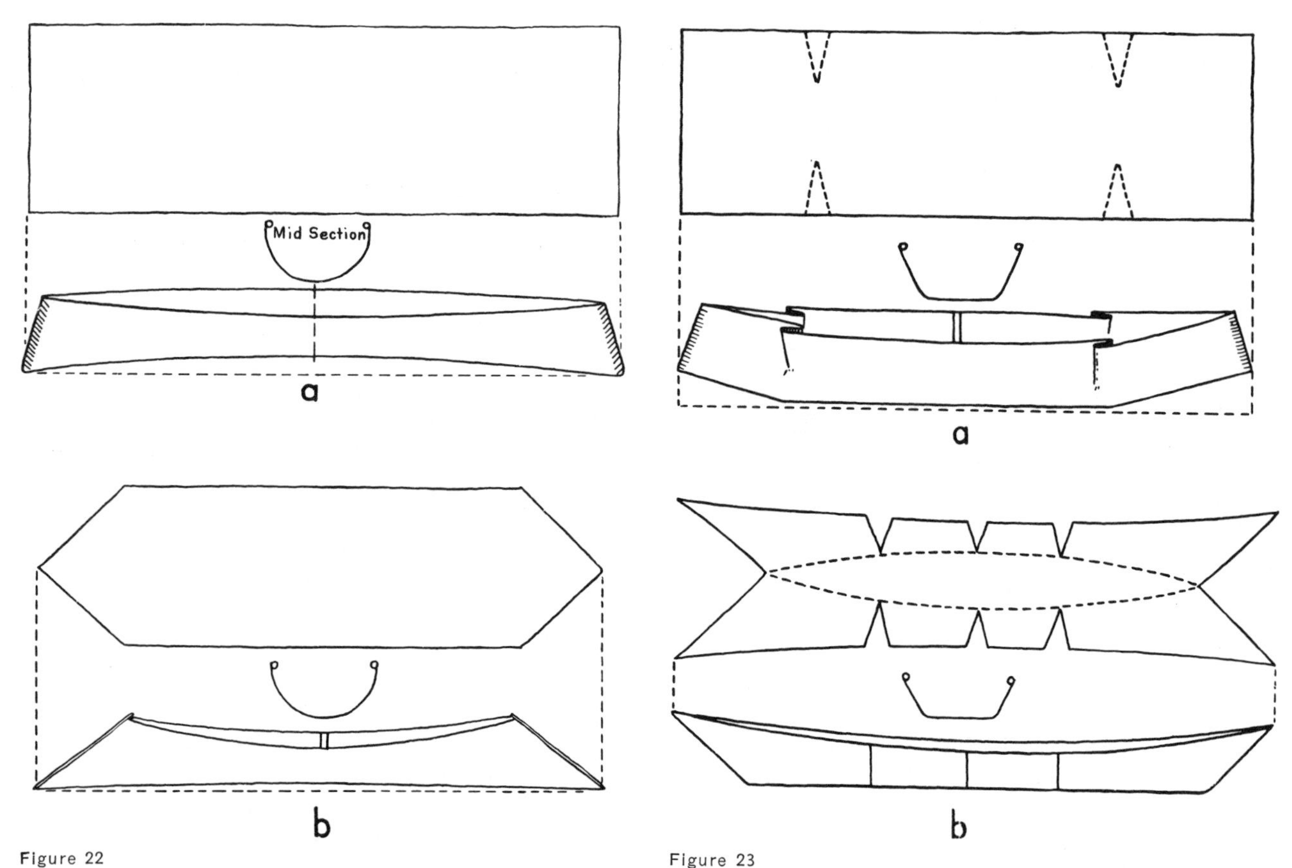


Figure 22

CANOE formed (a) without crimping or goring sides, showing hogged bottom; and (b) with ram ends to reduce hogging of bottom.

Figure 23

CANOE formed (a) by crimping sides, showing rockered bottom line, and (b) by simple gores in sides. The same effects are obtained by making bark cover of three pieces: sides and bottom.

together. The characteristics of these barks can readily be demonstrated with a sheet of paper: such a sheet can be made into a crude canoe-form by bending it lengthwise and joining the ends, but it will be obvious that the midsection takes a very unstable U-form. By forcing the ends inward to give a ram, or chin, effect to bow and stern, a somewhat flatter bottom can be obtained in the midsection. By crimping or folding the paper gore-fashion near each end of the canoe-form at the gunwale edge, some rocker is created in the bottom and the width of the gunwales is increased near the ends, giving more capacity. But without the crimping along the gunwale, when the midsection form is flattened on the bottom, the latter tends to hog. Many of these bark canoes utilized both the rams ends and crimping to obtain a more useful form. However, while a sheet of birch bark could be crimped or gored into a scow-form canoe such as the Asiatic birch-bark canoe, no example of this form from North America is known. On this continent all bark canoes were sharp at both ends, i.e., double-ended, although a number of North American dugouts were scow- (or punt-) shaped.

Birch bark gave much more freedom in the selection of form simply because it could be joined together in small odd-sized sheets to shape a hull, and because it was elastic enough to allow some "moulding" by pressure of the framework employed. Birch bark could be gored, or slashed, and rejoined without resort to folding or crimping; thus it permitted a smooth exterior surface to be achieved. The toughness of the bark was sufficient to allow some sewing in line with the grain, to add to the width of a sheet, if the proper technique were employed (this was also true to a lesser extent of spruce bark).

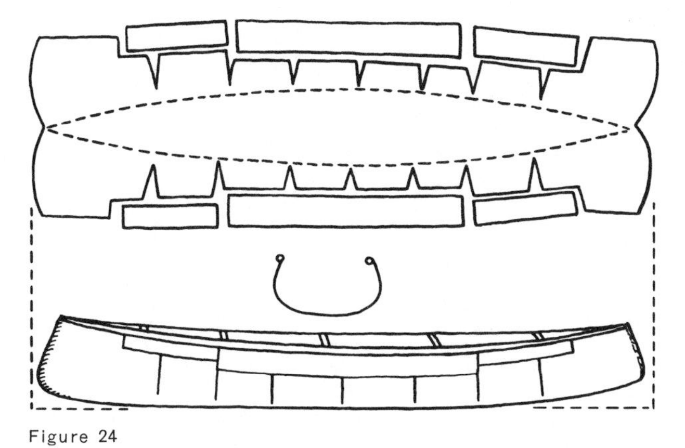

Figure 24

Canoe Formed by use of gores and panels.

The framework of most bark canoes depended upon the gunwale structure to give longitudinal strength to the hull; for this reason the structure was made sufficiently large in cross-section to be rather stiff, or was formed of more than one member. An inner and outer gunwale construction was employed in many bark canoes. The inner member was the strength member and was sometimes square, or nearly so, in cross-section. In some canoes bark was brought up on the outside of this gunwale member, lapped over the top, and lashed over it; in others the bark was lashed to both inner and outer gunwales. The outer gunwale, a rectangular-sectioned batten bent narrow-edge up, was applied like a guard, outside the bark, and was secured by pegs, by the lashings of the bark cover, or by widely spaced lashings. On top of the large inner gunwale and usually extending outward over the outer gunwale, a thin cap, pegged or lashed in the same manner as the outer gunwale, was sometimes added; this was intended to protect the lashing of the bark to the gunwale rather than to add longitudinal strength.

The corners of the inner gunwale, or of the single gunwale, were all rounded off to prevent them from cutting the sewing and lashings. The bottom outboard corner was sometimes rounded off more than the other, or beveled, in order to form between the outboard face of the gunwale and the bark a slot into which the heads of the ribs could be forced. An alternate method of accomplishing this was to notch or drill holes in the gunwales for the heads of the ribs.

The ends of the gunwales were fashioned in various ways. In some canoes the gunwales were sheered upward at the ends only slightly, the gunwale ends being secured to wide end boards in the stems or extended past them and secured to the stem-pieces. The apparent sheer in the latter might be formed by bending the outer gunwale, or outwale, and the cap (if one existed) to the required curve and then securing the ends to the stem-piece, or to the end boards, as the form of end profile dictated. If either the single gunwale or the outwale or both were sharply sheered, they were split, to a point near the end thwart, into two or four or even more laminations; even the rail cap, which was perhaps half an inch thick, might be split in the same manner to allow a sharp upward sweep at the stems. After being bent, the split members were temporarily wrapped to hold the laminations together. In no bark canoes did the ends of the gunwales curve back on themselves to form a hook just inboard of the bow and stern, despite the numerous pictures that show this feature. The gunwale ends sometimes projected almost perpendicularly upward, slightly above the top of the bow and stern, so that when the canoe was upside down its weight came on these rather than on the sewing of the ends of the craft.

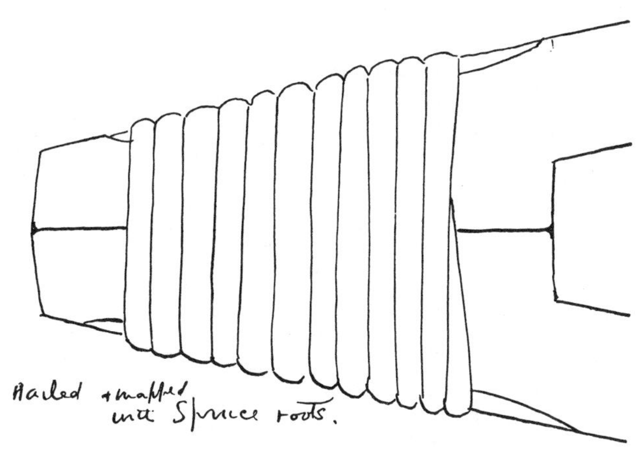


Figure 25

Gunwale Ends nailed and wrapped with spruce roots. (*Sketch by Adney.*)

The gunwale ends in some canoes were fastened together by means of one or more lashings, often widely spaced. After being lashed together, a narrow wedge was sometimes driven between the two gunwales from inboard to tighten the lashings. The ends were sometimes beveled on their bearing surfaces so as to make a neat appearance when joined. The various ways in which the gunwale ends at stem and

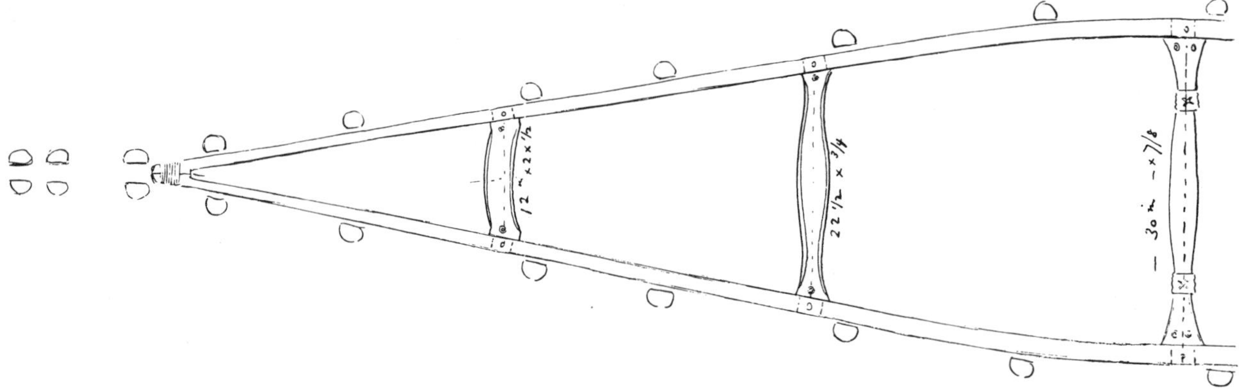

Figure 26

GUNWALES AND STAKES ON BUILDING BED, plan view. (*Sketch by Adney.*)

stern were finished can best be described when individual types are under examination. Some canoes had a small piece of bark over the ends of the gunwales but under the outwales that held it in place. Whether these pieces were employed to protect the lashing of the gunwales and adjoining work from the weather, or whether they were the vestigial remains of a decking once used, cannot be determined. In the Canadian Northwest the ends of bark canoes were sometimes decked with bark for a short distance inboard.

The bark was secured to the gunwales by a continuous spiral lashing all along the main gunwale or by separated lashing in series. In the first, the continuous lashing, where it passed through the bark, might show regularly spaced separations to avoid the tops of the ribs. In the second, the lashings were placed clear of the ribs. There were some slight variations in the lashings, but these were of minor importance so far as structural strength is concerned. In all cases, the bark was brought up to or over the top of the gunwale before being secured, so that the holes for the lashing were pierced at some distance from the edge of the bark to prevent it from splitting.

The ends of the thwarts were mortised into the gunwales and also secured by lashings. The number of thwarts varied with the tribal type, the size, and the purpose of the canoe. Usually an odd number, from three to nine, were used, though occasional canoes had two or four thwarts. Very small canoes for hunting might have only two or three thwarts, but most canoes 14 to 20 feet long had five. Canoes intended for portaging usually had one thwart at midlength to aid in lifting the canoe for the carry position. The distance between the thwarts might be determined by structural design, or might be fixed so as to divide the cargo space to allow proper trim. The thwarts might serve as backrests for passengers, but were never used as seats. There was no standard form for the shape of the thwarts, which varied not only to some degree by tribal classification, but even among builders in single tribe. They were usually thickest and widest over the centerline of the canoe, tapering outboard and then spreading again at the gunwales to form a marked shoulder at the mortise. The lashings to the gunwales often passed through two or more holes in this shoulder.

The ribs, or frames, of most canoes were very closely spaced and were wide, flat, and thin. They ran in a single length from gunwale to gunwale. In canoes having V-sections near the ends, the ribs were often so sharply bent as to be fractured slightly. Across the bottom they were wide but above the bilge they tapered in width toward the end, which was either a rounded point or a beveled or rounded chisel-edge. The ribs were forced under the gunwales so that the heads fitted into the bevel, or into notches or holes at the underside and outboard edge of the gunwale, between it and the bark cover. By canting the rib to bring its ends into the proper position and then forcing it nearly perpendicular, the builder brought enough pressure on the bark cover to mold it to the required form. Bulging of the bark at each frame was prevented by a thin plank sheathing. The ribs in many Eastern canoes were spaced so that on the bottom they were separated only by a space equal to the width of a rib.

Each piece of sheathing, better described as a "splint" than as "planking," was commonly of irreg-

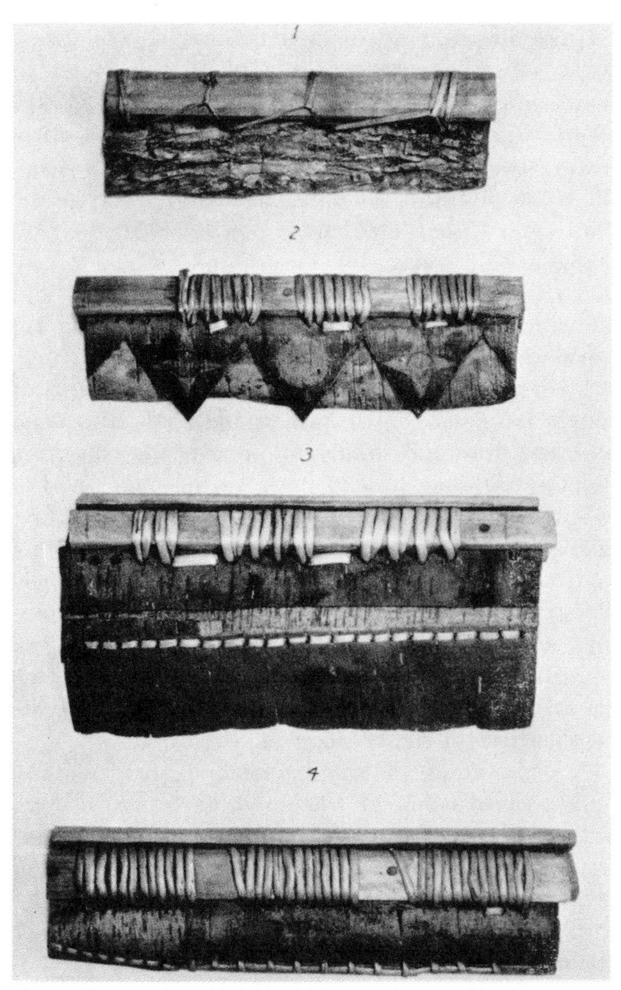


Figure 27

GUNWALE LASHINGS, examples made by Adney: 1, Elm-bark, Malecite; 2, St. Francis; 3, Algonkin; 4, Malecite.

ular form. The edges were often beveled to a marked thinness. While some builders laid the sheathing edge-to-edge in the bark cover, others overlapped the edges. Nearly all builders feathered the butts and overlapped them slightly. The sheathing was held in position by a number of light temporary ribs while the permanent frames, or ribs, were being installed. It is to be noted that the sheathing was neither lashed nor pegged; it remained fixed in place only through the pressure of the bent ribs and the restraint of the bark skin.

The exact method of fitting the sheathing varied somewhat from area to area, but not in every instance from tribe to tribe. The bottom sheathing used by some eastern Indians was in two lengths. The individual pieces were tapered toward the stems and the edges butted closely together. The sides were in three lengths, but otherwise similarly fitted. The butts lapped very slightly. In a second method, used to the westward, the sheathing was laid edge-to-edge in two lengths, with the butts slightly lapped. The center members of the bottom, usually five, were parallel-sided, but the outboard ends of those at the turn of the bilges were beveled, or snied, off. The members further outboard were in one length, with both ends snied off. The bottom thus appeared as an elongated diamond-form. The topside sheathing was fitted as in the first instance.

A variation in the second style used three lengths in the centerline sheathing. In still another varia-

Figure 28

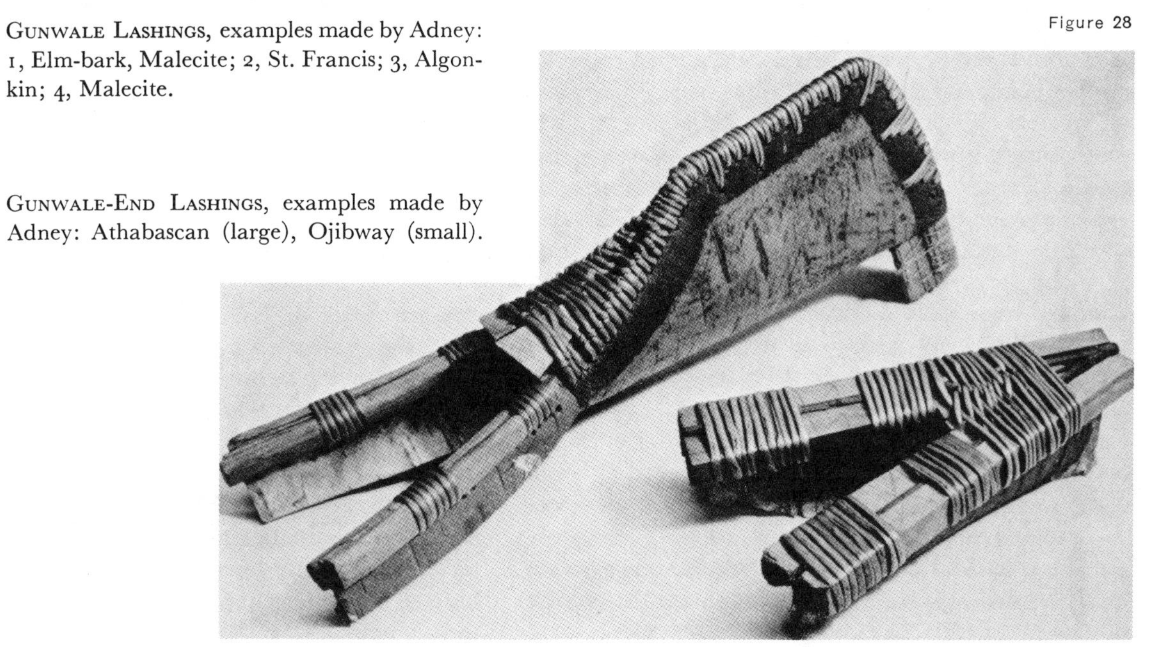

GUNWALE-END LASHINGS, examples made by Adney: Athabascan (large), Ojibway (small).

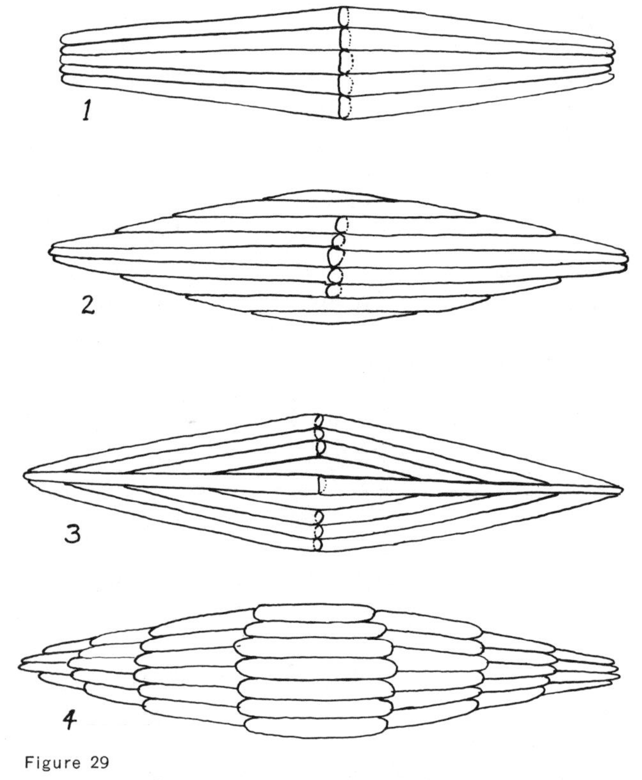


Figure 29

SPLINTS ARRANGED in various ways to sheath the bottom of a canoe: 1, Micmac, Malecite; 2, Central Cree, Têtes de Boule, etc.; 3, Montagnais; 4, Algonkin, Ojibway, etc.

tion a centerline piece was laid in two lengths without taper, the next outboard piece was then cut in the shape of a broad-based triangle, and the rest were laid in two lengths, with the sides parallel to the sides of the triangular strake and with their ends snied off against the centerline pieces. In a fourth style short pieces, roughly elongate-oval in shape, were overlapped on all sides and laid irregularly so that when in place they appeared "thrown in." With this style, the midship section was laid first and secured by a temporary rib, then the next toward the ends, with the butts shoved under the ends of the middle section. The next series was similarly laid so that the top member of each butt-lap faced toward the ends of the hull and was under a rib. The ends were not cut square across, but were either blunt-pointed or rounded. Five lengths of sheathing were often used, and the widths of the individual pieces of sheathing were rarely the same, so the seams were not lined up and presented an irregular appearance in the finished canoe. The sheathing was thin enough to allow it to take the curve of the bilge easily.

If the sheathing was lapped, the overlap was always slight. In some old canoes a small space was left between the edges of the sheathing, particularly in the topsides. In some northwestern bark canoes there was no sheathing; these used a batten system somewhat like that in the Eskimo kayak, except that in the bark canoes the battens were not lashed to the ribs, being held in place only by pressure. These kayak-like bark canoes had a bottom framework formed with chine members; some had a rigid bottom frame of this type, while others had bottom frames secured only by rib pressure. The purpose of the sheathing, it should be noted, was to protect the bark cover from abrasion from the inside, to prevent the ribs from bulging the bark, and to back up the bark so as to resist impacts; but in no case, even when battens were employed, as in the Northwest, did the sheathing add to the longitudinal strength of the bark canoe. The principle of the stressed rib and clamped sheathing, which is the most marked characteristic in the construction of the North American Indian bark canoe, is fundamentally different from that used in the construction of the Eskimos' skin craft.

A wide variety of framing methods are exhibited in the construction of the ends, or stems, of bark canoes. In the temporary types of the East, the bark was trimmed to a straight, slightly "ram" form and secured by sewing over two battens, one outboard on each side. Birch-bark canoes of the East usually had an inside stem-piece bent by the lamination method to the desired profile, the heel being left unsplit; as usual, the laminations were spirally wrapped, often with basswood-bark thongs. The stem-piece was then placed between the bark of the sides, and the bark and wood were lashed together with an over-and-over stitch. Sometimes variations of the short-and-long form of stitch were used here, and some builders also placed a halved-root batten over the ends of the bark before lashing to form a stem-band as protection to the seam. In some canoes the end lashing passed through holes drilled in the stem-pieces, often with the turns alternating in some regular manner through and around the stem-piece.

The stem-pieces were generally very light, and in some canoes the head was notched and sharply bent down and inboard, so that it could be secured to the ends of the gunwales. Some tribal types had no inner stem-piece, and the stem profiles were strengthened merely by the use of two split-root or halved-sapling battens, one on each side, outside the bark and under the sewing.

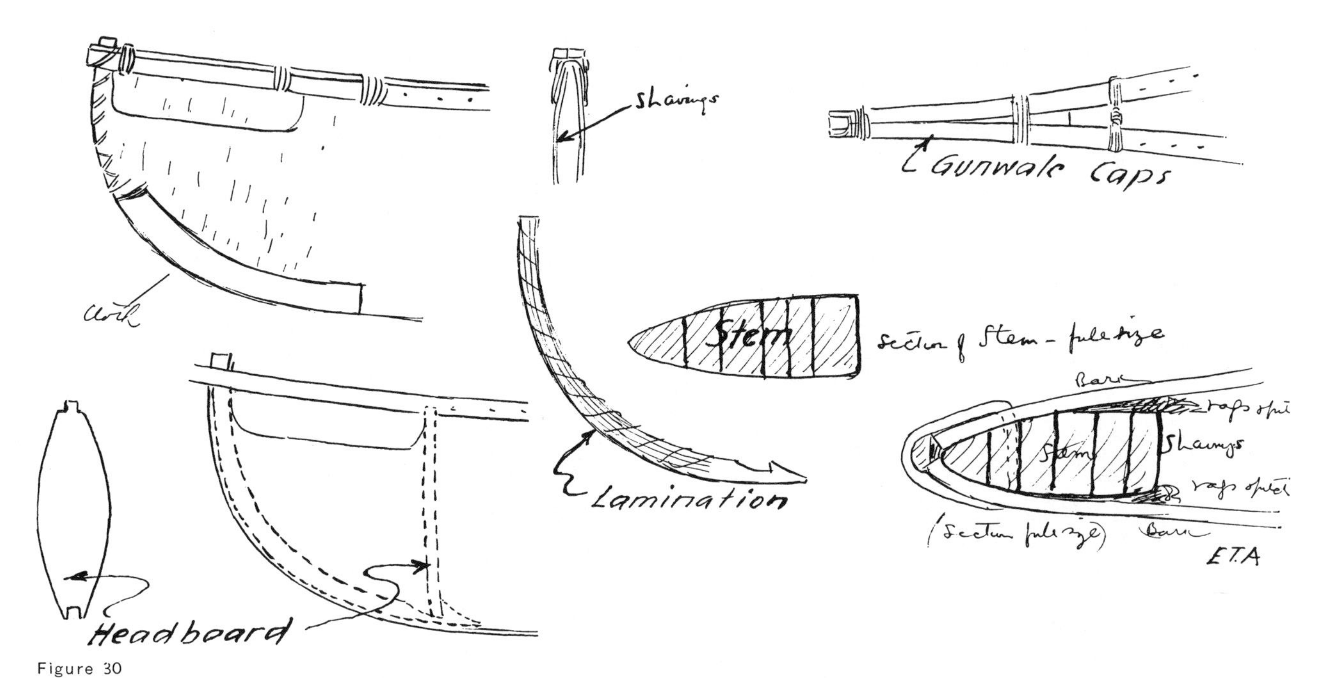


Figure 30

END DETAILS, INCLUDING CONSTRUCTION OF STEM-PIECES and fitting of bark over them, ending of gunwale caps at stem heads, and the headboard, with its location. Lamination of the stem pieces shows fewer laminae than is common. (*Sketches by Adney.*)

Birch-bark canoes to the westward used battens under the end lashing as well as rather complicated inside stem-pieces. In some parts of the West and Northwest, the ends were formed of boards set up on edge fore-and-aft, the bark being lashed through all, with the boards projecting slightly outboard of the ends of the bark cover to form a cutwater.

To support the inside stem-piece, some form of headboard was usually fitted near each end after the sheathing was in place. These were shaped to the cross-section of the canoe so as to form bulkheads. In some canoes, these miniature bulkheads stood vertical, but in others they were curved somewhat to follow the general curve of the end-profile, and this caused them to be shaped more like a batten than a bulkhead. Bent headboards were sometimes stepped so as to rake outboard. Sometimes the form of the headboard permitted the gunwale members to be lashed to it, and often there was a notch for the main gunwale on each side.

The headboards were sometimes stepped on the unsplit heel of the stem-piece; a notch was made in the bottom of the headboard to allow this. In two types of canoe in which there was no inner stem-piece, the headboards were stepped on short keel pieces, or "frogs," fore-and-aft on the bottom and extending slightly forward of the end of the sheathing to reinforce the forefoot. The purpose of the headboard was to strengthen the stem-piece, and in many cases it was an integral member of the end structure itself and helped to maintain its form. The headboard usually served to support the gunwale ends in some manner, it stretched the bark smooth near the stems, and it secured the ends of the sheathing where support from a rib would have been most difficult to obtain. Many canoes had the space between the headboard and the stem-piece stuffed with shavings, moss, or other dry material to help mold the bark to form beyond the sheathing in the ends. Some tribal groups decorated the headboards.

In a few canoes, the stem-piece was additionally supported by a short, horizontal member stepped in the forward face of the headboard and projecting forward to bear on the after side of the stem-piece. The latter was sometimes bent back onto itself above this member to form a loop around the top of the end-profile, and the gunwale ends or a part of the gunwale structure were secured to it. This complicated bending of the stem-piece, in conjunction with use of a headboard and a brace member, served to stiffen the end structure sufficiently to meet the requirements of service.

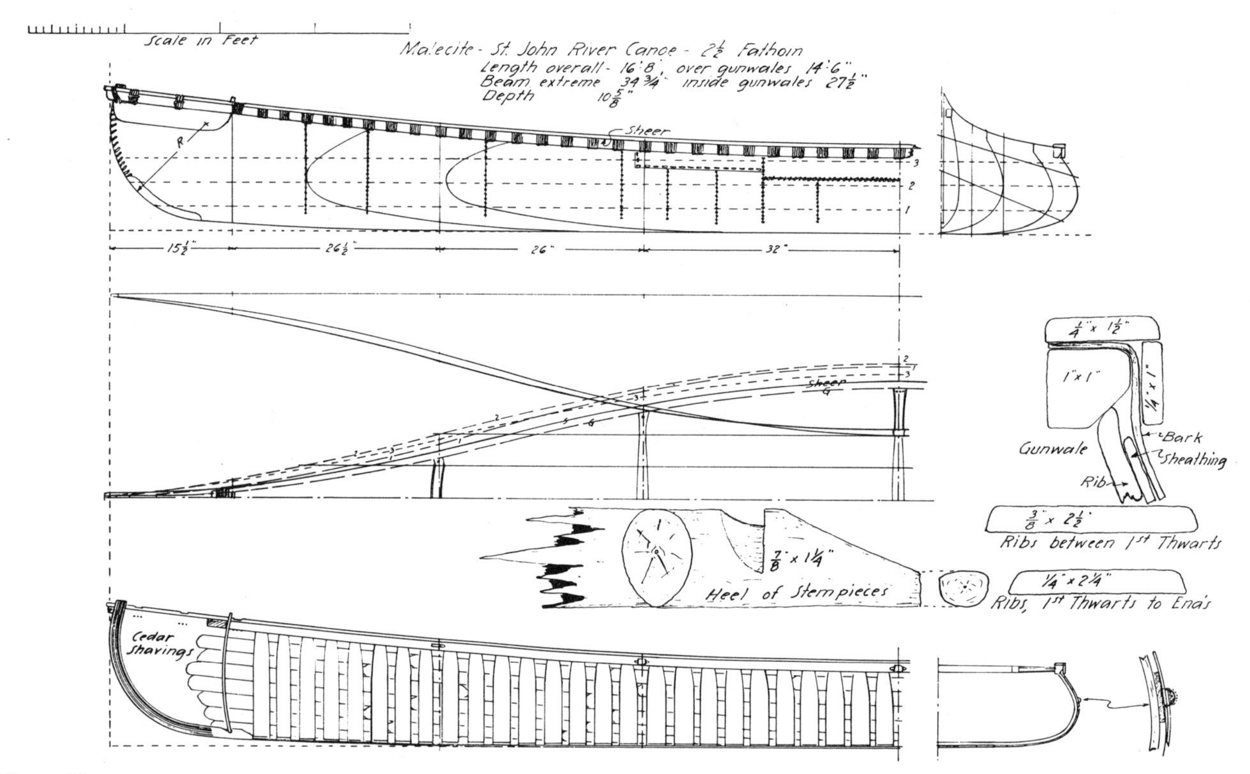


Figure 31

Malecite Canoe of the Type Described in This Chapter. This 2½-fathom St. John River canoe represents the last Malecite birch-bark model, and usually was fastened with tacks and nails, rather than with root lashings and pegs as described here.

The use of a bark cover over the gunwale ends has already been mentioned. In some eastern canoes, this was placed under the cap and outwale pieces and extended below the latter in a shallow flap on which the owner's mark or other decoration might appear; the flap was in fact a kind of name board. Such flaps do not appear on the partly decked bark canoes of the Northwest.

This general description of the structure of the bark canoes is sufficient to permit the explanation of the actual construction of a bark canoe to be more readily understood, and it also serves to illustrate the close connection between the method of construction and the formation of the lines, or model, of bark canoes. From the description, too, it can be seen that while the shape of a bark canoe was partially planned during the construction the control of every part of the model could not be maintained with the same degree of precision as in the building of an Eskimo skin boat or an Indian dugout.

## *Construction*

One aspect of canoe construction, the Indian method of making measurements, was briefly mentioned (p. 8) under a discussion of the origin of the measurement known in French Canada as the *brasse.* This was the distance from finger-tip to finger-tip of the arms out-stretched; in the fur trade in English times it was known as the fathom and it appears to have been about 64 inches, or less than the nautical fathom of 6 feet. Other measurements used were the greatest width of the ball of the thumb, which is very close to an English inch, and the width of the four fingers, each finger-breadth being close to three-fourths of an English inch. The length of the forearm, usually from the knuckles of the clenched hand to the elbow, was also employed by some Indians, as a convenient measurement.

Measurements in these units might be memorized and used in building, but many Indians used measuring sticks, and these served as "footrules." They were sometimes squared and were painted as well as notched.

A Malecite Indian, interviewed in 1925, had three such sticks for canoe building. One, for the length of the gunwale frame, was half the total length required; it was notched to show the distance at which the ends of the gunwales were lashed and also the position of the thwarts. Such a stick would be about 7 feet long for a 16-foot canoe, 8 feet for an 18-foot canoe. The second stick was notched to show half the length of each of the thwarts. The third stick had notches showing the height of the gunwale at each thwart and at the end, four notches in all for the half-length of the canoe. This stick measured from the surface of the building bed, not from a regular base line.

The method of measuring canoes appears to have been fairly well standardized, at least in historical times. As stated earlier, length was commonly taken over the gunwales only, and did not include the end profiles, which might extend up to a foot or slightly more beyond the gunwale ends, bow and stern. However, in certain old records the overall length is given, and in various areas other methods of measurement existed. Where a building frame was used, the given length of the canoe was the length of this frame; usually this approximated the length of the gunwales. The width of a canoe was measured by the Indian from inside to inside of the main gunwale members. The extreme beam might be only 2 or 3 inches greater than the inside measurement of the gunwales, but if the sides bulged out, the beam might actually be 6 or more inches greater. The depth was usually measured from the inside of the ribs to the top of the gunwale but in building it was measured from the surface of the building bed to the bottom of the main gunwales, as noted above in the description of the measuring sticks.

Thus it will be seen that the Indian measurements constituted a statement of dimensions primarily useful to the builder, for their main purpose was to fix the proportions rather than establish the actual length, width, and depth. Today we state the length of a canoe in terms of extreme overall measurement; the Indian was inclined to state the length in building terms, giving dimensions applicable to the woodwork only, just as the old-time shipbuilder gave the keel length of a vessel instead of the overall length on deck.

The building site was carefully selected. The space in which the canoe was to be set up had to be smooth, free of stones and roots or anything that might damage the bark, and the soil had to be such that stakes driven into it would stand firmly. A shady place was preferred, as the bark would not dry there as fast as in sunlight. Since the construction of a canoe required both time and the aid of the whole Indian family, the site had to be close to a suitable place for camping, where food and water could be obtained. It is not surprising, therefore, to find canoe building sites that apparently had been used by generations of Indians.

The preparation of the building bed was controlled by the intended form of the canoe to be built. If the bottom of the canoe was to be rockered, the cleared ground was brought to a flat surface for the length required for setting up the canoe. If the rocker was to be great, the middle of the bed would be slightly depressed. If the bottom was to be straight fore-and-aft, or very nearly so, the bed was crowned from 1½ to 2 inches higher in the middle than at the ends, so that the canoe was first set up with a hogged bottom. Very large canoes such as were used in the fur trade required as much as 4 inches crown in the building bed. Other dimensions being equal, the amount of crown was usually somewhat greater in canoes having bulging sides than in ones having more upright or flaring sides. Canoe factories such as were operated in certain fur-trading posts sometimes had a plank building bed suitably crowned and drilled for setting the stakes.

Two methods of setting up the canoe were used. In most of the eastern area, the gunwales were put together and used to establish the plan outline of the canoe on the building bed. But a building frame was used for constructing the various narrow-bottom canoes having flaring sides, and for some other tribal forms. The frame, made in the same general form as the gunwales when assembled, but less wide and sometimes much shorter, could be taken apart easily, allowing it to be removed after the canoe was built; hence it could be used to build as many canoes as desired to the same dimensions as the first, and was retained by the builder as a tool, or pattern, for future use.

The method of construction in which gunwales only were used in setting up the canoe will be explained first in order to show the general technique of construction. Use of the building frame will then be described. Important deviations from these methods will be described in later chapters under the individual tribal types in which they occur.

The Malecite canoe, a straight-bottomed craft about 19 feet long and 36 inches beam, is used as the example, hence the method of building to be described is that generally employed in the East, where variations in construction mainly involve the use or omission of structural elements.

The gunwales are the first members to be formed. In the Malecite canoe these are the inner gunwales, as the canoe will have outwales and caps. The gunwales are split from white cedar to produce battens that will square 1½ inches when shaped. The gunwales are tapered each way from midlength, where they are 1½ inches square, to a point 3 inches short of the ends, where they are ¾ by 1 to 1¼ inches. The edges of the gunwales are all rounded, and the outboard bottom edge is beveled almost ½ inch, at 45° to the bottom of the member. The last 3 inches at each end is formed like half a blunt arrowhead, as shown in the sketch of the member on page 31. The gunwales will be bent, side to side, on the flat as far as the ends are concerned, so the blunt arrowhead is formed on one of the wide faces of the ends as shown. The arrowhead form allows a neat joint when the gunwale ends are brought together, pegged athwartships, and then wrapped with a root lashing. In forming and finishing the gunwales, a good deal of care is required to get them to bend alike, so that the centerline of the finished frame will be straight and true.

To take the ends of the middle thwart, a mortise ¼ by 2 inches is cut in each gunwale member athwartships at exactly midlength, the length of the mortise being with the run of the gunwale. In it, the middle thwart, 33 inches long, is fitted. Made of a ⅞-inch by 3-inch piece of hard maple, the thwart tapers slightly in thickness each way from its center to within 5 inches of the shoulders, which are 30 inches apart. The thickness at a point 5 inches from the shoulder is ¾ inch; from there the taper is quick to the shoulder, which is 5⁄16 inch thick, with a drop to ¼ inch in the tenon. The width, 3 inches at the center, decreases in a graceful curve to within 5 inches of the shoulder, where it is 2 inches, then increases to about 3 inches at the shoulder. The width of the tenon is, of course, 2 inches, to fit the mortise hole in the gunwale. The edges of the outer 5 inches of the thwart are rounded off or beveled a good deal; inboard they are only slightly rounded.

The thwart is carefully fitted to the gunwale members and the ends are pegged. Some builders wedged the ends of this thwart from outside the gunwales, the wedge standing vertical in the thwart so that the gunwale would not split; however, it is not certain that wedging was used in prehistoric times, although it is seen in some existing old canoes. The pegs used in this canoe are driven from above, into holes bored through the gunwale and the tenon of the thwart to lock all firmly together. Three holes are then bored in the broad shoulders of the thwart about 1½ inches inboard of gunwale for the root lashing that is also used.

The ends of the gunwale members are now brought together, and to avoid an unfair curve appearing at the thwart in place, short pieces of split plank or of sapling, notched to hold them in place, are inserted between the gunwale members as temporary thwarts at points about 5 feet on each side of the middle thwart. After the ends are brought together and the final fitting is carried out, a peg is driven athwartships the ends and a single-part root lashing is carefully wrapped around the assembly.

Some canoe builders omitted the blunted half-arrowhead form at the gunwale end. Instead, the inside faces were tapered to allow the two parts to bear on one another for some distance. The gunwales were then pinched together and lashed with one or more wrappings. Finally, a thin wedge was sometimes driven from inboard between the two gunwale ends to tighten the wrappings. The wedges were usually so carefully fitted as to be difficult to identify. It is probable that this wedged gunwale ending represents the prehistoric form, and the blunted half-arrowhead ending is a result of the use of steel tools.

After the ends of the gunwales have been securely fastened together, the first pair of permanent thwarts is fitted. These are located 36 inches, center to center, on each side of the middle thwart, a distance that determines the centers of the mortises in each gunwale member. Each thwart, made from a ¾-inch by 3-inch piece, tapers smoothly in thickness from the ¾-inch center to the 5⁄16-inch shoulder. The tenon is of the same dimensions as that of the middle thwart, the width takes the same form as that of the middle thwart, and the edges are similarly beveled and rounded. The distance between the shoulders, taken along the centerline, is 22½ inches, and the centerline length of the thwart 25½ inches. However, the shoulders and ends of the tenons must be bevelled to follow the curve of the gunwales hence the extreme length of the thwart is actually very close to 26 inches. The worker determines the bevel of the shoulders by fitting the thwart to the run of the gunwales, the

Figure 32

MALECITE CANOE BUILDING, 1910. (*Canadian Geological Survey photos.*)

Weighting gunwales on bark cover on building bed.

Resetting stakes.

Shaping bark cover and securing it to stakes.

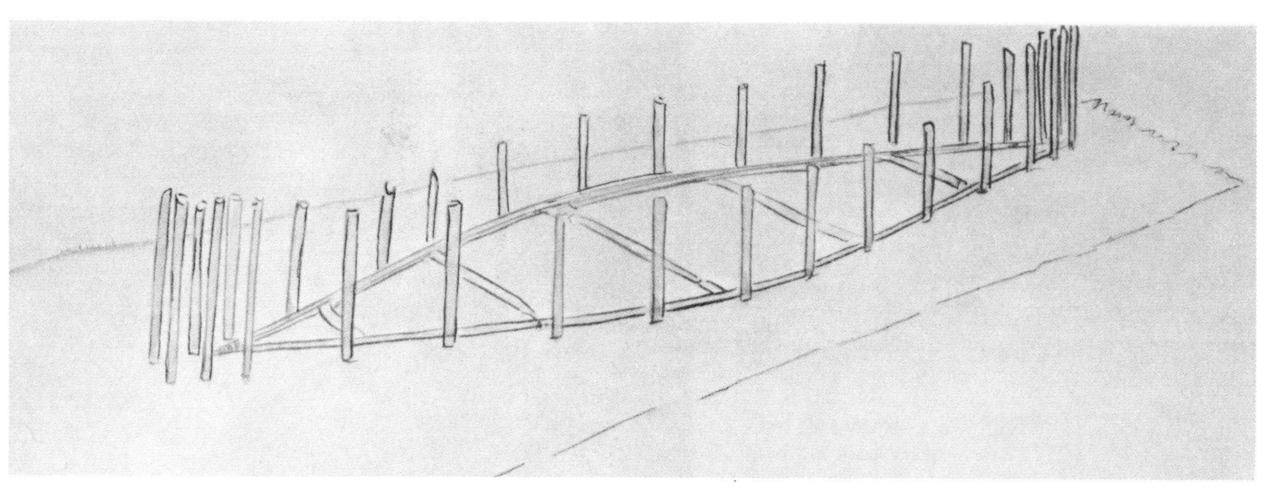

Figure 33

FIRST STAGE OF CANOE CONSTRUCTION: assembled gunwale frame is used to locate stakes temporarily on building bed. Instead of the gunwales, a building frame was used in some areas. (*Sketch by Adney.*)

temporary thwarts being shifted so that the distance between the gunwales equals that set by the measuring stick. These two thwarts having been fitted, the tenons are pegged as before, but in the shoulders only one lashing hole is bored instead of the three employed in the middle thwart.

The second pair of thwarts is placed 30 inches, center to center, from the first pair, one at each end, and on the basis of this measurement the tenons are cut as for the others. These two thwarts are made of ⅝- by 4-inch pieces tapering in thickness each way from the center to the shoulder, where they are a scant 5/16 inch thick, the tenons having the same dimensions as in the other thwarts. In width the thwarts are worked to an even 3 inches from shoulder to shoulder, but in the form of a curve so that when each thwart is in place its center will be bowed toward the ends of the canoe, viewed from above. As in the first pair, the shoulders and ends are cut to a bevel to fit the gunwale; at the centerline they each measure 12 inches shoulder-to-shoulder in a straight line athwartships and 15 inches end-to-end. Allowing for bevel, the maximum length is just over 15 5/16 inches. These thwarts are drilled for single gunwale lashings and the corner edges are well rounded from shoulder to shoulder. The distance from the centerlines of these last thwarts at the bow and stern to the extreme ends of the joined gunwales is 33 inches, so the finished gunwale length is 16 feet.

After the endmost thwarts are pegged into place, the temporary stays are removed. At each step of construction the alignment of the gunwales is checked by measuring with the measuring sticks and by sighting, since the shape of the assembled gunwales, in this case of the inner gunwales, is very important in determining the sharpness of the completed canoe and the fairness of its general form.

The assembled gunwales are now ready to be laid on the building bed which, for the Malecite canoe, is 20 feet long, about 3½ feet wide, and is raised about 1½ inches at midlength so that the canoe bottom will be straight when the craft is in the water. The gunwale frame having been carefully centered on this bed, with the middle thwart exactly over the highest point in the surface of the bed, some scrap splitplanking is laid across the gunwales and the whole weighted down with a few flat stones. Next, 34 stakes from 30 to 50 inches long are prepared, each made of a halved length of sapling. Around the outside of the gunwale frame 26 of these are driven in pairs opposite one another across the frame, about 24 inches apart and placed so that none is opposite a thwart, except for the stakes at the extreme ends of the gunwale frame, which are spaced about a foot from their nearest neighbors and are face-to-face, about 1½ inches apart. All the stakes are driven with the flat face about an inch from the gunwale frame and parallel to its outside edge. Finally two more pairs of stakes are driven at each end, the

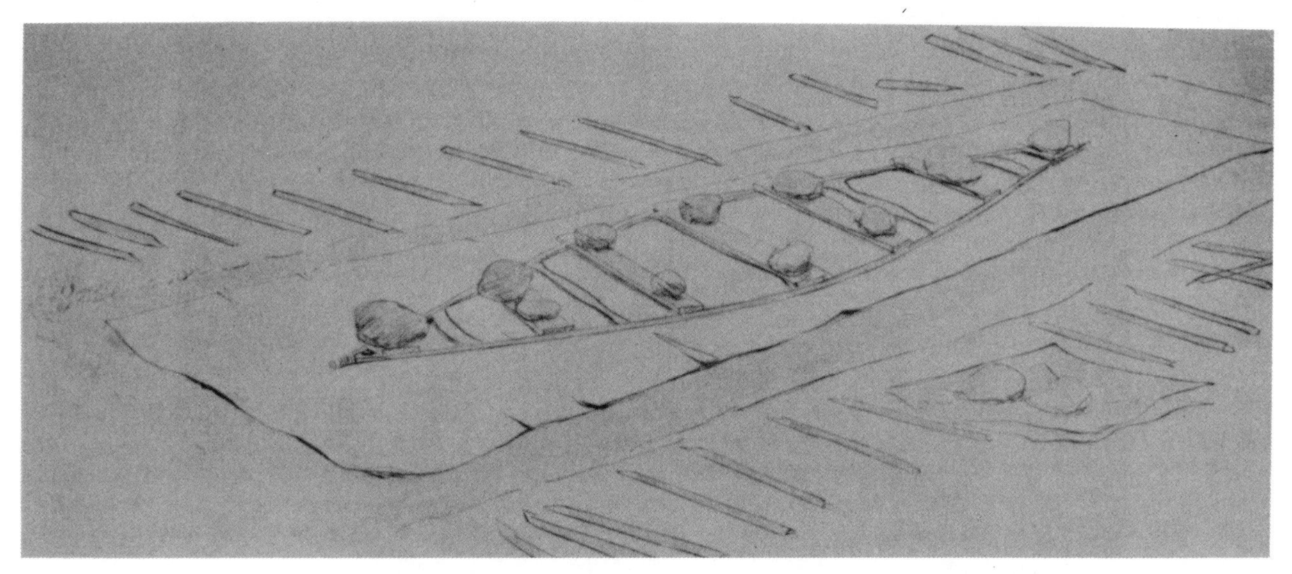

Figure 34

SECOND STAGE OF CANOE CONSTRUCTION: stakes have been removed and laid aside, and the gunwales shown in first stage have been removed from the building bed. The bark cover is laid out on the building bed, and the gunwales are in place upon it, weighted down with stones. (*Sketch by Adney.*)

first pair about a foot beyond the end of the gunwale frame and 1½ inches apart, the second about 6 inches beyond these and similarly spaced. The length between the outermost stakes, measured over the gunwale frame, is about 18½ feet. Great care is taken to line up the last pairs of stakes with the centerline of the gunwale frame.

If the canoe is to have a slight rocker near the ends and is to be straight over the rest of the bottom, the ends of the gunwale frame will be blocked above the building bed so that the frame is not hogged on the bed.

After the builder is satisfied with the staking, each stake is carefully pulled up and laid to one side, off the bed but near its hole. The weights are then removed from the gunwale frame, which is lifted from the bed and laid aside, and the bed, if disturbed is repaired and re-leveled.

The roll of birch bark is now removed from storage, perhaps in a nearby pool where it has been placed to keep it flexible, and unrolled white side up on the building bed. As the bark dries, it will become more and more stiff, so it will be necessary to moisten it frequently during construction to maintain its flexibility.

The bark is usually long enough, but often it is not wide enough. If the bark is too short, it may be pieced out at this time, or later. If it is not wide enough it is centered on the bed; the piecing out will be done later. The gunwale frame is now laid on the bark, care being taken to place it as nearly as possible in its former position on the bed.

The bark outside the frame is then slashed from the edge to a point close to the end of each thwart, and also to points along the frame halfway between the thwarts, so that the edges can be turned up. While it is being slashed, the bark cover is bent slightly, so that is is cut under tension. Later, when the required shape can be determined, these slashes will be made into gores, the Malecite canoes having flush seams, not overlaps, in the topsides and bottom. If a fault is noted along the outer edge of the bark, a slash may be placed so as to allow the fault to be cut out in the later goring; irregularity in the position of the cuts does no great harm to the progress of building these canoes. The slashes are usually carried to within an inch of the gunwales on the bed. It is not customary to slash the bark close to the end, there the bark can usually be brought up unbroken, depending upon the form of the end.

When the bark has been cut as described, it can be turned up smoothly all around the frame so that the stake holes can be seen and a few of the stakes can be replaced. The frame and the bark are then

Figure 35

Malecite Canoe Builders Near Fredericton, N.B., using wooden plank building bed with stakes set in holes in the platform. This was a late method of construction, which probably originated in the early French canoe factory at Trois Rivières, Que.

realigned so that all stakes may be replaced in their holes without difficulty. When the frame and bark are aligned, the frame is weighted as before and the bark is turned up all around it, the stakes being firmly driven, as this is done, in their original holes. The longest stakes are at the ends of the frame, as the depth of the hull is to be greatest there. The tops of each pair of opposite stakes are now tied together with a thong of basswood or cedar bark, to hold them rigid and upright.

After the bark is turned up around the frame, its lack of width becomes fully apparent. At this stage, some builders fitted the additional pieces to gain the necessary width; others did it later. The method of piecing the bark cover and the sewing technique, however, is explained here.

The bark is pieced out with regard to the danger of abrasion that would occur when the canoe is moving through obstructions in the water, or when it is rolled or hauled ashore and unloaded. If the bark is to be lapped below the waterline, the thickness of the bark of both pieces in the lap is scraped thin so a ridge will not be formed athwart the bottom; here, however, most tribes used edge-to-edge joining. If there are laps in the topsides, the exposed edge is toward the stern; if in the midlength, upward toward the gunwale; and if it is in the end the lap may be toward the bottom, because this makes it easier to sew, and because in the ends of the canoe there is less danger of serious abrasion. Many tribes used edge-to-edge joining everywhere in the topsides so that the direction of lapping was not a matter of consideration. The type of goring, whether by slash and lap or by cutting out a V-shaped gore, will, of course, have much to do with the selection cf the method of sewing to be used.

It is to be recalled that in canoe building no needle was used in sewing the bark; the ends of the root strands were sharpened and used to thread the strand through the awl holes. Much of the topside sewing in a bark canoe was done with small strands made by splitting small roots in half and then flattening the halves by scraping. Large root strands quartered and prepared in the same manner, or the cores of these, were sometimes used in heavy sewing or lashing at the gunwale or in the ends of a canoe.

As noted previously, root thongs were used well water-soaked or quite green, for they became very stiff and rather brittle as they dried out. Once in place, however, the drying did not seem to destroy their strength. Rawhide was also used for such sewing by some tribes.

The sewing was done by Indian women, if their help was available, and the forms of stitching used in canoe building varied greatly. The root sewing at the ends of the canoes ranged from a simple over-and-over spiral form to elaborate and decorative styles. Long-and-short stitching in a sequence that usually followed some formal pattern was widely used. Among the patterns were such arrangements as one long, four short, and one long; or two longs, two or three shorts, and two longs; or one short, five of progressively increasing length, and then one short; or six progressively longer followed by six progressively shorter. Cross-stitching, employing the two ends of the sewing root as in the lacing of a shoe was also common. Sometimes this was combined with a straight-across double-strand pass to join the ends of the X. The harness stitch, in which both ends of the sewing root were passed in opposite directions through the same holes, was often used, as was the 2-thong in-and-out lacing from each side used in northwestern canoes having plank stem-pieces.

If the root strand was too short to complete a seam, instead of being spliced or knotted the end was tucked back under the last turns or stitches, on the inside of the bark cover. In starting, the tail was placed under the first turn of the stitch, so that it could not be pulled through. To finish sewing with double-ended strands, as in the harness stitch, both ends were tucked under the last turn or two.

Commonly two or more turns were taken through a single hole in the bark; this might be done to clear some obstruction such as a frame head at the gunwale, or to provide a stronger stitch, or turn, as in the harness stitch and others, or to allow for greater spacing between awl holes in the bark. (Since the awl blade was tapered, the size of the hole it made in the bark could be regulated by the depth of penetration of the blade as it was turned in the hole.)

The length of stitches varied with the need for strength and watertightness. Long stitches were about 1 inch, short stitches from about ⅜ to ½ inch in length. The run of the grain, of course, was a consideration in the length of stitch used.

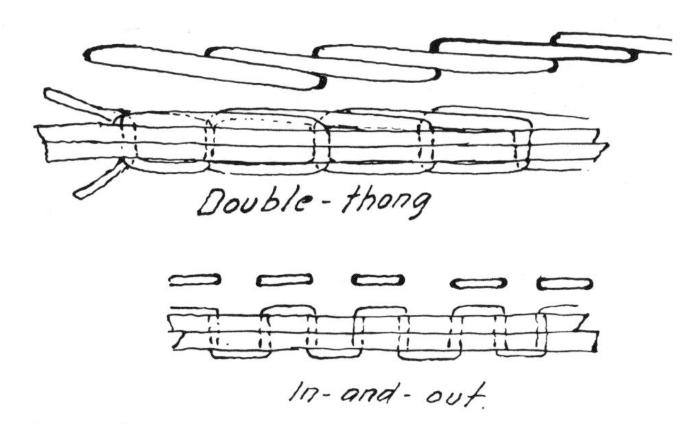


Figure 36

SEWING: two common styles of root stitching used in bark canoes.

The piecing of the side panels was done with a great variety of sewing styles, according to strength requirements. The strain put upon the bark in molding it by rib pressure was greater in the midlength than in the ends; and the sewing differed accordingly. The over-and-over spiral, with a batten under the sewing, was used for sewing in the midlength, as was back-stitching, a variety of basting stitch in which a new pass is started about half way between stitches, thus forming overlapped passes or turns. Back-stitching was usually done in a direction slightly diagonal to the line of sewing, so as to cross the grain of the bark at an angle with each pass. The double-thong in-and-out stitch, in which each thong goes through the same hole from opposite sides, was frequently used. The simple, spiral over-and-over stitch was used in sewing panels in the ends of canoes, as was the simple, in-and-out basting stitch using either a single or double strand.

When the sides were pieced out edge-to-edge, the sewing was usually done spirally, over and over a narrow, thin batten placed outside the bark cover. This batten might be either a thin split sapling or, more commonly, a split and thinned piece of root. If the pieced-out sides were lapped, then the harness stitch was commonly used. The lap might be some inches wide to decrease the danger of splitting while the bark was being punched with the awl, afterward the surplus was cut away leaving about a half inch of overlap. On rare occasions the strength of a lapped-edge seam was increased by the use of a parallel row of stitching.

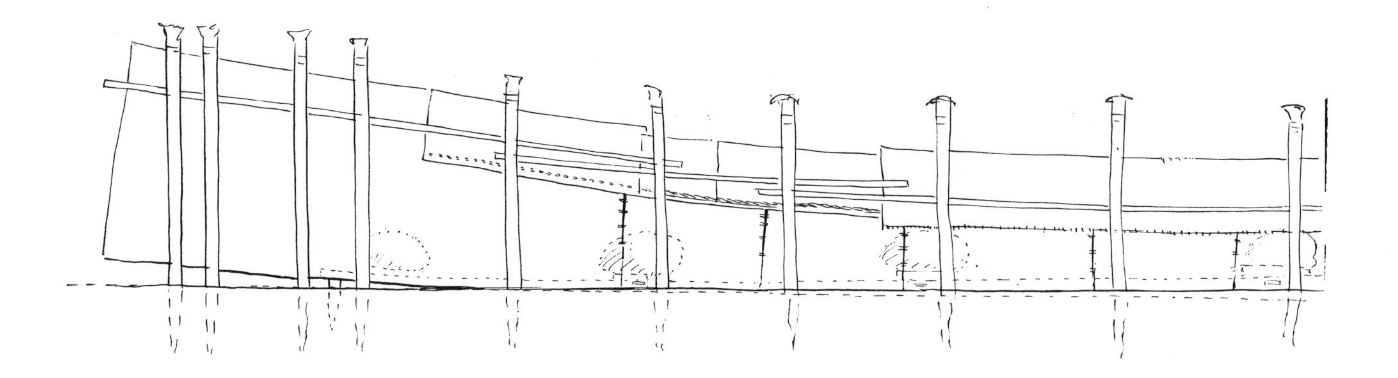

Figure 37

COMPARISON OF CANOE ON THE BUILDING BED (above), with gunwales or building frame weighted down by stones inside bark cover, and (below) canoe when first removed from building bed during fifth stage of construction. (*Sketches by Adney.*)

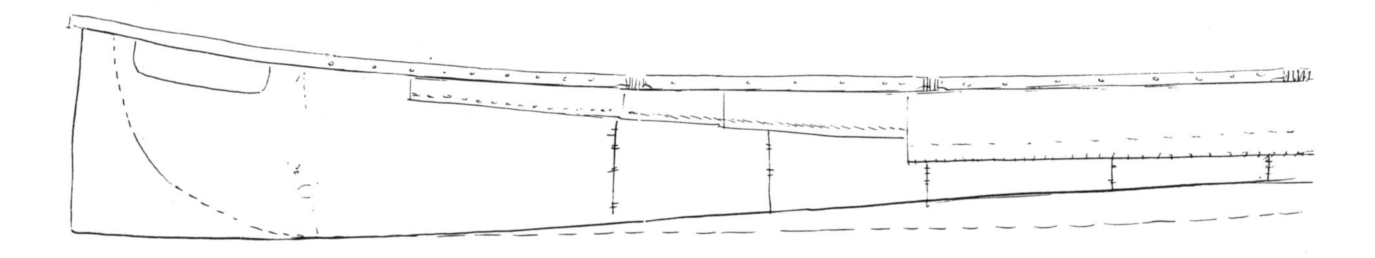

In making the canoe watertight, it is to be remembered that some forms of stitch make the bark lie up tight all along its edges while others bind only where the stitch crosses the seam. The in-and-out stitch, which was used only above the waterline, cannot be pulled up hard without causing the bark to pucker and split and cannot be made very watertight with gum. The over-and-over stitch, in either a spiral form or square across the seam on the outside and diagonally on the inside, is very strong; when a batten is used under the stitches it can be pulled up hard and allows a very watertight gumming. When this style of sewing is used without a batten across the run of the grain, as in the gore seams, it cannot be pulled up as hard, but will serve. Backstitching, which was much used in the topsides, can be pulled up quite hard and makes a tight seam when gummed, as do the harness stitch and cross-stitch. The ends, regardless of the style of sewing used, were more readily made tight by gumming than the other seams in a bark canoe.

Two basic methods, with some slight and unimportant variations, were used to fasten the bark to the gunwales. One employed a continuous over-and-over stitch, the other employed groups of lashings. On a canoe with the lashing continuous along the gunwales, the turns were made two or more times through the same hole on each side of each rib head to allow space for them. This might also be done where the lashing was in groups, as described above. Usually, a measuring stick was used to space the groups between thwart ends so that each group came between the rib heads. The groupings could be independent lashings, or the strand could be carried from one group to another. If the latter, it was passed along under the gunwale in a number of in-and-out stitches or in a single lone stitch either inside or out, or else it was brought around over the gunwale from the last full turn. Some tribes use both ends of the lashing, passing them through the same hole in the bark from opposite directions below the gunwales; the ends might be carried in the same manner in a

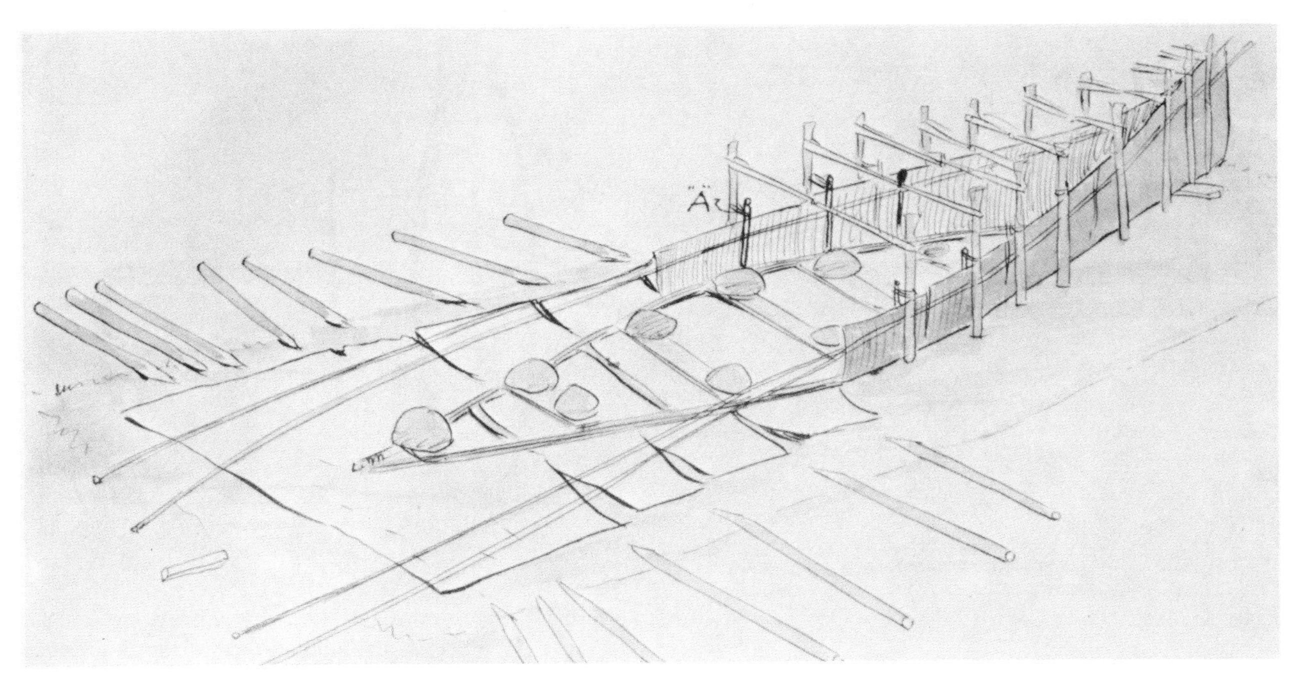


Figure 38

THIRD STAGE OF CANOE CONSTRUCTION: the bark cover is shaped on the building bed. The gores have been cut; part of the cover is shaped and secured by stakes and battens. "A" shows battens secured by sticks lashed to stakes. (*Sketch by Adney.*)

long stitch to the next group. In some elm and other bark canoes employing basswood or cedar-bark lashings the bark was tied with a single turn at wide intervals; when roots were used in these, however, small groupings of stitches were customary. When group lashings were used with birch bark, the intervals between groups was usually relatively short, though in a few canoes the groups and intervals were of nearly equal length.

In an independent group, the ends of the strand were treated as in whipping, the tail being under the first turns made and the end tucked back under the last—usually on the inside of the gunwales. Where there were inner and outer gunwales the lashing was always around both, and the tail might be jammed between them. If a cap was used on the gunwales, the lashings were always under it. The use of a knotted turn to start a lashing occurred only in the old Têtes de Boule canoes.

On the Malecite canoe, the sides are pieced out in one to three panels rather than in one long, narrow panel on each side. The panel for the midlength requires the greatest strength and is usually lapped inside the bottom bark. The latter is first trimmed straight along its edge, and the panel inserted behind it with a couple of inches of lap. Then the two pieces of bark are sewn together over a halved-root batten with an over-and-over stitch. (Other tribes used some form of the harness stitch, or a similar style, allowing great strength.) The middle panel does not extend much beyond the ends of the first pair of thwarts on each side of the middle. The next panels toward the ends are lapped outside the bottom bark and are sewn with the back-stitch. Then, if still another panel is required at each end, this too is lapped outside and is sewn in the lap with an in-and-out stitch. The ends of the panels are usually sewn with an over-and-over stitch that runs square with the seam outside and diagonally to it inside the bark. (The harness stitch was used here by some tribes, as were many forms of the cross-stitch.) The ends of the canoe and the gores have already been sewn during an earlier stage of the building process.

Once the sides are pieced out, the bark is ready to be turned up and around the gunwale frame and clamped perpendicularly. To effect this, small stakes are made by halving saplings, so that each half is about a half inch thick. The butt of each half is cut chisel-shaped, with the bevel on the flat side; the rounded face is smoothed off, and it may be tapered

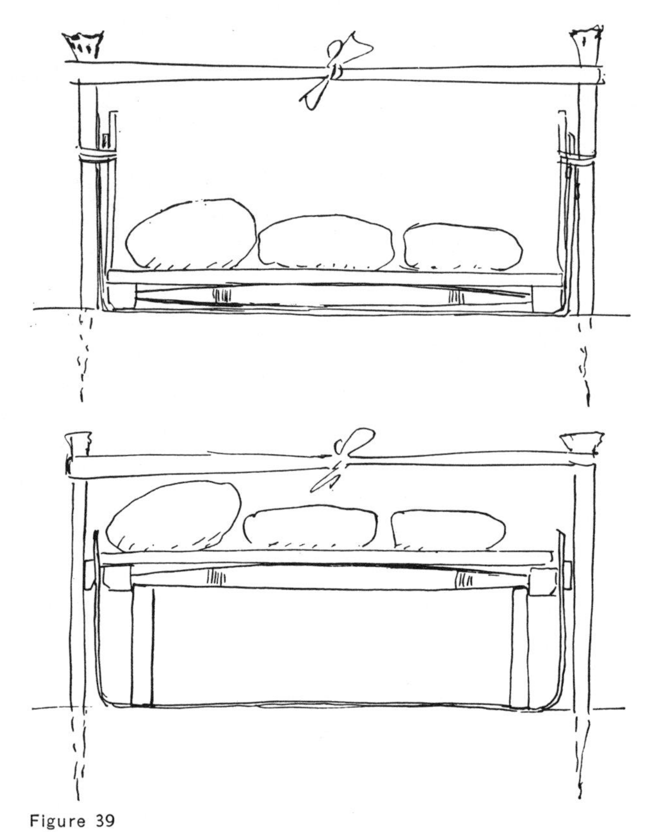

Figure 39

Cross Section of canoe on building bed during third stage of construction (above) and fourth stage. (*Sketch by Adney.*)

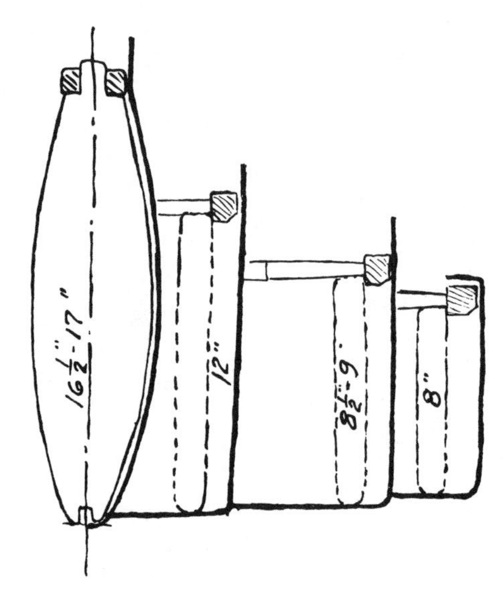


Figure 40

Multiple Cross Section through one side of a canoe on the building bed: at the headboard, middle, first, and second thwarts. Gunwale is raised and supported on sheering posts set under thwarts. Crown of the building bed is shown by varying heights of bottoms of the four sections.

toward the head of the stake. Between two of the slashes a length of bark is now brought up against the outer stakes; against the bark the small, inside stake is placed with the round face of the chisel-pointed butt wedged against the outer face of the gunwale. The top is then levered against the outside stake, so that the flat face of each clamps the bark in place. The top of the inner stake is then bound to the outer.

In setting the inside stakes, care is taken that their points do not pierce the bark. No inside stakes are required at the ends, as here the outside stakes are so close together in opposing pairs as to hold the bark in a sharp fold along the centerline of the cover. This of course is also true of the stakes beyond the ends of the gunwales.

After a few lengths of bark have been thus secured, they are faired between the stakes by inserting thin strips of split sapling, or battens of wood or root, along each side of the bark, under the inside and outside stakes. These battens are placed about half-way up the upturned bark. Some builders used long wooden battens, as this gave a very fair side when enough lengths were secured upright; others got the same results with short battens, the ends of which were overlapped between a pair of stakes on each side.

When the bark has been turned up and clamped, the gores may be trimmed to allow it to be sewn with edge-to-edge seams at each slash. This is usually done after the sides are faired, by moving the battens up and down as the cuts are made, then replacing them in their original position. The gores or slashes, if overlapped, are not usually sewn at this stage of construction.

With the inside stakes in place, the longitudinal battens secured, and the gores cut or the overlaps properly arranged, all is ready for sheering the gunwales. First the weights are removed from the gunwale frame so that it can be lifted. If the inside stakes have been properly made and fitted this can be done without disturbing the sides, though the ties across each pair of outside stakes may have to be slacked off somewhat. Before lifting the frame, some short posts, usually of sapling or of waste from splitting out the gunwales and thwarts, are cut in

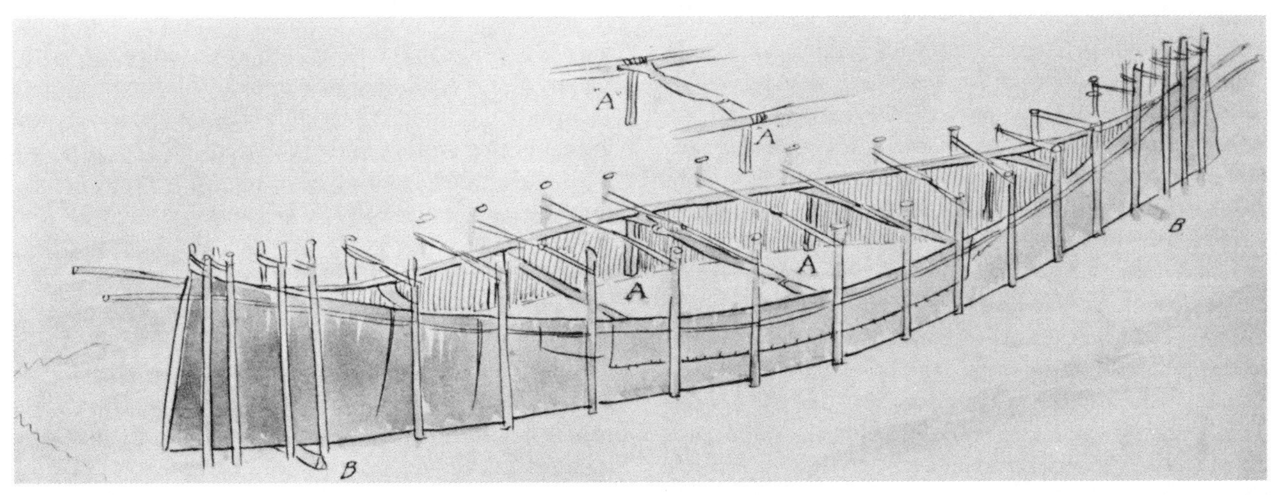


Figure 41

FOURTH STAGE OF CANOE CONSTRUCTION: bark cover has been shaped and all stakes placed. The gunwales have been raised to sheer height; "A" indicates the sticks which fix the sheer of the gunwales; "B" indicates blocks placed under ends to form rocker. Side panels are shown in place, and cover is being sewn to gunwales. (*Sketch by Adney.*)

lengths determined by the measuring stick or from memory, one for each end of each thwart, and one for each end of the gunwale frame. Those under the middle thwart ends in this canoe are 7½ inches long, those under the next thwarts out from the middle will be 9 inches, those under the end thwarts will be 12 inches, and those at the gunwale ends will be 17 inches long. These posts, cut with squared butts, are laid alongside the bed. The gunwale frame is now lifted and the pair of posts to go under the middle thwart are stepped on the bark cover, the gunwale is lowered onto them, and while the frame and posts are held steady, stones are laid on a plank over the middle thwart. Next, the ends of the gunwales are held and lifted so that a pair of posts can be placed at the thwarts next out from the middle. More weights are placed over these, the operation is repeated for the end thwarts and, finally at the gunwale ends, so that the gunwales now stand on posts on the bark cover, sprung to the correct fore-and-aft sheer and steadied by the bearing of the outside of the gunwale frame on the rounded faces of the inside stakes. Now the sheer has been established and the depth of the canoe is approximated.

To protect the bark cover from the thrust of the weights used to ballast the frame, some builders inserted small bark or wood shields for padding under the heels of the posts. By some tribes the posts were notched on one face, to fit inside the gunwales near the thwarts, and there were also other ways of assembling the gunwales themselves.

It should be apparent that the operations just described would serve only for canoes in which the sheer had a gentle, fair sweep. For canoes in which the sheer turned up sharply at the ends, the gunwale members might have to be split into laminations and pre-bent to the required sheer before being assembled into the gunwale frame. To accomplish this, the laminations were scalded with boiling water until saturated and then the gunwale members were staked out on the ground or tied with cords to set the wood in the desired curves as it dried out. The laminations were then wrapped with cord and the gunwale was ready to assemble. To produce a hogged sheer, the gunwales were made of green spruce and then staked out to season in the form desired; a hogged sheer was also formed by steaming or boiling the gunwale members at midlength.

The canoe, as now erected on the building bed, has a double-ended, flat-bottomed, wall-sided form. The gunwales are sprung to the proper breadth and sheer, and the bark is standing irregularly above them. At this point, on canoes not having outwales, the bark cover was laced or lashed to the gunwales. Since the Malecite canoe has outwales, these are now made and fitted. They consist of two white cedar battens about 19½ feet long, perhaps 1 inch wide, and ½ inch thick. The face that will be the outboard side is usually somewhat rounded, as are all the corners, and the corner that will be on the inside and bottom of each batten when it is in place is somewhat beveled. The outwales are placed between the bark and the outside stakes, the inside stakes being removed one by one as this is done. The removal of the inside stakes allows room for the outwale to be inserted in their place, between the outside stakes and the inner gunwale face, and it allows the bark to be brought against the outside face of the inner gunwales. In the process of fitting the outwales, the battens along the sides may have to be removed and replaced, or shifted, and the cross-ties of each pair of outside stakes may require adjustment. Beginning at midlength, the outwale is pegged through the bark cover to the inner gunwales at intervals of 6 to 9 inches. The pegging is not carried much beyond the end thwarts in any canoe and could not be in canoes having laminated gunwales near the ends.

The Malecite canoe has bark covers over the ends of the inner gunwales, and these are now fitted so that they can be passed under the outwales and clamped in place. The ends of the outwales are forced inside the stakes at and beyond the ends of the gunwales, assuming a pinched-in appearance there, and they may reach a few inches beyond the ends of the bark cover; they will be cut and shaped to the length of the finished canoe later.

The outwale pegs are made by splitting from a balk of birch, larch, or fir roughly squared dowels about ¼ inch square and 6 to 9 inches long. Each dowel is then tapered and rounded each way from the middle to form two shanks that are between ⅛ and 3/16 inch in diameter over 2 to 3 inches of length. The ends may be sharpened by fire. The dowels are then cut in two, providing a pair of pegs with large heads. These are driven in holes drilled through the outwales, bark cover, and gunwales, and when well home, the protruding ends are cut off flush. Toward the ends of the gunwales, the spaces between the pegs increase, and at the extreme ends, the outwale will be lashed to the gunwale by widely spaced groupings of root strand. These are usually temporary, as the final lashing of the bark to the gunwales will secure the outwales.

After the outwales are secured in place, the bark is fastened to the assembled gunwales with group lashings. In the Malecite canoe being built, these are independent, each grouping consisting of eight to ten complete turns of the root strand. The intervals between, roughly 2 inches, are usually spaced by means of a special measuring-stick to insure evenness. Before the lashing is actually begun, however, the excess bark standing above the gunwales is cut away. The bark either is trimmed flush with the top of the gunwale, or enough is left for a flap that will fully cover the top of the inner gunwale, to be turned down under the lashing. The latter method, the stronger, was used by many builders. In making the turns in the group lashings, two or three turns may be taken through a single hole in the bark; the Malecites did this to avoid having the holes too close together. The result is that the group when seen from outboard appears as a W-form, with only two or three holes in the bark for an entire group. Care is taken to lay up the turns over the gunwales neatly, turn against turn without open spacing or overlaps and crossings.

When this is completed, the ends of the thwarts can be lashed, the strand passing through the holes in the shoulders, around the two gunwale members, and through one or two holes in the bark cover. The groupings for the bark cover are spaced so that these lashings do not overlap them, and thus the lashings serve a dual purpose.

Next, the gores are usually sewn and the ends of the side panels closed. To do this, the temporary side battens outside the bark are removed. Since this is a Malecite canoe, the gores are sewn edge-to-edge with an over-and-over stitch, the strand crossing the seam square outside and diagonally inside. When these seams and those remaining in the upper panels are sewn, the rather stiff bark holds the shape formed on the building bed to a remarkable degree.

The canoe can now be raised from the building bed. To set it up at a most convenient working height, the weights are first removed from the gunwales and the remaining stakes are pulled up. The canoe is then lifted from its bed and turned upside down over a couple of logs, or crude horses. Traditionally, logs or sapling were rested across two pairs of boulders or the logs were tied between two pairs of trees at convenient distances apart. More recently, horses, formed by sticking four legs into auger holes drilled in the bottom of a 4-foot length of timber, were used. After the canoe is on its supports the ends are ready to be closed in.

The stem-pieces customarily used by the Malecite builder are formed from two clear white cedar billets a full 36 inches long and in the rough nearly 1½ inches square. The billets are first shaped so that

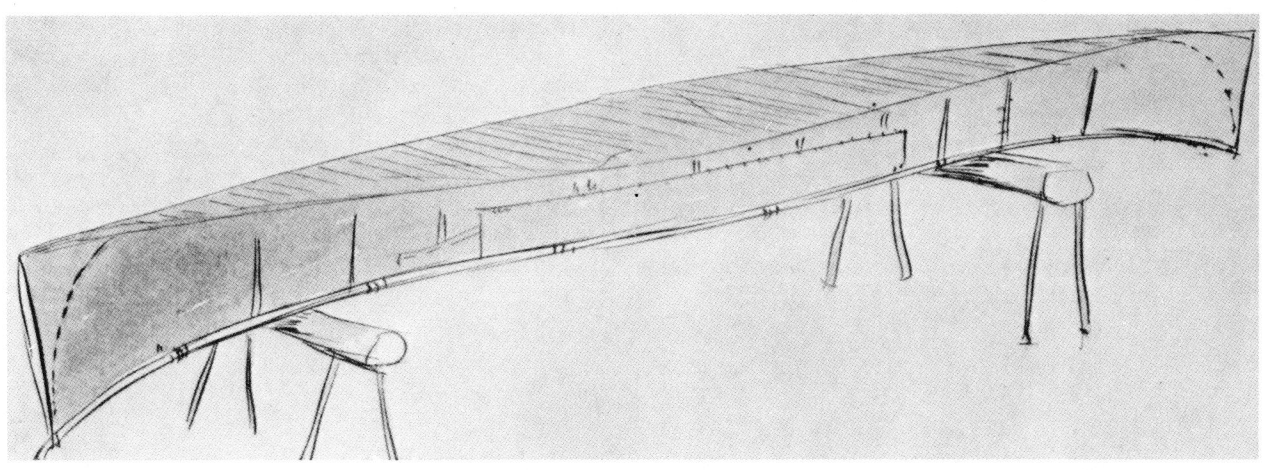

Figure 42

FIFTH STAGE OF CANOE CONSTRUCTION: canoe is removed from building bed and set on horse in order to shape ends and complete sewing. Bark cover has dried out in a flat-bottomed and wall-sided form. (*Sketch by Adney.*)

the outboard face of each stem-piece is about ¾ inch wide, making it a truncated triangle in cross-section. Then, along lines parallel to the base of the truncated triangle, it is split into six laminations which are carried to within 6 or 7 inches of the end selected to be the heel of the stem-piece. Just clear of the laminations a notch is cut into the top side of the heel, to hold the headboard, as will be seen. The piece is then treated with boiling water until the laminations are flexible, and the curve of the stem-piece can be formed and either pegged out or tied with cords until it dries in the desired shape. When dry the laminations are tightly wrapped with basswood bark cord, leaving the form of the stem-piece a quarter arc of a circle, with short tangents at each end, as shown in the illustration (p. 35).

Next, the ends of the outwales are cut to a length determined by the quality of the bark already in place; if the bark in one end is not very good, it may be cut away somewhat and the canoe made shorter by this amount at both ends in finishing. After the ends of the outwales have been cut, both are notched on the inside at the extreme ends to take the head of the stem-piece. The outwales may or may not project ¼ or ½ inch beyond the stem and the stem head may project ½ or 1 inch above the top of the outwales of the canoe; these matters, at the builder's option, decide the length of the notch and the fitting of the stem-pieces.

The stem-piece is now placed between the folded bark end of the canoe with the heel resting for a small distance along its length on the bark bottom; the head must come to the right height above the outwales, as noted. While one worker holds the stem-piece in place, another trims away the excess bark at the end to the profile of the outboard face of the stem-piece. Thus the profile of each end is cut and the rake of the ends is established. The bark is next lashed to the stem-piece. In this canoe it is done with a spiral over-and-over stitch, a batten made of a large split root being placed over the edges of the bark, as the lashing proceeds, to form a stem band. The turns pass alternately from outboard around the inboard face of the stem-piece and through it; the awl inserted in the laminations from one side opens them enough to allow the strand to be forced through. Care is taken to pull up the strand very hard each time. As the outwale is approached, the bark is cut away at the notching in each so that the outwales can be brought snugly against the sides of the stem-piece. Here the strand is brought up one or two times over the outwales, abaft the stem head, before the bitter end is tucked, thus locking the outwales to the stem-piece and the bark. Then a lashing is placed around the outwales just inboard of the stem-piece, passing through a hole in the flap of the end deck-piece of bark and through the side bark. This lashing holds the outboard end of the deck piece flap. At the inboard end of the flap, another lashing is required, but the pinched-in outwales require additional securing outboard of this point; hence a lashing is passed just inboard of the middle

Figure 43

RIBS BEING DRIED AND SHAPED FOR OJIBWAY CANOE. (*Canadian Geological Survey photo.*)

of the flap, a little outboard of the ends of the inwales, and about six inches inboard from this lashing another is passed through the side bark and around the gunwale and outwale on each side. These three lashings hold the outwales snug to the ends of the gunwales and against the projecting bark ends in the pinched-in form of projecting outwales.

The heels of the stem-pieces rest on the bottom bark and the sewing is carried down to where the cutting of the profile makes an end to the seam, the solid part of the heels extending about 6 to 8 inches inboard of this. Next, any sewing required on the bottom is done. When the bark cover has been given a final inspection on the outside and all sewing has been completed, the canoe is lifted from its supports, righted, and set on the bed or on a smooth grassy place.

All seams are now payed with gum on the inside of the bark while this can still be done without interference from the sheathing or those parts of the structure remaining to be installed. The Malecites used only spruce gum tempered with animal fat. The gum, heated until it is sufficiently soft to pour like heavy syrup, is spread with a small wooden paddle or spoon, and is then worked into the seam and smoothed by rubbing with the thumb dipped in water to prevent the gum from sticking and burning. It is first worked into the ends, between the bark and each side of the stem-pieces, particularly near the heel below the waterline. When the crevices are filled, a piece of bark (in later times a piece of cloth was used) wide enough to cover the gum alongside is well smeared with warm gum and pressed down along the inside of the stem-pieces. On each seam, at gores, and on side panels a thin narrow strip of bark is smeared with gum and pressed over the seam after the latter had been well payed. The bark is now carefully scrutinized for small splits, holes, or thin spots since these can be easily patched from the inside at this stage of construction. In fitting bark strips and in gumming, great care is taken to obtain a flat surface; the edges of the strips inside are faired to the inside face of the bark by smearing gum along the edges. The canoe is now ready to be sheathed and ribbed out.

The sheathing for this canoe has been split in advance out of clear white cedar in splints about 5 to 9 feet long, 3 to 4¼ inches wide, and ⅛ inch thick.

The butts of each piece have been whittled to a feather edge, the bevel extending back about 2 inches. Also, some pieces of basket ash have been split out of saplings for temporary ribs to hold the sheathing in place.

A total of 50 or more ribs in five lengths, the longest about 5 feet, have been made up from white cedar heartwood and bent to the desired shape.

In deciding the rough lengths of the ribs, the builder can resort to various methods. He can prebend ribs in pairs to a number of arbitrarily chosen shapes: the first set of six pairs to the desired midsection form; a second set of five pairs to the form of the section between the middle and first pair of thwarts; a third, of five pairs, to the section at the first thwarts each way from the middle; a fourth, of four pairs, to the section between the end and the first pair of thwarts each way from the middle; a fifth, of three pairs, to the section at the end thwarts; and a sixth, of two or three pairs, for the section at or near the headboards. This makes from 50 to 52 frames in a canoe measuring 18 or 19 feet overall.

Each frame piece is treated with boiling water and then bent, over the knee or around a tree, to a slightly greater degree than is needed. While thus bent, each pair is wrapped lengthwise over the end with a strip of basswood or cedar bark to hold the ribs in shape. Sometimes a strut is placed under the bark strips to maintain the desired form, or a cross-tie of bark may be employed. The ribs are then allowed to season in this position.

Another method, which will be illustrated later (p. 53), involves placing ribs of green spruce in their approximate position and forcing them against the bark. In this method, a number of long battens are placed over the roughly bent ribs laid loosely inside the bark cover, and are spread by forcing a series of short crosspieces, or stays, between them athwartships. The bark is given a good wetting with boiling water to make it flexible and elastic, so that the pressure applied to the battens by the temporary crosspieces brings the bark to the shape desired for the canoe. The rough lengths of the ribs are determined by use of a measuring stick or by measurements made around the bark with a piece of flexible root or a batten of basket ash. The ribs, in any case, are made somewhat longer than required to allow a final fitting when being placed over the sheathing.

It can be seen that the exact form the canoe takes is largely a matter of judgment and of the flexibility and elasticity of the bark, rather than of precise molding on a predetermined model, or lines.

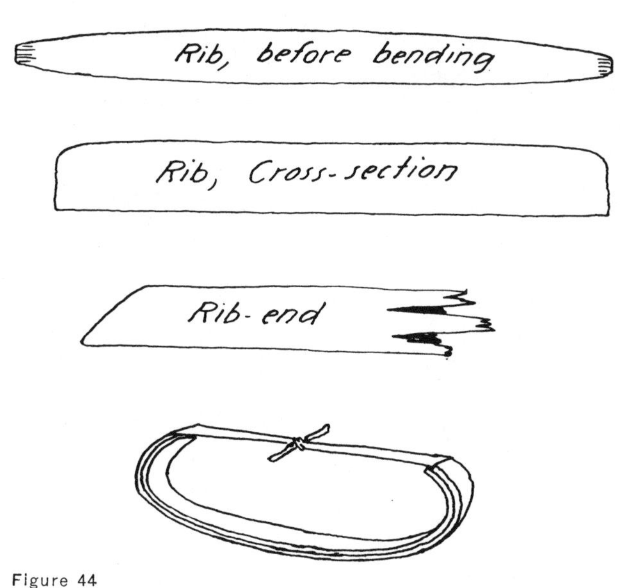


Figure 44

DETAILS OF RIBS and method of shaping them in pairs in a bark strap or thong so that they take a "set" while drying out.

In the Malecite canoe the ribs are wide amidships, 3 or 4 inches, and narrow to 2½ or 2 inches toward the ends. The thickness is an even ⅜ inch. Most birch-bark canoes have ribs of even thickness their full length, but in a few the thickness is tapered slightly above the turn of the bilge, usually when the tumblehome is high on the sides and rather great. The width, as previously explained, is usually carried all across the bottom; above the bilges there is a moderate taper.

The sheathing of the canoe is now first to be put in place. In the Malecite canoe the center pieces are the longest; they are tapered each way from their butts, which overlap about 2 inches amidships. The ends are made narrow enough to fit readily into the sharp transverse curve of the bottom and are long enough to pass under the heels of the stem pieces for an inch or two. The pieces of sheathing on each side of the center pieces are fitted in the same manner, and by the time two or three courses are in place they must be held in some manner at the ends. This is accomplished by means of the rough temporary ribs mentioned earlier. The sheathing is laid edge-to-edge, with the butts overlapping, and, if there are not enough long pieces to complete the bottom amidships, three or four lengths, with overlapped butts, will be used. As the sheathing progresses, more temporary ribs will have to be added. At the turn of the bilge, the sheathing will bend transversely as pressure is applied by the temporary ribs; the bark must be again wetted so that the angular bilge can be forced into a roughly rounded form. Particular care is required in finishing the sheathing below the gunwale to be certain that the top strake will be close up against the sewing of the bark at gunwales, but no particular attempt is made to make the edges of the sheathing in the topsides maintain edge-to-edge contact.

The pressure of the temporary ribs, the heads of which are forced under the gunwales, and the elasticity of the bark due to treating it with boiling water are enough to rough-shape the canoe.

Before the permanent ribs are placed the sheer is checked. If it appears to have straightened, the ends of the gunwales are supported by means of short posts placed under them, with the heels standing on the heels of the stem pieces or on the sheathing. Then some stakes, each having a projecting limb or root, are cut and are driven into the ground with the limb hooked over the gunwale to force it down.

After measurements have been made for the first rib with a strand of root or an ash batten, it is now cut to a length slightly more than would permit the rib to be forced upright when in place. The ends of the rib are set in place in the bevel, or notch, on the underside of the gunwales, against the bark cover, and with the bottom part of the rib standing inboard of the head. Then, with one end of a short batten placed against its inboard side, the rib is driven toward the end of the canoe with blows from a club on the head of the batten. If the rib drives too easily it is removed and laid aside; if too hard, it is shortened. It must go home tightly enough to stretch slightly the bark cover by bringing pressure to bear on the whole width of the sheathing. Care is taken, in this operation, to keep moist not only the bark but also the sewing, particularly along the gunwales, so that all possible elasticity is obtained. The ribs are set, one by one, working to within two or three frames of the midship thwart; then the other end of the canoe is begun. The last three or four ribs to be placed are thus amidships. In every rib driven, the tension is great, but no rib is driven so that it stands perpendicular to the base. Those first driven stand with their bottoms nearer the midship thwart than the ends, and this angle, or slant, continues to amidships; the ribs in the other end of the canoe slant in the opposite direction.

It will be evident that skill is required to estimate how much pressure the bark will stand before bursting under the strain of the driven ribs. It is also apparent that the shape of the canoe is controlled by the shaping given the ribs in the prebending, for this fixes the amount of tumble-home and the amount of round, or rounded-V, given to the bottom athwartships. No fixed rules appear to exist; the eye and judgment of the builder are his only guides. To show how much strain is placed on the bark, however, it may be noted that inspection of two old canoes showed that the gunwale pegs had been noticeably bent between the inner and outer gunwales.

It appears to have been a rather common practice, after all the ribs had been driven into place, to allow the canoe to stand a few days and then again to set the frames (where unevenness appears in the topsides) with driving batten and maul, the bark cover and the root sewing or lashings having been again thoroughly wetted.

The headboards are now to be made. These are shaped in the form of an elongate-oval from a wide splint of white cedar about 4 inches wide at mid-length and ¼ inch thick. The narrow end is first cut off square or nearly so; the bottom end is notched to fit in the notch in the heel of the stem-piece and the top has a small tenon at the centerline that will be fitted into a hole drilled or gouged in the underside of the inner gunwales where they join at the ends. The length of the headboards in the canoe being built is 15¾ inches over all, and when they have been made for each end, they are checked as to width and height to see that they can be fitted. Next, the extreme ends of the canoe between the stem and the headboards are stuffed with dry cedar shavings or dry moss so that the sides stand firm on each side of the bow outboard of the ends of the sheathing, which ends rather unevenly, just outboard of where the headboards will stand. This completed, the headboards are forced into position by first stepping the heel notch in the stem-piece notch and then bending the board by placing one hand against its middle and pulling the top toward the worker. This shortens the height of the board enough so the tenon projecting on its head can be sprung into the small hole under the inner gunwales, where it becomes rigidly fixed. Its sprung shape pushes up the gunwales and makes the side bark of the ends very taut and smooth, while supporting the gunwale ends.

Two thin strips about 19 feet long are next split out of white cedar to form the gunwale caps; these are ¼ to ⅜ inch thick, and taper each way from about 2 inches wide in the middle to 1 inch wide at the ends.

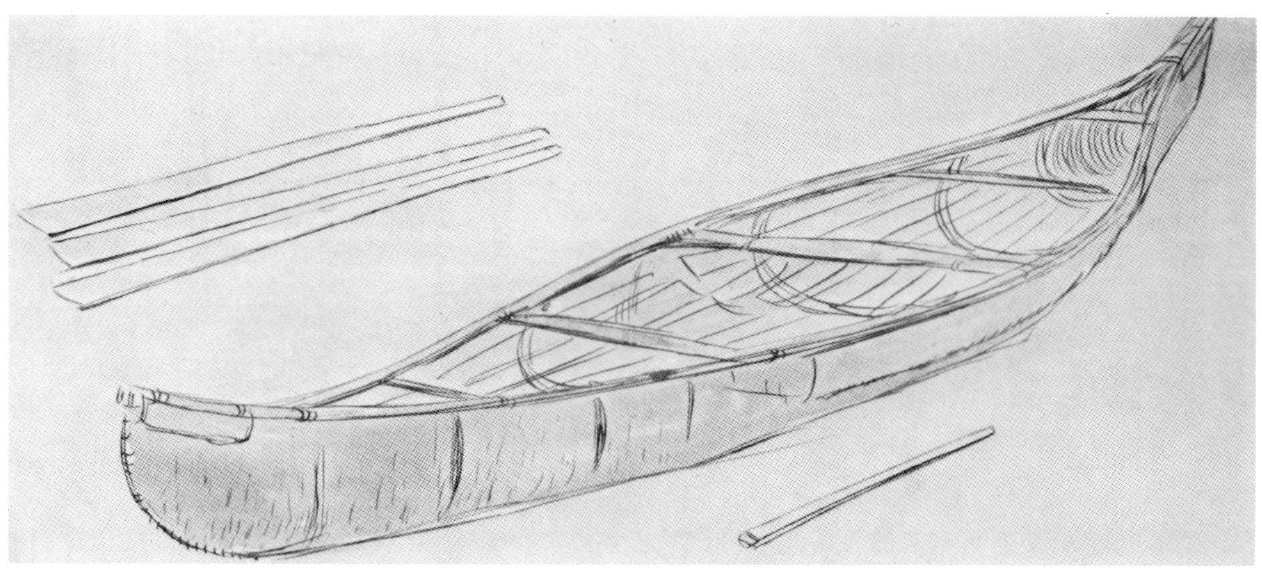

Figure 45

Sixth Stage of Canoe Construction: canoe has been righted and placed on a grassy or sandy spot. In this stage splints for sheathing (upper left) are fixed in place and held by temporary ribs (lower right) under the gunwales. The bark cover has been completely sewn and the shape of the canoe is set by the temporary ribs. (*Sketch by Adney.*)

These are laid along the top of the inner gunwales and fastened down with pegs placed clear of the gunwale lashings. The ends of the strips are usually secured by two or three small lashings; the caps thus formed often stop short of the ends of the inner gunwale members. If the caps are carried right out to the stems, as was the practice of some Malecite builders, the lashings of the outwale are not turned in until after the caps are in place, in which case the bark deck pieces, or flaps, are put in just before the final lashing is made.

Next, the canoe is turned upside-down and all seams are gummed smoothly on the outside. The ends, from the beginning of the seam to above the waterline, may be heavily gummed and then covered with a narrow strip of thin bark, heavily enough smeared with gum to cause it to adhere over the seam. In more recent times a piece of gummed cloth was used here. Above this protective strip, the end seams are filled with gum so that the outside can be smoothed off flush on the face of the cutwater between the stitches. All seams in the side and bottom are gummed smooth and any holes or patches remaining to be gummed are taken care of in this final inspection.

If the canoe is to be decorated (not many types were) the outside of the bark is moistened and the rough, reddish winter bark, or inner rind, is scraped away, leaving only enough to form the desired decorations. When paints of various colors could be obtained, these were also employed, but the use of the inner rind was apparently the older and more common method of decorating.

The paddles are made from splints of spruce or maple, ash, white cedar, or larch. Two forms of blade were used by the Malecite. The older form is long and narrow, with the blade wide near the top and the taper straight along each edge to a narrow, rounded point. Above the greatest width, the blade tapers almost straight along the edge, coming into an oval handle very quickly. At the head, the handle is widened and it ends squared off, but the taper toward the handle is straight, not flared as in modern canoe paddles; there is no swelling. Paddles of a shape similar to this, some without a wide handle, were used by other eastern Indians. The more recent form of Malecite paddle has a long leaf-shaped, or beaver-tail, blade, much like that of the modern canoe paddle, except that it ends in a dull point; the handle is as in the old form but the head is swelled to form the upper grip. The face of the blade, in both old and new form, has a noticeable ridge down the centerline.

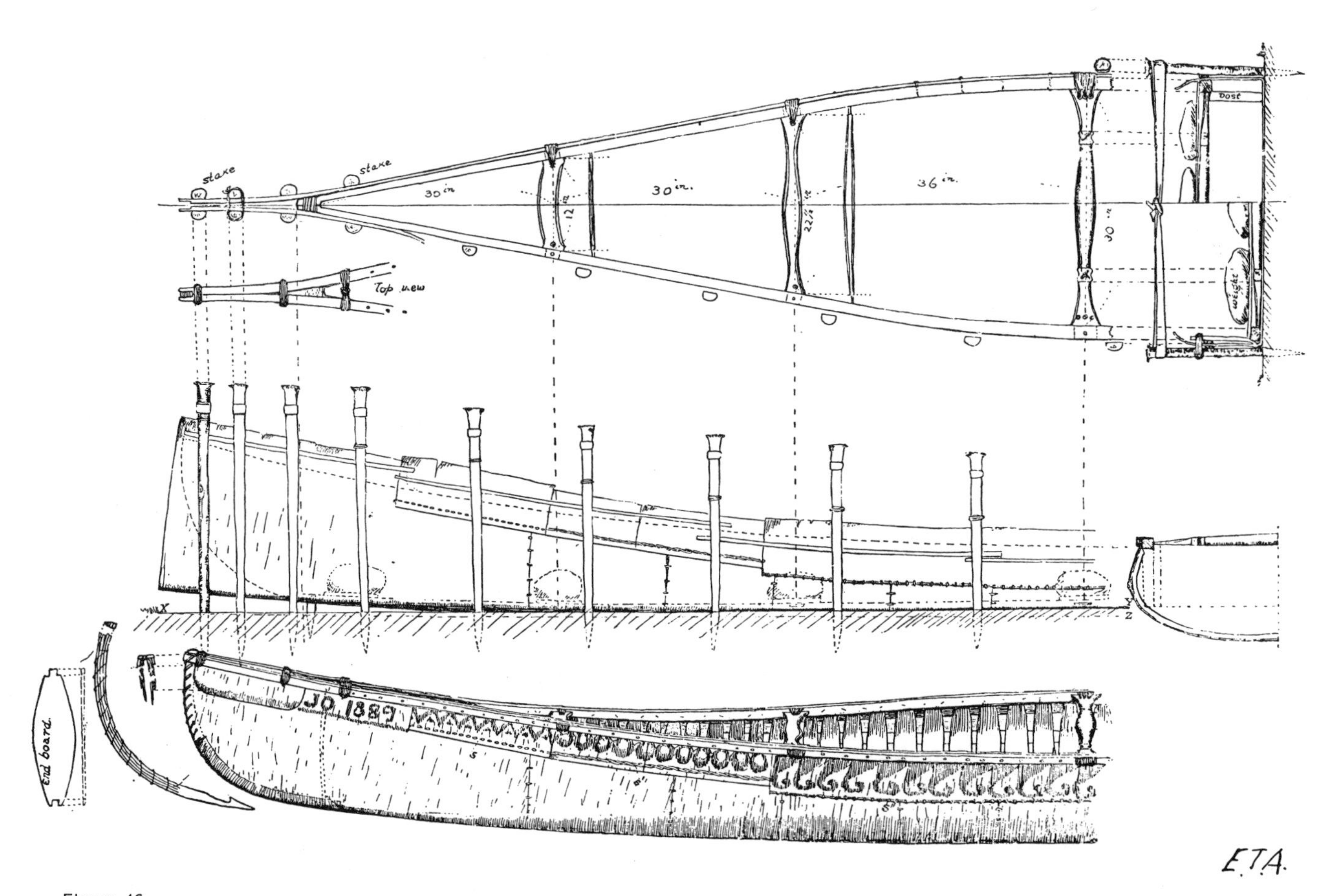


Figure 46

General Details of Birch-Bark Canoe Construction, in a drawing by Adney. (From *Harper's Young People*, supplement, July 29, 1890.)

The eastern style of construction described here produced what might be called a wide-bottom canoe with some tumble-home above the turn of the bilge, but a different method of construction was used to produce canoes having a narrow bottom and flaring sides. These canoes were not set up on the building bed, in the first steps of shaping the hull, with the gunwale frame on the cover bark. Instead, a special building frame, mentioned earlier, was used. Each tribe using the building frame had its own style, but the variations were confined to minor matters or to proportion of width to length.

In general, the building frame is made of two squared battens, about 1¼ inch square for an 18-foot canoe. These, sometimes tapered slightly toward each end, are fitted with crosspieces with halved notches in each end to fit over the top of the battens. There may be as many as nine or as few as three of these crosspieces, with seven apparently a common number. Where ends of the long battens join they are beveled slightly on the inside face and notches are cut on the outside face to take the end lashings. Each crosspiece end is lashed around the long battens, a hole being made in each end of the crosspiece for this purpose. The lashings, commonly bark or rawhide thongs, are all temporary, as the building frame has to be dismantled to remove it from the canoe. Sometimes holes are drilled in the ends of the crosspieces, or in the long battens, and in them are stepped the posts used to fix the sheer of the gunwales.

The methods of construction, using the building frame, varied somewhat among the tribes. Since the gunwale was both longer and wider across than the building frame, the posts for sheering were set with outboard flare. However, some builders made the gunwales hogged by staking them out when green, and then set them above the building frame with vertical posts. These gunwales would not be fitted with thwarts nor would the thwart tenons always be cut at this stage. The bark was lashed to the gunwales while they were in the hogged position with the ends secured; the gunwales were then spread by inserting spreaders, or stays, between them, after which the thwarts were fitted. This method required knowledge of just how much hog should be given to the gunwales, and it must be stated that not all builders guessed right enough to produce a good-looking sheer. Judging the hogging required in the gunwales was complicated by the fact that most of these canoes had laminated ends in the gunwales at bow and stern, and a quick upturn there as well. This method of construction persisted, however, because the straight sides made easy the sewing of gores and side panels. In some Alaskan birch-bark canoes the building frame was, in fact, part of the hull structure and remained in the canoe. In these, the building frame was hogged and then flattened by the ribs in construction so as to smooth the bottom bark by placing it under tension. In some canoes the posts for sheering the canoe rested under the thwarts rather than under the gunwales. In most canoes the building frame was taken apart and removed from the canoe when the gunwale structure was complete and in place, sheered.

Where large sheets of bark were available, the setting up with the building frame or gunwale was made easier than where the bark had to be pieced out for both length and width. If large pieces of bark could be obtained there was little or no sewing on the bottom; only the gores or laps, and the panels, in the side required attention after the bark had been lashed to the gunwales. In such instances, the set-up did not require perpendicular sides, as the sides could be completed after the canoe was removed from the building bed and the building frame had been removed from the hull. There were many minor variations in the set-up and in the sequence of the sewing. In view of the slight opportunities that now exist for examining the old building methods and construction sequences, it is impossible to be certain that the one used by a tribe in recent times was that employed in prehistoric times by their ancestors.

Instead of a laminated stem-piece, a large root whittled to the desired cross section was sometimes used by builders among the Malecites and other eastern tribes. This was bent into the ends while green and to it was lashed the bark, so that the stem dried in place to the desired profile curve. No inner stem-piece was used by the Micmacs, who formed the end structure by placing a split-root batten on each outside face of the bark and passing the lashing around both. When a plank-on-edge was used to form the stem-piece, as mentioned earlier, no headboard was required, as the gunwales ends could be brought to the plank structure. In canoes having the complicated stem structure seen in the large fur-trade canoes and some others, the headboard became an integral part of the stem structure, rather than an independent unit, and was placed in the canoe during building with the stem-pieces.

There was much variation in the form of gunwale structure employed in bark canoes. A strip of bark was added all along the outwale by some tribes, so that between the gunwale members and for a short distance below the sewing the bark was doubled; the bottom of this strip was, in fact, a flap not secured and thus was much like the flaps at the ends of the Malecite canoe, but without covering the top of the main gunwales. The outwale and inwale cross sections of some canoes were almost round. The use of a single gunwale member is commonly followed by continuous lashing of the bark along it. On some northwestern canoes having continuous lashing, the ends of the ribs were made in sharp points that could penetrate between the turns of root sewing, under the gunwales. The ends of the ribs in some of these were secured more firmly by tying them to long battens placed between the ribs and the bark cover just below the gunwales. The northwestern canoes built in this manner had double gunwales, an outwale and an inwale, but no bevel or notch for the rib heads. The ends of the gunwales, inner and outer, were secured in many ways. Some, instead of being pegged and lashed, were simply tied together; others were fastened by a rather elaborate lashing through the bark and around the gunwales. Caps were sometimes allowed to overlap at the ends and were pinned together with pegs or lashed. In some canoes the outwales were lashed, rather than pegged, to the inwales, and for this and for the caps rawhide appears to have once been widely used. In some canoes the head of the stem-piece was bent inboard sharply and lashed to the ends of the inwales or outwales. In many canoes the gunwales, instead of stopping short of the stem-piece, ran to it and were lashed there.

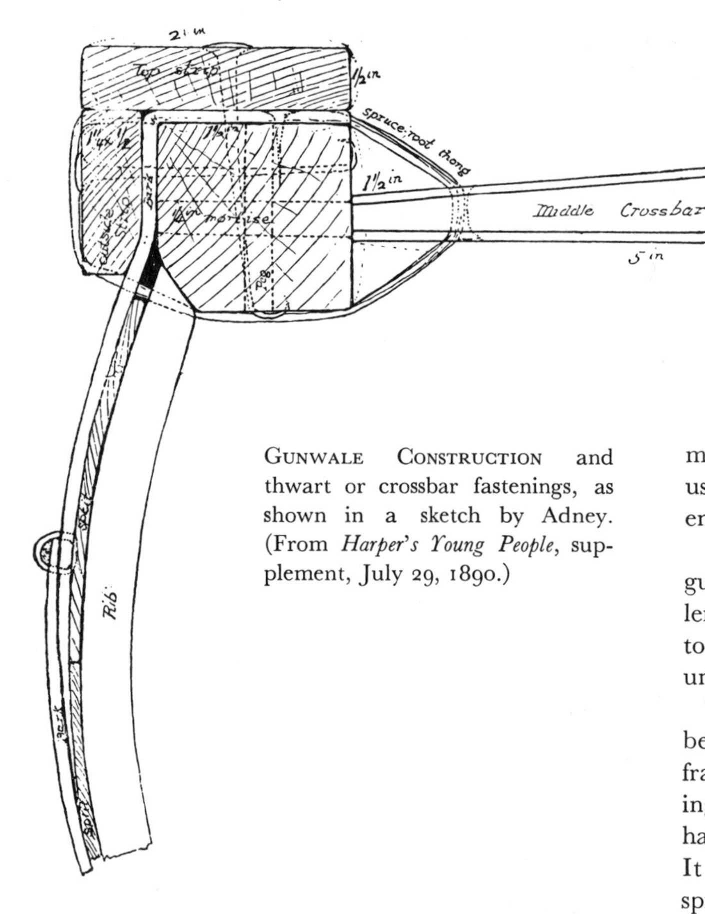


Figure 47

GUNWALE CONSTRUCTION and thwart or crossbar fastenings, as shown in a sketch by Adney. (From *Harper's Young People*, supplement, July 29, 1890.)

At the start of ribbing out a canoe, the first two or three ribs might not be put at each end until after the headboards had been fitted, and sometimes a rib was placed on each side of the middle thwart, apparently to hold securely the sheathing butted amidships while the ribbing progressed toward them from the ends. When a canoe was short and rather wide, the ribs usually were bent by placing them inside the faired bark cover before the sheathing was installed, there to dry and set or to season, depending on whether they were steamed or green. Prebending the ribs, as described in the building of a Malecite canoe, worked well only when the canoe was long, narrow, and sharp. The spacing of the ribs was done by eye, not by precise measurement, and was never exactly the same over the length of the canoe. Ribs near the ends were usually spaced at greater intervals than those in the middle third of the length.

The extension of the bark beyond the ends of the inner gunwale in an eastern canoe was often about one foot on each end, but this distance was actually determined by the length of the bark available and by the usual reluctance of the builder to add a panel at the end.

For the height of the end posts, in sheering the gunwales, a common Malecite measurement was the length of the forearm from knuckles of clenched fist to back of elbow. These posts were often left in place until the stems were fitted.

The use of a building frame is known to have been common in areas where, normally, the gunwale frame would be employed in the initial steps in building. In a few instances this occurred when a builder had a number of canoes of the same size to construct. It seems probable that the use of the building frame spread into Eastern areas comparatively recently as a result of the influence of the fur-trade canoes on construction methods. The employment of the plank building bed in the East is known to have occurred among individual canoe builders late in the nineteenth century as a result of this influence.

The use of nails and tacks instead of pegs and root lashing or sewing in bark canoe construction became quite widespread early in the nineteenth century; it is to be seen in many old canoes preserved in museums. The bark in these is often secured to the gunwales with carpet or flat-headed tacks, and both the outwale and the cap are nailed to the inner gunwales with cut or wire nails. Various combinations of lashings and nailing can be seen in these canoes, although such combinations are sometimes the result of comparatively recent repairs or restorations rather than evidence of the original construction. No date can be placed on the introduction of nails into Indian canoe building, although it may be said that nailing was used in many eastern areas before 1850.

Among the many published descriptions of the method of building bark canoes the earliest give very incomplete information on the building sequence and usually contain obvious errors as to proportions and materials. (An example is that of Nicolas Denys, who, sometime between 1632 and 1650, saw bark canoes being built in what is now New Brunswick and Cape Breton.) The best descriptions are relatively recent and, as a result, may describe methods of construction that are not aboriginal.

The description given here is based upon notes made by Adney in 1889–90 and upon inspection of old canoes from the various tribal areas. It was noted that, although among canoes of the same approximate length there was some variation in dimensions and some variety in end form, the construction appeared to vary remarkably little, and it is apparent that the Malecites held very closely to a fixed sequence in the building process. There was, however, great variation in detail. The number of gore slashes in canoes 18 to 19 feet long varied from 10 to 23 on a side. The number was not always the same on both sides of a canoe nor were the gores always opposite one another. Canoes with long, sharp ends often had a large number of closely spaced gores in the middle third of the length, with widely spaced gores toward the ends. Full-ended canoes, on the other hand, had rather equally spaced gores their full length. The amount and form of rocker was also a factor in spacing the gores, and when the rocker was confined to short distances close to the ends there would naturally be rather closely spaced gores in these portions of the sides.

A number of the building practices remain to be described, but these will be best understood when the individual tribal canoe forms are examined. No written description of building canoes can be understood without reference to drawings, and to promote this understanding construction details have been shown on many of those of individual canoes of each tribal type.

Figure 48

"Peter Joe at Work." Drawing by Adney for his article "How an Indian Birch-Bark Canoe is Made" (*Harper's Young People*, supplement, July 29, 1890).

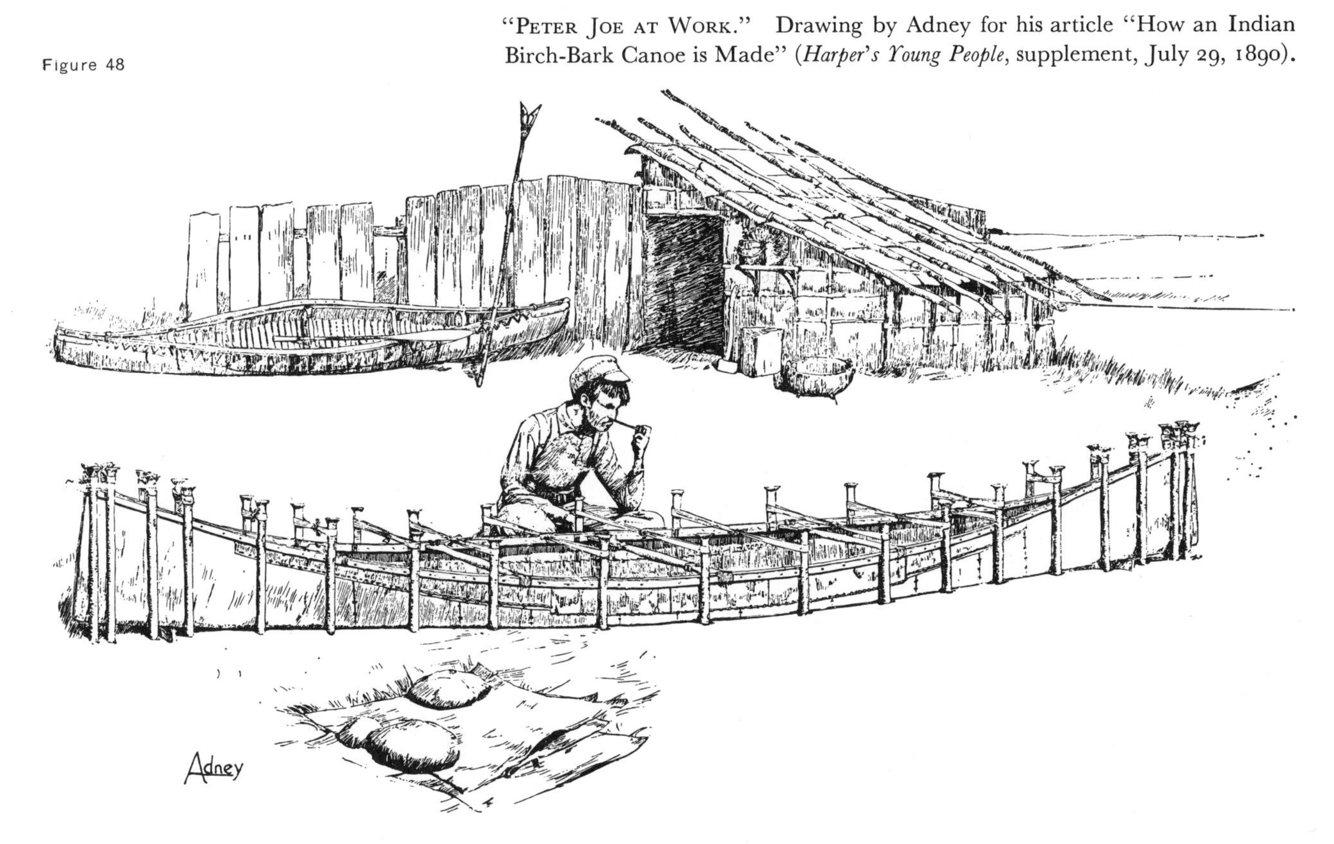

# *Chapter Four*

# EASTERN MARITIME REGION

STUDY OF THE TRIBAL FORMS of bark canoes might well be started with the canoes of the eastern coastal Indians, whose craft were the first seen by white men. These were the canoes of the Indians inhabiting what are now the Maritime Provinces and part of Quebec, on the shores of the St. Lawrence River and in Newfoundland, in Canada, and of the Indians of Maine and New Hampshire, in New England. Within this area were the Micmac, the Malecite, and the mixture of tribal groups known as the Abnaki in modern times, as well as the Beothuk of Newfoundland. All these groups were expert canoe builders and it was their work that first impressed the white men with the virtues of the birch-bark canoe in forest travel.

## *Micmac*

The Micmac Indians appear to have occupied the Gaspé Peninsula, most of the north shore of New Brunswick and nearly all the shores of the Bay of Fundy as well as all of Nova Scotia, Prince Edward Island, and Cape Breton. They may have also occupied much of southern and central New Brunswick as well, but if so they had been driven from these sections by the Malecites before the white men came. The Micmacs were known to the early French invaders under a variety of names; "Gaspesians," "Canadiens," "Sourikois," or "Souriquois," while the English colonists of New England called them merely "Eastern Indians." The name Micmac is said to mean "allies" and not known, but this name was in use early in the 18th century, if not before 1700.

The Micmac were a hunting people with warlike characteristics; they aided the Malecite and other New England Indians in warfare against the early New England colonists and in later times aided the French against the English in Nova Scotia and New Brunswick. These Indians lived in an area where water transport represented the easiest method of travel and so they became expert builders and users of birch-bark canoes, which they employed in hunting, fishing, general travel, and warfare.

The area in which they lived produced fine birch bark and suitable wood for the framework. Through experience, they had become able to design canoes for specific purposes and had produced a variety of models and sizes. The hunting canoe was the smallest, being usually somewhere between 9 and 14 feet long, with an occasional canoe as long as 15 feet. This light craft, known as a "woods canoe" and sometimes as a "portage canoe," was intended for navigating very small streams and for portaging. Another model, the "big-river canoe," somewhat longer than the woods canoe, was usually between 15 and 20 feet long. A third model, the "open water canoe," was for hunting seal and porpoise in salt water and ranged from about 18 feet to a little over 24 feet in length. The fourth model, the "war canoe," about which little is known, appears to have been built in either the "big-river" or "open-water" form, and to the same length, but sharper and with less beam so as to be faster.

The tribal characteristics of the Micmac birch-bark canoes were to be seen in the form of the midsection, in certain structural details, and in their generally sharp, torpedo-shaped lines. The construction was very light and marked by good workmanship. The distinctive profiles of bow and stern, which do not appear in the canoes of other tribes in so radical a form, were almost circular, fairing from the bottom around into the sheer in a series of curves. The break in the profile of the ends at the sheer, a break that marks in more or less degree, the end profile of other tribal forms, never occurs in the Micmac canoe. At most, a slight break in the "streamlined"

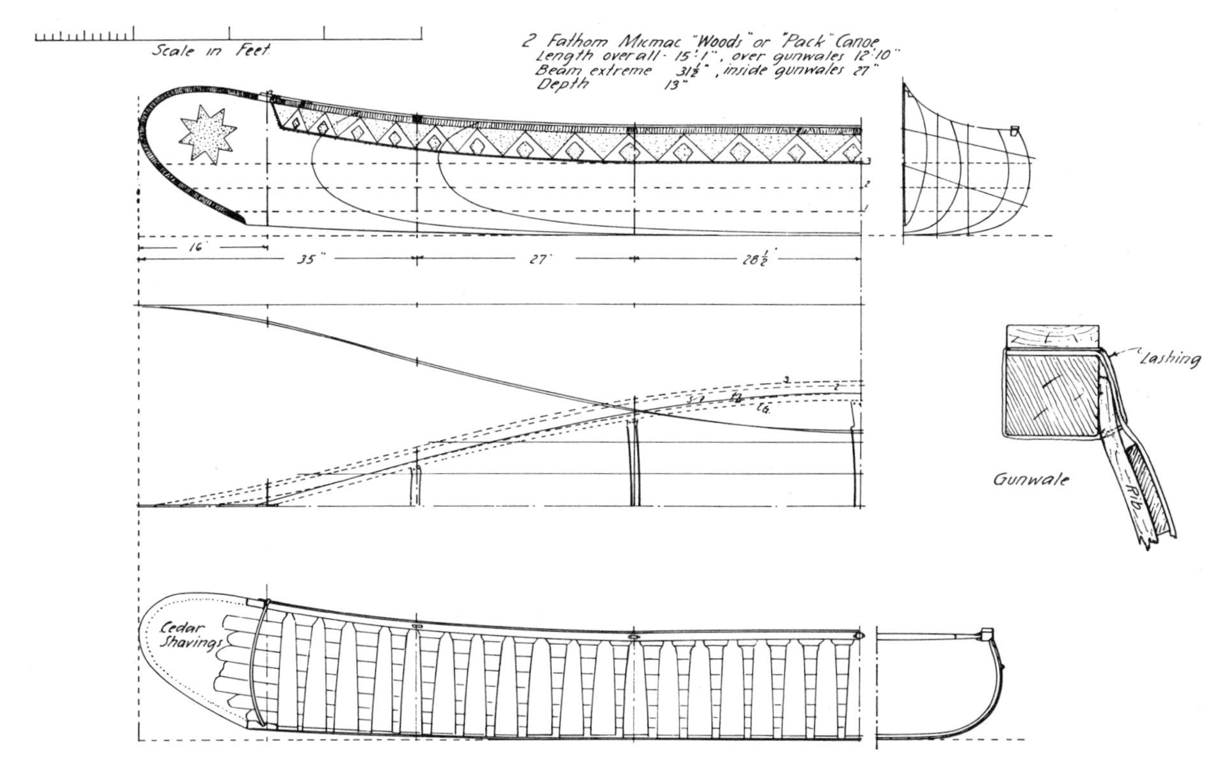


Figure 49

MICMAC 2-FATHOM PACK, OR WOODS, CANOE for woods travel with light loads, used by the Nova Scotia Micmacs.

curve might occur at the point where the profile was started in the bottom, at which point there might be a short, hard curve.

The form of the sheer line of the Micmac canoes apparently varied with the model: the woods canoe had the usual curved sheer with the point of lowest freeboard about amidships, the big river canoe had either a nearly straight sheer or one very slightly hogged, while the open-water canoe had a strongly hogged sheer in which the midship portion was often as much as 3 or 4 inches above that just inboard of the ends. However, there is a possibility that, at one time, the sheer of all Micmac canoes was more or less hogged. The little that is known of the war canoes of colonial times indicate that they had the strongly hogged sheer that now marks the open-water model, through it is also known that some of these were really of the big-river model, which in later times had usually no more than a vestige of the hogged sheer.

The hull-forms of the Micmac canoes were marked in the topsides by a strong tumble-home, carried the full length of the hull, that gave these canoes more beam below than at the gunwale. The form of the midsection varied with the model; the woods canoe usually had a rather flat bottom athwartships, the big river canoe a slightly rounded bottom, and the open water canoe either a well-rounded bottom or one in the form of a slightly rounded V. The fore-and-aft rocker in the bottom was always moderate, usually occurring in the last few feet near the ends; however, many of the canoes were straight along the bottom. This condition will be again referred to in discussing the building beds used in this type. The ends were usually fine-lined; in plan view the gunwales came into the ends in straight or slightly hollow lines. The level lines below the gunwales might also be straight as they came into the ends, but were commonly somewhat hollow; a few examples show marked hollowness there. Predominantly, the Micmac canoes were very sharp in the ends and paddled swiftly. Early Micmac canoes seem to have been narrower than more recent examples, which are usually rather broad as compared to the types used by some other tribes.

Structurally, the Micmac canoes were distinguished by the construction of the ends and by their light

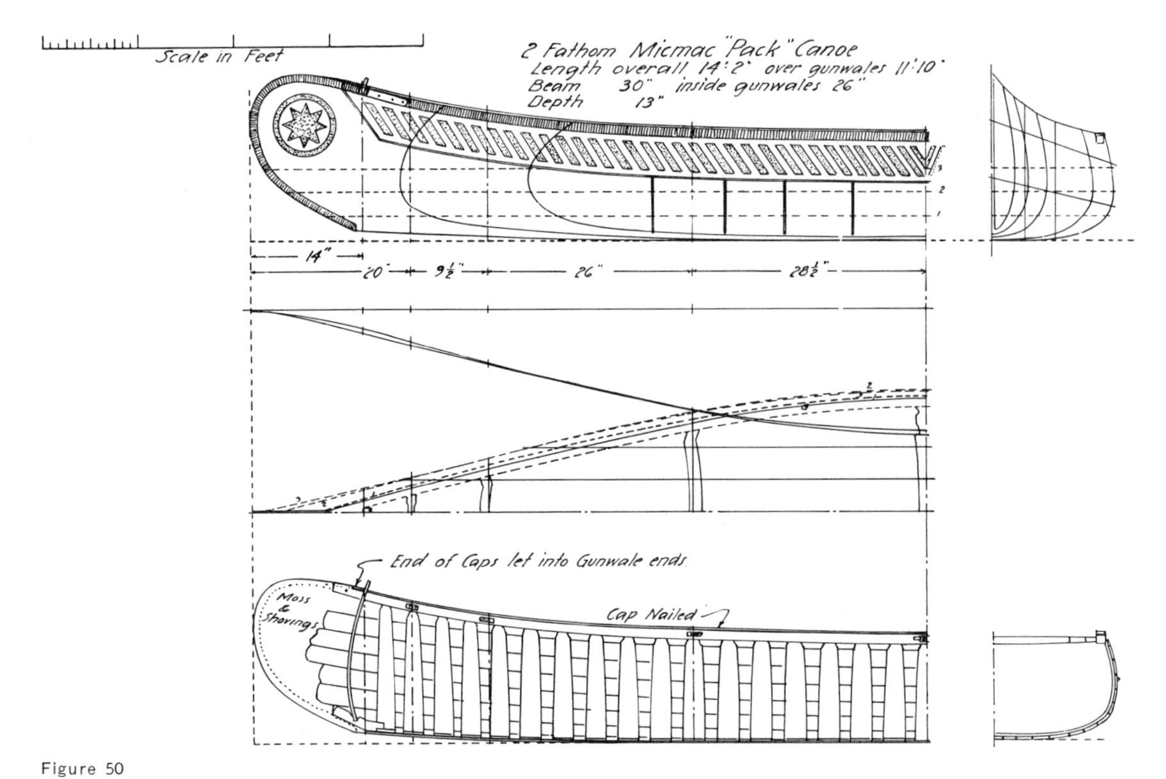


Figure 50

MICMAC 2-FATHOM PACK, OR WOODS, CANOE with Northern Lights decoration on bow, and seven thwarts.

build throughout. The canoes had no inner framework to shape the ends; stiffness there was obtained by placing battens outside the bark, one on each side of the hull, that ran from the bottom of the cut in the bark required to shape the ends to somewhat inboard of the ends of the gunwales at the sheer. These two battens, as well as a split-root stem-band covering the raw ends of the cut bark, were held in place by passing a spiral over-and-over lashing around all three. Sometimes thicker battens reaching from the high point of the ends inboard to the end thwarts were added, in which case the side battens were stopped at the high point of the ends and there faired into the thick battens.

The gunwale structure was rather light, the maximum cross section of the main gunwale in large canoes being rarely in excess of 1¼ inches square. These members usually tapered slightly toward the ends of the canoe and had a half-arrowhead form where they were joined. Old canoes had no guard or outwale, but some more recent Micmac canoes have had a short guard along the middle third of the length. Often there was no bevel to take the rib ends on the lower outboard corner of the main gunwales, and the gunwales were not fitted so that their outboard faces stood vertically. Instead, the tenons in the gunwales were cut to slant upward from the inside, so that installation of the thwarts would cause the outboard face to flare outward at the top. Between this face and the inside of the bark cover were forced the beveled ends of the ribs, which were cut chisel-shape. However, some builders beveled or rounded the lower outboard corner of the main gunwale, as described under Malecite canoe building (p. 38). The bark cover in the Micmac canoe was always brought up over the gunwales, gored to prevent unevenness, and folded down on top of them before being lashed. The gunwale lashing was a continuous one in which the turns practically touched one another outboard, though they were sometimes separated under the gunwale to clear the ribs, which widened near their ends, so the intervals between them were very small.

The other member of the gunwale structure was the cap; its thickness was usually ¼ to ⅜ inch, reduced slightly toward the ends. Its inboard face and the

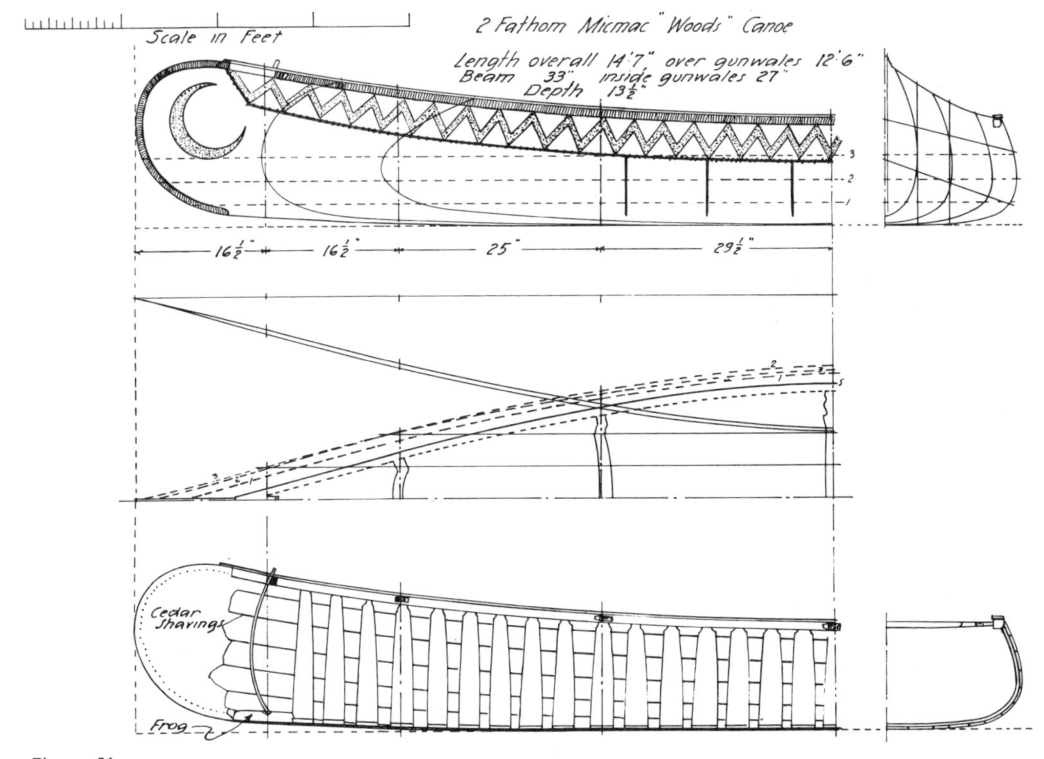


Figure 51

MICMAC 2-FATHOM PACK, OR WOODS, CANOE with normal sheer and flat bottom.

bottom were flat, but the top was somewhat rounded, with the thickness reduced toward the outboard edge. The cap was fastened to the main gunwales with pegs and with short lashing groups near the ends, but in late examples nails were used. The ends of the caps were bevelled off on the inboard side, so that they came together in pointed form. The cap usually ended near the end of the gunwale but in some canoes, particularly those that were nail-fastened, the cap was let into the gunwale (see p. 50) so that the top was flush with end of the gunwale.

The ends of the gunwales were supported by headboards that were bellied outboard to bring tension vertically on the bark cover. The heel of the board stood on a short frog, laid on the bottom with the inboard end touching or slightly lapping over the endmost rib. The frog supported the heels of the headboard and also the forefoot of the stem-piece, which otherwise would have but partial support from the sewing battens outside the ends at these points. The headboard was rather oval-shaped and the top was notched on each side to fit under the gunwale; the narrow central tenon stood slightly above the top of the main gunwales when the headboard was sprung into place and was held in position by a lashing across the gunwales inboard of the top of the headboard. The heel was held by the notch in the frog. Cedar shavings were stuffed into the ends of the canoe between the stempiece and the headboard to mold the ends properly, as no ribs could be inserted there. All woodwork in these canoes was white cedar, except the headboards and thwarts, which were maple, and the stem battens, which were usually basket ash but sometimes were split spruce roots.

The more recent Micmac canoes usually had no more than five thwarts; this number was found even on small woods canoes. However, old records indicate that canoes 20 to 28 feet long on the gunwales were once built with seven thwarts. The shape of the thwarts varied, apparently in accordance with the builder's fancy. The most common form was nearly rectangular in cross-section; in elevation, it was thick at the hull centerline and tapered smoothly to the outboard ends; and in plan it was narrowest at the hull centerline and increased in width toward the ends, the increase being rather sharp at the shoulders

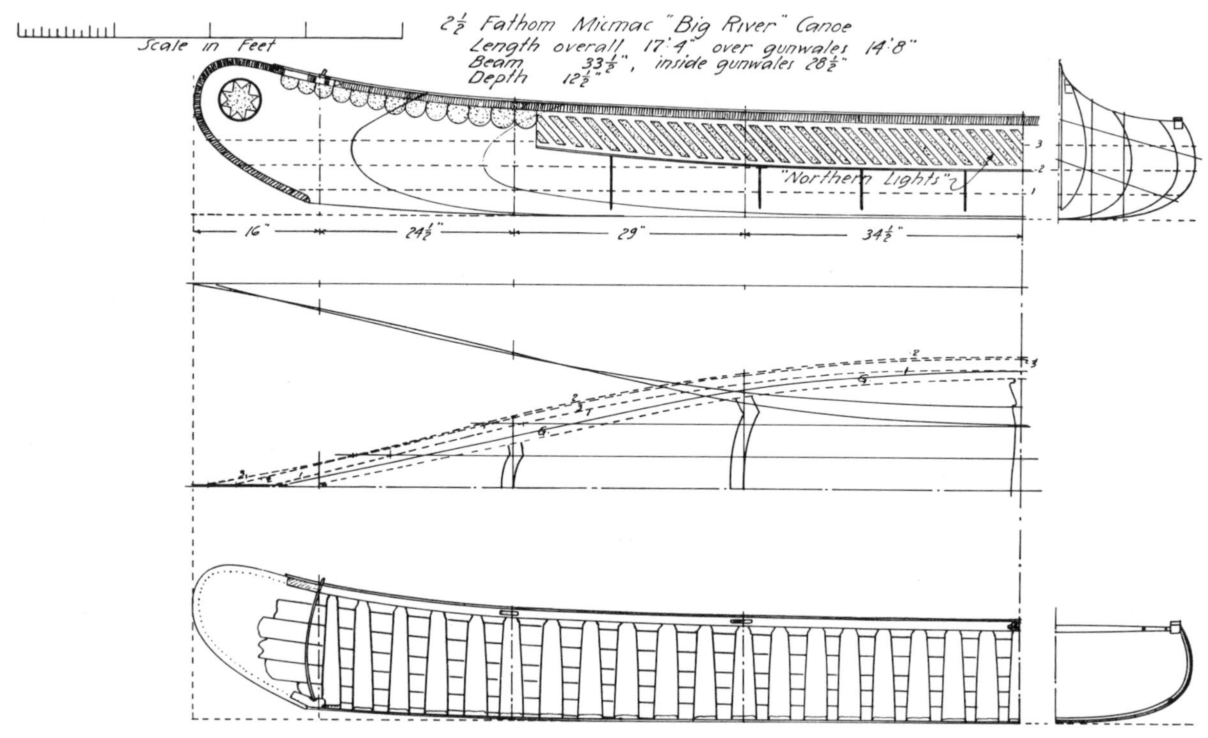


Figure 52

MICMAC 2½-FATHOM BIG-RIVER CANOE, built for fast paddling and of moderate capacity.

of the tenon. In some, the tenon went through the main gunwales and touched the inside of the bark cover; in others the ends of the thwarts were pointed in elevation, square in plan, and were inserted in shallow, blind tenons on the inboard side of the main gunwales. A single 3-turn lashing through a hole in the shoulder and around the main gunwale was used in every case.

Sometimes the thwarts just described were straight (in plan view) on the side toward the middle of the canoe, and only the middle thwart was alike on both sides. In others the straight side of the end thwart and of that next inboard were toward the bow and stern of the canoe. In still others, the middle thwart had a rounded barb form in plan, with the barb located within 6 or 7 inches of the shoulder and pointed toward the tenon; the next thwarts out on each side of the middle thwart were shaped like a cupid's bow but slightly angular and aimed toward the ends of the canoe, and the end thwarts were of similar plan. In one known example having such thwarts, there were two very short thwarts at the ends of the canoe, of the usual plain form described earlier, each a few inches inboard of the headboard. Thus this canoe had seven thwarts in the old fashion.

The ribs, or frames, were thin, about ¼ or 5/16 inch thick, and across the bottom of the canoe they were often 3 inches wide. In the topsides the ribs were tapered to about 2 inches in width; when the bottom and outboard corner of the main gunwales were not beveled, the rib ends were cut square across on the wide face and chisel-shaped. When the gunwale corner was beveled, the ribs were formed with a sharply tapered dull point at the ends. From the middle of the canoe to the first thwarts each way from the middle, the ribs were spaced 1 inch edge-to-edge. From the first thwarts to the ends, the spacing was about 1½ inches. Most builders made the ribs narrower toward the ends; if those in the middle of the canoe were 3 inches wide, those near the ends might be 2½. They were shaped and placed as described for the Malecite canoe in Chapter 3.

In the construction of a Micmac canoe, the gunwales were first formed, assembled, and used as a building frame. If the sheer was to be hogged, this was done by treating the main gunwales with boiling

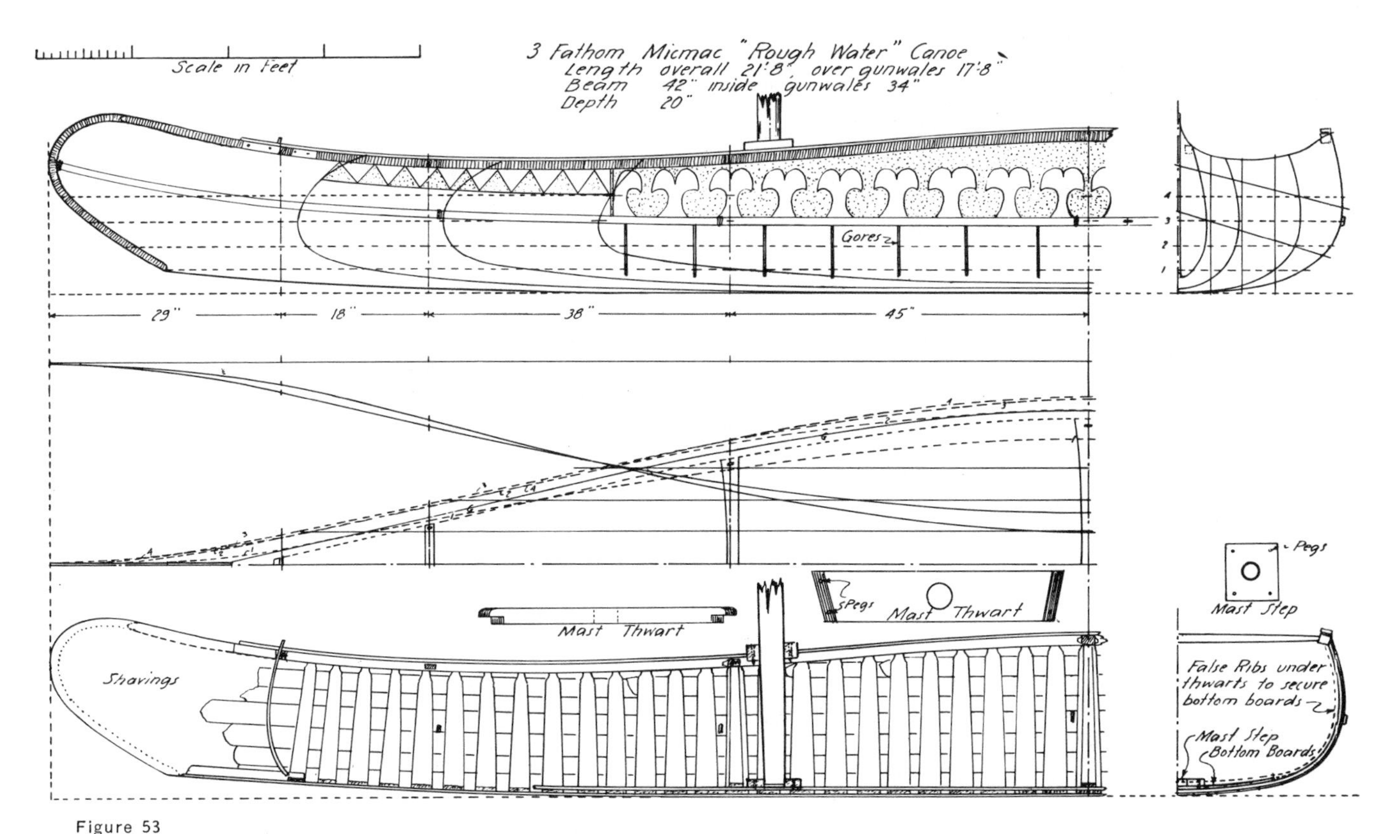


Figure 53

MICMAC 3-FATHOM OCEAN CANOE FITTED FOR SAILING. Short outwales or battens project gunwales to strengthen the ends of the canoe. Some specimens of this type of canoe had almost no rocker in the bottom.

water before assembly and then staking them out to dry in the required sheer curves. The building bed was well crowned, usually 2 to 2½ inches because of the very wide bottom and the tumble-home of these canoes. Most Micmac canoes appear to have had only slight fore-and-aft rocker in the bottom; the bottoms of the seagoing type were often quite straight, and the other two types had a slight rocker of perhaps 1½ inches, most of it near the ends. When the sheer was hogged, the amount of hog was probably close to the amount of crown in the building bed. The ends of the gunwales, when laid on the bed, were blocked up to about the desired amount of rocker to be given the bottom.

The bark cover was selected with great care from the fine stand of paper birch available to the Micmac. Except in emergencies, only winter bark was used. The cover was gored six to eight times on each side, and most of these cuts were grouped amidships, owing to the sharpness of the ends. The gores were trimmed edge-to-edge, without overlap, as the Micmac preferred a smooth surfaced canoe, and the sewing was the common spiral, over and over. The width of the bark cover was usually pieced out amidships on each side (at least in existing models) by the addition of narrow panels. These may not have been necessary in the very old canoes, which appear to have been much narrower than more recent examples. The horizontal seams of the panels were straight, or nearly so, and did not follow the sheer. The closely spaced spiral over-and-over stitch was sewn over a batten, the lap being toward the gunwale. As has been said, a continuous over-and-over gunwale lashing was used. The thwart lashings were through single holes in the thwart shoulders, three turns being usual, and two turns around the gunwale on each side were added, all passing through the bark cover, of course. The sewing was neat and the stitches were even.

The wood lining, or sheathing, of the Micmac canoe was like that described for the Malecite canoe in the last chapter. The sheathing was a full ⅛ to

Figure 54

Micmac Rough-Water Canoe, Bathurst, N.B. (*Canadian Geological Survey photo.*)

about 3/16 inch thick. The strakes were laid edge-to-edge longitudinally, with slightly overlapping butts amidships, and were tapered toward the ends of the canoe. The maximum width of any strake at the butts was about 4 inches.

In some of the rough-water canoes fitted to sail, a guard strip running the full length of the canoe and located some 6 or 7 inches below the gunwale was placed along both sides to protect the strongly tumble-home sides from abrasion from the paddles, particularly when the craft was steered under sail. These strips, about 5/16 inch thick and 3/4 inch wide, were butted on each side, a little abaft amidships, and were held together by a single stitch. The guards were secured in place by rather widely spaced stitches around them that passed through the bark cover and ceiling, between the ribs in the topsides. At bow and stern, the ends of the guards butted against the battens outside the bark at the end profiles and were secured there by a through-all lashing.

The proportions and measurements of the Micmac canoes appear to have changed between the colonial period and the late 19th century. From early refer-

Figure 55

Micmac Woods Canoe, built by Malecite Jim Paul at St. Mary's Reserve in 1911, under the direction of Joe Pictou, old canoe builder of Bear River, N.S. Modern nailed type. (*Canadian Geological Survey photo.*)

ences, it is apparent that the early canoes were much narrower than later ones, in proportion to length, as mentioned earlier. An 18-foot rough-water canoe of the 18th century appears to have had an extreme beam of between 30 and 34 inches and a gunwale beam, measured inside the members, of 24 to 28 inches, the depth amidships being about 18 to 20 inches. A similar canoe late in the 19th century would have had an extreme beam of nearly 40 inches, a beam inside the gunwales of 33 or 34 inches, and a depth of about 18 inches or less. An early woods canoe, about 14 feet long overall, appears to have had an extreme beam of only 29 inches and a beam inside the gunwales of about 25 or 26 inches. A woods canoe of 1890 was 15 feet long, 36½ inches extreme beam, and 30 inches inside the gunwales, with the depth amidships about 11 inches. A big-river canoe of this same date was a little over 20 feet in extreme length, 18 feet over the gunwales, 41 inches extreme beam, and 34 inches gunwale width inside, with a depth amidships of about 12½ inches. An 18-foot big-river canoe of an earlier time was reported as being 37 inches extreme beam, 30½ inches inside the gunwales, and 13 inches depth amidships. The maximum size of the rough-water seagoing canoe, in early times, may have been as great as 28 feet but with a narrow beam of roughly 29 or 30 inches over the gunwales, and say 24 inches inside, with a depth amidships as much as 20 or 22 inches due to the strongly hogged sheer there. In modern times, such canoes were rarely over 21 feet in overall length and had a maximum beam of about 42 inches, a beam inside the gunwales of 36 or 37 inches, and a depth amidships of 16 or 17 inches.

In early colonial times, and well into the 18th century, apparently, the Micmac type of canoe was used as far south as New England, probably having been brought there by the Micmac war parties aiding the Malecite and the Kennebec in their wars against the English. The canoe in the illustration on page 12 is obviously a Micmac canoe and apparently one used by a war party. As it was brought to England in 1749 in the ship *America*, which was built in Portsmouth, New Hampshire, and probably sailed from there, it seems highly probable that the canoe had been obtained nearby, perhaps in eastern Maine.

The small woods canoe, most commonly about 12 feet long, appears first to have been used by all the Micmac. By the middle of the 19th century, however, this type was to be found only in Nova Scotia, owing to the movement of most of the tribe toward the north

Figure 56

MICMAC ROUGH-WATER CANOE fitted for sailing. (*Photo W. H. Mechling, 1913.*)

shore in New Brunswick, where their inland navigation was confined to large rivers and the coast. Hence the Micmac in New Brunswick used the big-river model and the seagoing type. The latter was last used in the vicinity of the head of Bay Chaleur and was often called the Restigouche canoe, after the Micmac village of that name. It was replaced by a 3-board skiff-canoe and finally by a large wooden canoe of the "Peterborough" type with peaked ends and lapstrake planking; some of the latter may still be seen on the Gaspé Peninsula.

The use of sail in the Micmac canoes cannot be traced prior to the arrival of the white men. The use probably resulted from the influence of Europeans, but it is possible that the prehistoric Indians may have set up a leafy bush in the bow of their canoes to act as a sail with favorable winds. The old Nova Scotia expression "carrying too much bush," meaning over-canvassing a boat, is thought by some to have originated from an Indian practice observed there by the first settlers. In early colonial times, the Micmac used a simple square sail in their canoes and this, by the last decade of the 19th century, was replaced by a spritsail probably inspired by the dory-sail of the fishermen. The Indian rig was unusual in several respects. The sheet, for example, was double-ended; one end was made fast to the clew of the sail and the other to the head of the sprit, so that it served also as a vang. The bight was secured

Figure 57

Micmac Rough-Water Canoe, Bay Chaleur. *(Photo H. V. Henderson, West Bathurst, N.B.)*

within reach of the steersman by a half hitch to a crossbar fixed well aft across the gunwales. The sail, nearly rectangular and with little or no peak, was laced to the mast, and the sprit was supported by a "snotter" lanyard tied low on the mast. A sprit boom was also carried by some canoes; this was secured to the clew of the sail and to the mast, a snotter lanyard being used at the latter position.

The mast was secured by a thwart pegged, or nailed, across the gunwale caps. Sometimes, the thwart was also notched over the caps, so that the side-thrust caused by the leverage of the mast would not shear the fastenings. The crossbar for the sheet was sometimes similarly fastened and fitted, with its ends projecting outboard of the gunwales. The heel of the mast was sometimes stepped into a block, which was usually about 5 inches square and 1½ inches thick, nailed or pegged to the center bottom board, or sometimes it was merely stepped into a hole in the center bottom board. The bottom boards, usually

Figure 58

Micmac Rough-Water Sailing Canoe, Bay Chaleur. *(Canadian Geological Survey photo.)*

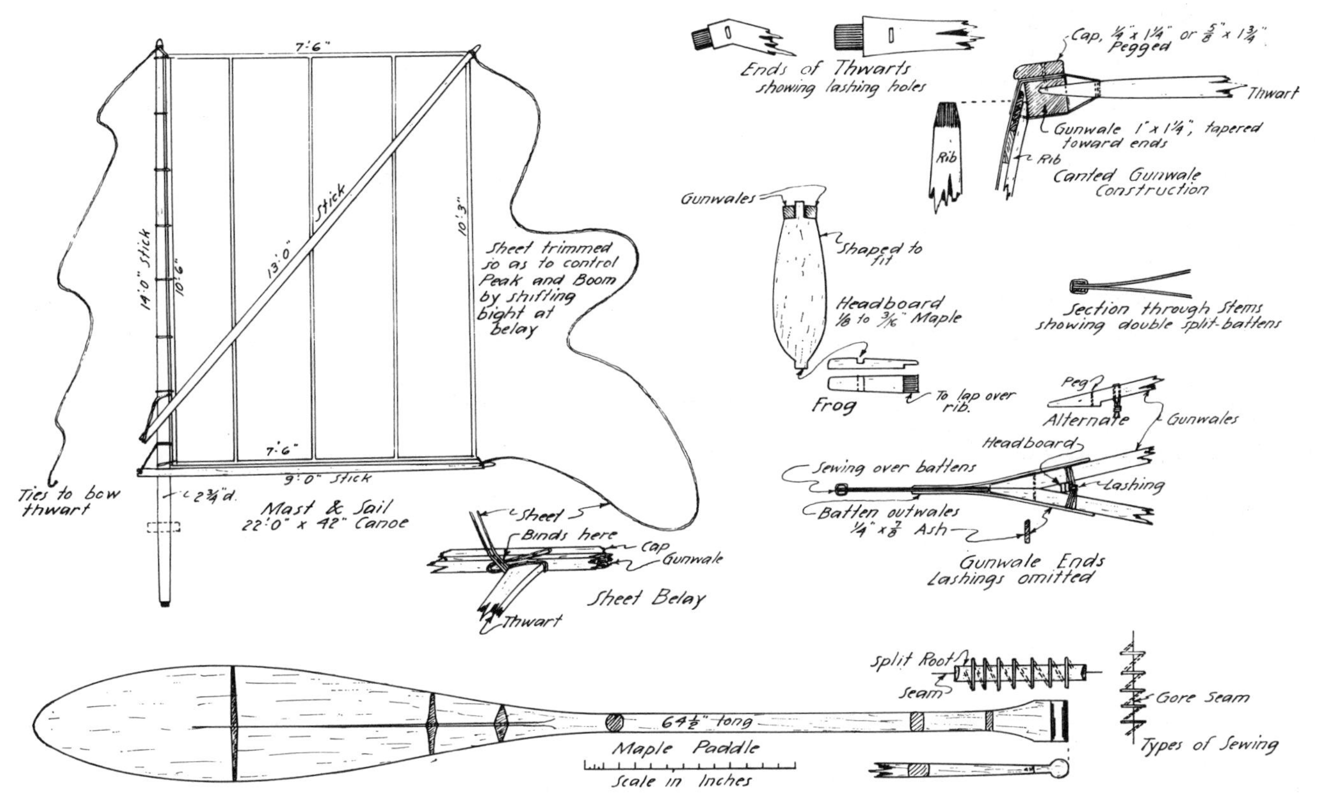


Figure 59

Details of Micmac Canoes, Including Mast and Sail.

three in number were of wide, thin stock and were clamped in place over the ribs by three or four false frames driven under the thwarts, just as were the canoe ribs under the gunwales.

The canoes could not sail close-hauled, as a rule, though some Indians learned to use a leeboard in the form of a short plank hung vertically over the lee side and secured by a lanyard to a thwart, the board being shifted in tacking. An alternate was to have a passenger hold a paddle vertically on the lee side. There seems to have been no fixed proportions to the area of sail used; the actual areas appear to have been somewhere between 50 and 100 square feet, depending upon the size of the canoe. Joseph Dadaham, a Micmac, stated in 1925 that he used "24 yards" in the sail of a "rough-water canoe" 20 feet long and about 44 inches beam, while one 18 feet long and about 36 inches extreme beam carried "16 to 18 yards"; it is obvious that the "yards" are of narrow sail cloth and not square yards of finished sail. In the last days of sailing bark canoes, mast hoops and a halyard block were fitted so that the sail could be lowered instead of having to be furled around the mast (to accomplish this the "crew" had to stand). Dadaham also stated that for his sheet belay he used a jamb-hitch which could be released quickly when the canoe was found to be overpowered by the wind. It appears that during the last era of these bark canoes the rig had been improved to fit it for open-water sailing.

The paddles used by the Micmac appear to have varied in shape. If the canoe shown in Chapter 1 (p. 12) was indeed a Micmac canoe as supposed, the paddle shown there is quite different from the later tribal forms illustrated above, and it is possible that the top grips shown in the more modern forms were never used in prehistoric times, when the pole handle shown with the old canoe may have been standard.

The Micmac canoes were decorated by scraping away part of the inner rind of the birch bark, leaving portions of it in a formal design. It seems very probable that the Micmac seldom used this form of decoration in early times, but later they used it a great deal in their rough-water canoes, perhaps as a result of contact with the Malecite. The formal designs used as decoration by the Micmac did not have any particular significance as a totem or religious symbol; they were used purely as decoration or to identify the owner. Such forms as the half-moon, a star in various shapes, or some other figure might be used by the builder, but these were apparently only his canoe mark, not a family insignia or his usual signature, and could be altered at will.

The usual method of decoration was to place the canoe mark on both sides of the canoe at the ends and to have along the gunwales amidships a long narrow panel of decoration, usually of some simple form. The panel decorations are said by Micmacs to have been selected by the builder merely as pleasing designs. One design used was much like the fleur-de-lis, another was a series of triangles supposed to represent camps, still another was the northern lights design, a series of closely spaced, sloping, parallel lines (or very narrow panels) that seem to represent a design much used in the quill decoration for which the Micmac were noted. Canoes are recorded as having stylized representations of a salmon, a moose, a cross, or a very simple star form; these may have been canoe marks or may once have been a tribal mark in a certain locality. A series of half-circles were sometimes used in the gunwale panels, which were rarely alike on both sides of the canoe, and it is probable that use was made of other forms that have not been recorded. Colored quills in northern lights pattern were used in some model or toy canoes but not in any surviving example of a full-size canoe. It is quite possible, however, that such quill-work was once used in Micmac canoe decoration. Painting of the bark cover for decorative purposes in Micmac canoes has not been recorded.

Historical references to the canoes of the Micmac are frequent in the French records of Canada; it must have been Micmac canoes that Cartier saw in 1534 at Prince Edward Island and in Bay Chaleur. The most complete description of such canoes is in the account of Nicolas Denys, who came to the Micmac country in 1633 and remained there almost continuously until his death at 90, in 1688. His travels during this period took him into Maine as far as the Penobscot and throughout what are now New Brunswick and Nova Scotia. While his descriptions are primarily concerned with the Malecite dress, houses, and hunting and fishing techniques, his notes on birch-bark canoes seem to indicate very clearly that he is describing a hogged-sheer Micmac rough-water canoe. He says, for example, that the length of these canoes was between 3 and 4½ fathoms, the fathom being the French *brasse*, so that they ranged in length from 16 to 24 feet over the gunwales. This gunwale length seems reasonable, since Denys gives the beam as only about 2 English feet, obviously a gunwale measurement in view of the great tumblehome in these canoes. That the Micmac rough-water canoe is the subject of Denys' observations is further indicated by his statement that the depth was such that the gunwales came to the armpits of a man seated on the bottom. This could only be true in a canoe having a hogged sheer in the lengths given, and is, in fact, a slight exaggeration unless the man referred to was of less than average height. The

Figure 60

Micmac Canoe, Bathurst, N.B. (*Canadian Geological Survey photo.*)

depth would be about 22 English inches, great even for a 24-foot canoe. Denys states that the inside sheathing of these canoes was split from cedar. He also states that the splints were about 4 inches wide, were tapered toward the ends, and ran the full length of the canoe. It is probable that they were butted amidships, as in known examples; this, however, would have been covered by a rib and might not have been noticed.

Denys says that the Indians "bent the cedar ribs in half-circles to form ribs and shaped them in the fire." Adney believed this meant by use of hot water. However, this bending could have been done by what was known in 17th-century shipbuilding practice as stoving, in which green lumber was roasted over an open fire until the sap and wood became hot enough to allow a strong bend to be made without breakage. Wood thus treated, when cooled and seasoned somewhat, would hold the set. While it is certain that later Indians knew how to employ hot water, it does not follow that all tribes used this method, particularly in early times.

Denys also states that the roots of "fir," split into three or four parts, were used in sewing. He apparently used "fir" as a general name for an evergreen. It is probable that the roots used were of the black spruce. The technique of building he describes is about the same as that outlined in the last chapter. He says that the gunwales were round and that seven beech thwarts were employed, practices that differ from those in more recent Micmac canoe building, and he notes the goring of the bark cover. Denys states the paddles were made of beech (instead of maple as was perhaps the case) with blades about 6 inches wide and their length that of an arm (about 27 inches), with the handle a little longer than the blade. He also says that four, five, or six paddlers might be aboard a canoe and that a sail was often used. "Formerly of bark," the sail was made of a well-dressed hide of a young moose. Since it could carry eight or ten persons, the canoe Denys is referring to is obviously a large one. In his building description he does not mention headboards, rail caps, or the end forms. It may be assumed that he was then describing a canoe he had seen during construction but whose building he did not follow step by step.

De la Poterie, in his book published in 1722, gives a profile and top view of what must have been a Micmac canoe. The probable length indicated must have been about 22 English feet overall and about 32 inches extreme beam; seven thwarts are shown.

Late in the 19th century there appears to have been some fusion of Micmac and Malecite methods of construction, as Malecite built to Micmac forms and vice versa. This apparently did not produce a hybrid form so far as appearance was concerned but it did affect construction, in that inner end-frames were used and other details of the Micmac design were altered. The Micmac, having early come into close contact with the Europeans, were among the first Indians to employ nails in the construction of bark canoes, and this resulted in an early decadence in their building methods. Hence, some examples of their canoes show what the Indians termed broken gunwales, in which the ends of the thwarts were not tenoned into the gunwales, but rather were let flush into the top by use of a dovetail cut or, less securely, by a rectangular recess across the gunwale, and were held in place with a nail through the thwart end and the gunwale member.

From scanty references by early writers, it appears that a spiral over-and-over lashing was originally used by the Micmac on the ends and gunwales. The lower edges of the side panels were sewn over-and-over a split-root batten. In some extant examples the gores are sewn with a harness stitch; in others a simple spiral stich is used. The cross-stich does not appear to have been used by the Micmac. The gunwale caps were certainly pegged and the ends lashed; the bark cover was folded over the gunwale tops and clamped by the caps as well as secured by the gunwale lashings. Tacking the bark cover to the top of the gunwales, with the cap nailed over all, marks the later Micmac canoes. The use of nails and tacks seems to have begun earlier than 1850.

In spite of decadent construction methods used in the last Micmac birch-bark canoes, the model re-

Figure 61

Micmac Woman gumming seams of canoe, Bathurst, N.B., 1913. (*Canadian Geological Survey photo.*)

mained a very good one in each type. The half-circular ends, sharp lines, and standard midsectional forms were unaltered; the hogged sheer was retained in some degree in at least two of the canoe types, the rough water and the big river, right down to the end of bark-canoe building by this tribe. The very fine design and attractive appearance of the Micmac canoe may have contributed to the early acceptance by the early explorers and traders of the birch-bark canoe as the best mode of water transport for forest travel.

## *Malecite*

Another tribe expert in canoe building and use was the Malecite. These Indians were known to the early French explorers as the "Etchimins" or "Tarratines" (or Tarytines). Many explanations have been given for the name Malecite. One is that it was applied to these people by the Micmac and is from their word meaning "broken talkers," since the Micmac had difficulty in understanding them. When the Europeans came, these people inhabited central and southern New Brunswick and the shore of Passamaquoddy Bay, with small groups or tribal subdivisions in the area of the Penobscot to the Kennebec. These were early affected by the retreat of the New England Indians before the whites into eastern and northern Maine and southeastern Quebec. As a result, the Penobscot and Kennebec Indians became part of the group later known as Abnaki, while the Passamaquoddy Indians remained wholly Malecite and closely attached to those living along the St. John River in New Brunswick. Like their neighbors the Micmac, the Malecite were hunters and warlike; during the colonial period they were usually friendly to the French and enemies of the English settlers in their vicinity. It is not certain that the tribe now called by that name were actually of a single tribal stock; it is possible that this designation really covers a loose federation of small tribal groups who eventually achieved a common language. In addition, the tribal designation cannot be wholly accurate because of the fact that much of the original group living in New England were absorbed in the Abnaki in the 17th and 18th centuries. Therefore, the Malecite are considered here to be those Indians formerly inhabiting valleys of the St. John and the St. Croix Rivers, and the Passamaquoddy Bay area. The remaining portions, the Kennebec and Penobscot Indians, must now be classed as Abnaki, of whom more later (see p. 88).

In considering the birch-bark canoes of the Malecite, it is important to understand that this tribal form includes not only the types used in more recent times in New Brunswick and on Passamaquoddy Bay, but also an overlapping type related to the later Abnaki models. The old form of Malecite canoe used on the large rivers and along the coast appears to have had rather high-peaked ends, with a marked overhang fore and aft. The end profiles had a sloping outline, strongly curved into the bottom, and a rather sharply lifting sheer toward each end. This form was also to be seen in old canoes from the St. John River (the lower valley), the Passamaquoddy, the Penobscot, and the upper St. Lawrence. By late in the 19th century, however, this style of canoe had been replaced by canoes having rounded ends, the profiles being practically quarter-circles and sometimes with such small radii that a slight tumble-home appeared near the sheer. The small radius of the end curves is particularly marked in some of the seagoing porpoise-hunting canoes of the Passamaquoddy. In modern forms, the amount of sheer is moderate and the quick lift in the sheer to the ends is practically nonexistent. On the St. Lawrence, the radii of the end curves are very short and the upper part of the stems stands vertical and straight; the sheer, too, is usually rather straight. The older type, with high-peaked ends, was also marked by very sharp lines forward and aft, and had a midsection with tumble-home less extreme than in the Micmac canoes. The bottom, athwartships, was usually somewhat rounded (in coastal canoes the form might be a rounded V) and the bilges were rather slack, with a reverse curve above, to form the tumble-home rather close to the gunwales. The river model probably had lower ends and less rake than the coastal type, but surviving examples of both give confusing evidence. The river canoes usually had a flatter bottom than the coastal type, the latter having somewhat more rocker fore-and-aft. The sections near the ends were rather V-shaped in the coastal canoes, U-shaped in the river canoes.

The old form of small hunting canoe is represented by but one poor model (see p. 72) in which the ends are lower and with much less rake than those of the river type. From this very scant evidence, it seems probable that the small woods canoes were patterned on the river canoe in all respects but the profile of the ends.

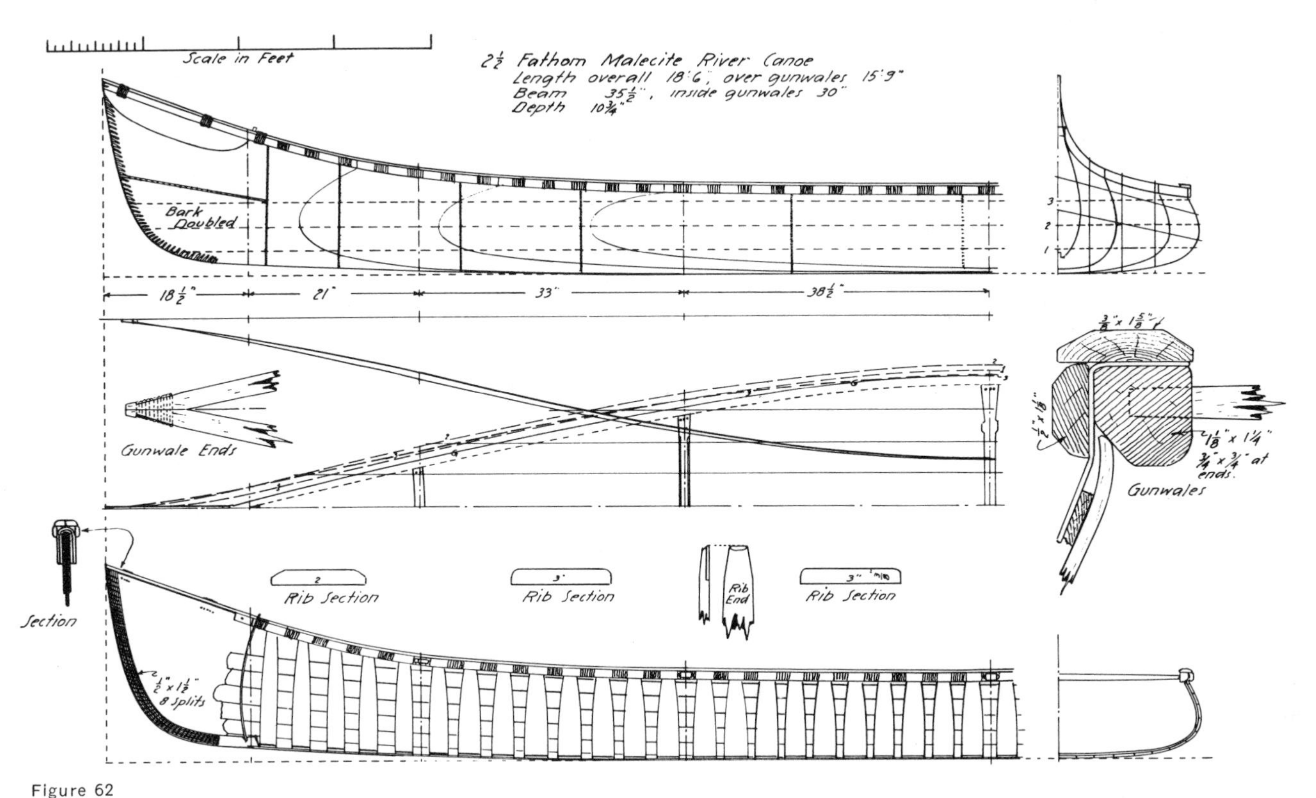


Figure 62

MALECITE 2½-FATHOM RIVER CANOE, 19TH CENTURY. Old form with raking ends and much sheer.

From the early English and French accounts, it is evident that none of the maritime Indians used very large or long war canoes, capable of holding many men. The old war canoes of the Malecite appear to have been either of the coastal or river types as the circumstances of their place of building and use dictated. The slight information available in these accounts suggests that the war canoe did not differ in appearance from the other types of Malecite canoes, and that they were not of greater size. The Malecite appear to have followed the same practices as the Micmac, using for war purposes canoes of standard size and appearance but narrower and built for speed, since a war party sought to travel rapidly to and from its objective in order to surprise the enemy and escape before organized pursuit could be formed. The Malecite placed four warriors in each canoe, two to paddle and two to watch and use weapons while afloat. However, only on rare occasions were bows and arrows used from canoes afloat; most fighting was done on land. Each canoe carried the personal mark of each of the four warriors, apparently one mark on each flap, or *wulegessis*, under the gunwales near the ends. When a war leader was carried however, only his mark was on his canoe. After a successful raid, the Malecite used to race for the last mile or so of the return journey, and the winning canoe was given, as a distinction, some mark or picture, often something humorous such as a caricature of an animal. This practice, however, was not confined to war canoes; in rather recent times it has been noted that such pictures were placed on any canoe that had shown outstanding qualities in racing competition or in exhibitions of skill.

When making long canoe trips, the Malecite followed the widespread Indian practice of using the canoe as a shelter at night. When a camping place was reached, the canoe was unloaded, carried ashore, and turned upside down so that the tops of the ends and one gunwale rested on the ground. If the ends were high enough, as in the old Malecite type, one gunwale was raised off the ground far enough to permit a man to crawl under. If, as in the Micmac canoes, the ends were too low to allow this, they were raised off the ground by short forked sticks, with the forks resting against the end thwarts and

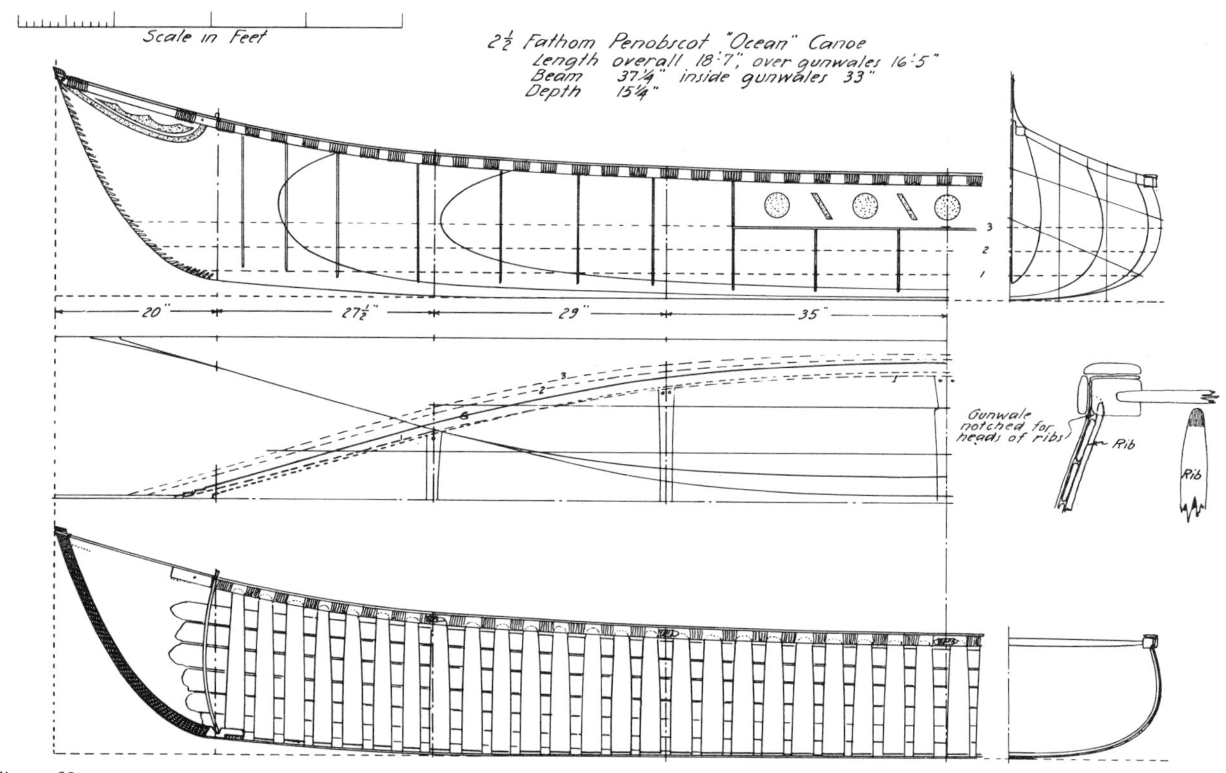


Figure 63

Old Form of Malecite-Abnaki 2½-Fathom Ocean Canoe of the Penobscots. In the Peabody Museum, Salem, Mass.

the upper gunwale and the heels stuck into the earth. The dunnage (provisions or other cargo) was then stowed on the ground under the ends of the canoe and the two men would sleep under a single blanket with their feet pointed in opposite directions, each with his head on a pile of dunnage. If there were too many men aboard to do this, in bad weather a crude shelter was made by resting some poles on the upturned bilge and covering them with sheets of bark; under such a shelter meals could be cooked.

As did many of the eastern Indians, the old Malecite tribesmen built canoes of materials other than birch bark. When a canoe was required for a temporary use such as in hunting, it could be made of spruce bark. (As the designs of such canoes were rather standardized, they will be dealt with in Chapter 8.) When bark was unobtainable, the Malecite built canoes covered with moosehide, or, in rare instances, they built wooden dugouts.

The old Malecite river canoe shown on page 71 will serve to illustrate a description of the details of construction that were used. These canoes were obviously built with their gunwales (which were the length of the bottom only) serving as a building frame. The ends of the gunwales were supported by headboards stepped on the heels of the inner stem-pieces, and the stems raked outward from their heels. The gunwale ends were joined to the head of the stem-piece by the outwales and the gunwale caps. Bark was used to the ends of the canoe. One side of the bark cover was cut so that it stood well above the sheer line from the gunwale end outboard, and the opposite side was cut to the level of the sheer. The first piece was then folded over the opposite side and down, so that it covered both the extreme ends of the gunwales and the top of the inner stem-piece. Another piece of bark was then fitted over this fold, and this new piece formed the flaps below the outwales on each side, the *wulegessis*. The outwales ran past the gunwale ends and were cut off flush with the outboard face of the stem; the caps ran likewise and covered the bark over the head of the inner stem piece. The characteristic sheer of these canoes, where the rise toward the ends began, showed a quick curve that faired into a rising straight line at the gunwale and then continued straight and rising to the stem head.

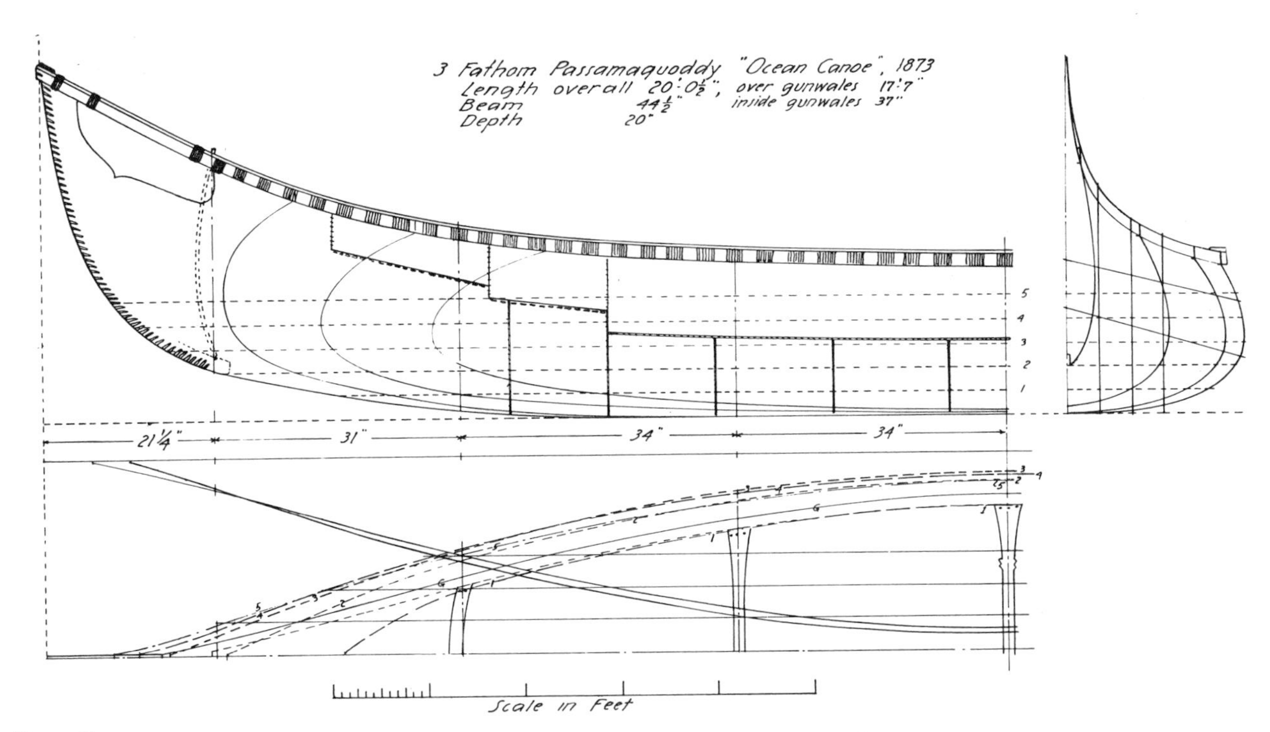


Figure 64

LARGE 3-FATHOM OCEAN CANOE OF THE PASSAMAQUODDY porpoise hunters. These canoes were sometimes fitted to sail or outrigged for rowing. The last of this type had much lower ends.

The *wulegessis* was therefore quite long. The ends of the gunwales were not of the half-arrowhead shape, but were snied off on their inboard sides so that they met on a rather long bevel; the lashing was slightly let in to the outboard faces to keep it from slipping over the gunwale ends. The caps of the gunwales were similarly reduced in width, where they came together over the ends of the canoe.

The main gunwale members were about 1¼ inches square amidships, tapering to ¾ inch at the ends. The lower outboard corner was beveled to take the ends of the ribs, as shown on page 71, and the lower inboard corner was also beveled or rounded, but to a lesser degree. The upper inboard corner, shown beveled in the drawing of figure 62, was sometimes slightly rounded, as were the outwales. Amidships the outwale was about 1 inch deep, and it tapered toward the ends, where its depth was about ⅝ inch, the thickness being ½ inch amidships and a scant ⅜ inch at the ends. On the canoe shown, the cap was ⅜ inch thick, tapering to about 5/16 inch at the ends, and 1¾ inches wide amidships, tapering to about ⅝ or ½ inch where the caps came together at the ends. The top corners of the cap were beveled in the example.

The sheathing appears to have been about 3/16 inch thick on the average. On the bottom and sides it was in two lengths, overlapping slightly amidships. Toward the ends of the canoe the sheathing was tapered, maximum width of the splints being about 4 inches amidships.

The canoe, which was 18 feet 6 inches long overall, had 46 ribs. These were about 3 inches wide and ⅜ inch thick from the center to the first thwart outboard on each side, and 2 inches wide from these thwarts to the ends, except for the endmost five ribs, which were roughly 1¾ inches wide. The drawing on page 71 shows the shape of the thwarts. The ends were tenoned through the gunwales, and there were three lacing holes in the ends of the middle and first thwarts and two in the end thwarts. The beam of the canoe inside the gunwales was 30 inches and outside, 31¼ inches; the tumble-home made the extreme beam 35½ inches. The canoe was rather flat bottomed athwartships and quite shallow, the depth amidships being 10¾ inches.

The building bed must have had about a 1½ inch crown at midlength. It is probable that the stem pieces were not fixed in place until after the gun-

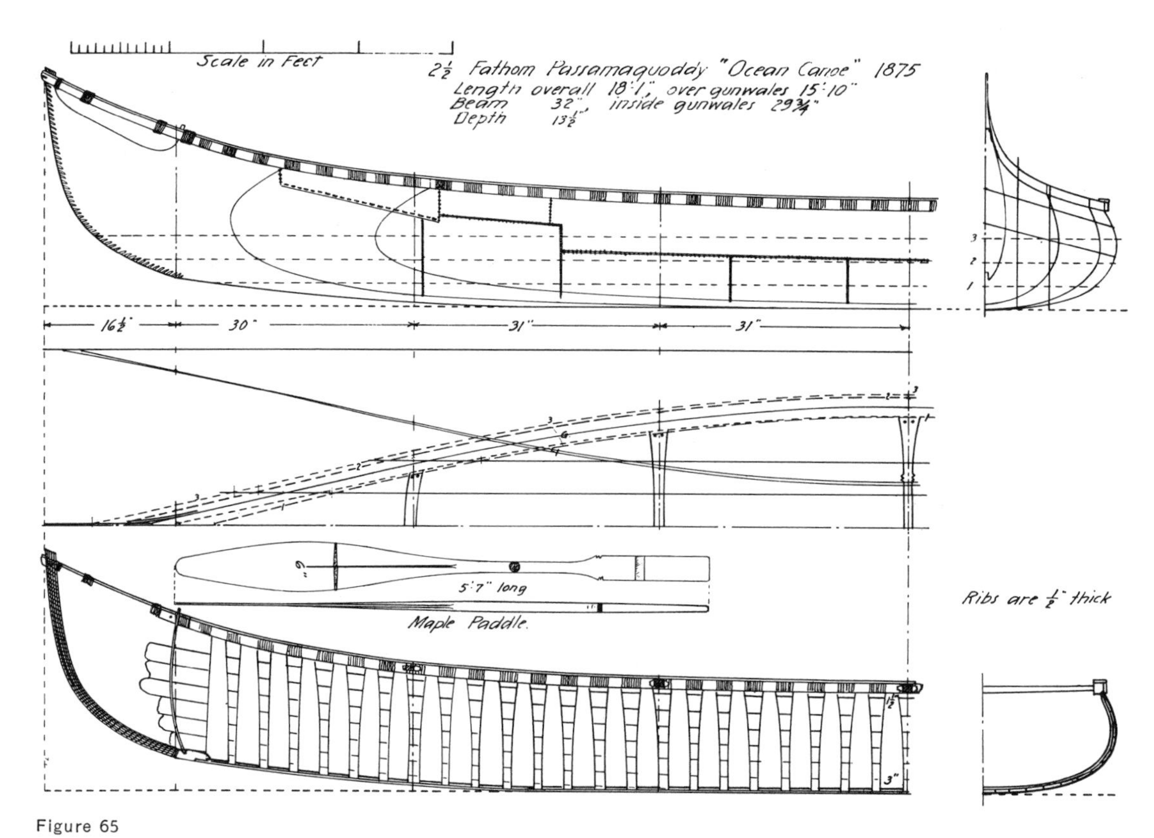


Figure 65

OLD FORM OF PASSAMAQUODDY 2½-FATHOM OCEAN CANOE with characteristic bottom rocker and sheer. This rather small, fast canoe for coastal hunting and fishing was common in the 19th century.

wales had been raised to sheer height. The gunwales were lashed with the Malecite group lashings, each of four turns through the bark and spaced at 3 to 3½ inches apart in the midlength and at 2 inches from the end thwarts to the headboards. Two auxiliary lashings were placed over the outwales and caps outboard of the gunwale ends, one about 6 inches beyond the ends of the gunwales and the other against the inboard side of the stem-piece. The end closure was accomplished by the usual spiral lashing passed through the laminated stem pieces. The latter were split (to within about 4 inches of the heel), into six or more laminae that were closely wrapped with bark cord. The headboards were bellied toward the ends to keep the bark cover under tension, and the ends outboard of the headboards were stuffed with shavings or moss.

A canoe from the Penobscot River, obtained in 1826 by the Peabody Museum, Salem, Massachusetts, and described in *The American Neptune* for October 1948, shows that the Penobscot built their canoes on the old Malecite model. The canoe is apparently a coastal type. It has some round in the bottom amidships and V-sections toward the ends; it is 18 feet 7 inches long overall, 37¼ inches maximum beam, 15¼ inches deep amidships, and the ends stand 26 to 28 inches above the base line, the bow being slightly higher and with more rake than the stern. The rocker takes place within 4 feet of the ends, with the bottom straight for about 8 feet along the midlength. The bilges amidship are slack, and the reverse curve to form the tumble-home starts within 6 inches of the gunwales (see drawing, p. 72.)

A much later coastal canoe of the Passamaquoddy, a porpoise- and seal-hunting canoe built in 1873, will also serve to show the old type (see p. 73). This style of canoe was usually built in lengths ranging from 18 to 20 feet overall, the maximum beam was between 25 and 44 inches, and the beam inside the gunwales was between 29½ and 36 inches. The depth amidships ranged from about 18 to 21 inches, and the height of the ends above the base was from 28 or 30 inches to as much as 45 inches. The ribs numbered from 42 to 48 and were 3 inches wide and

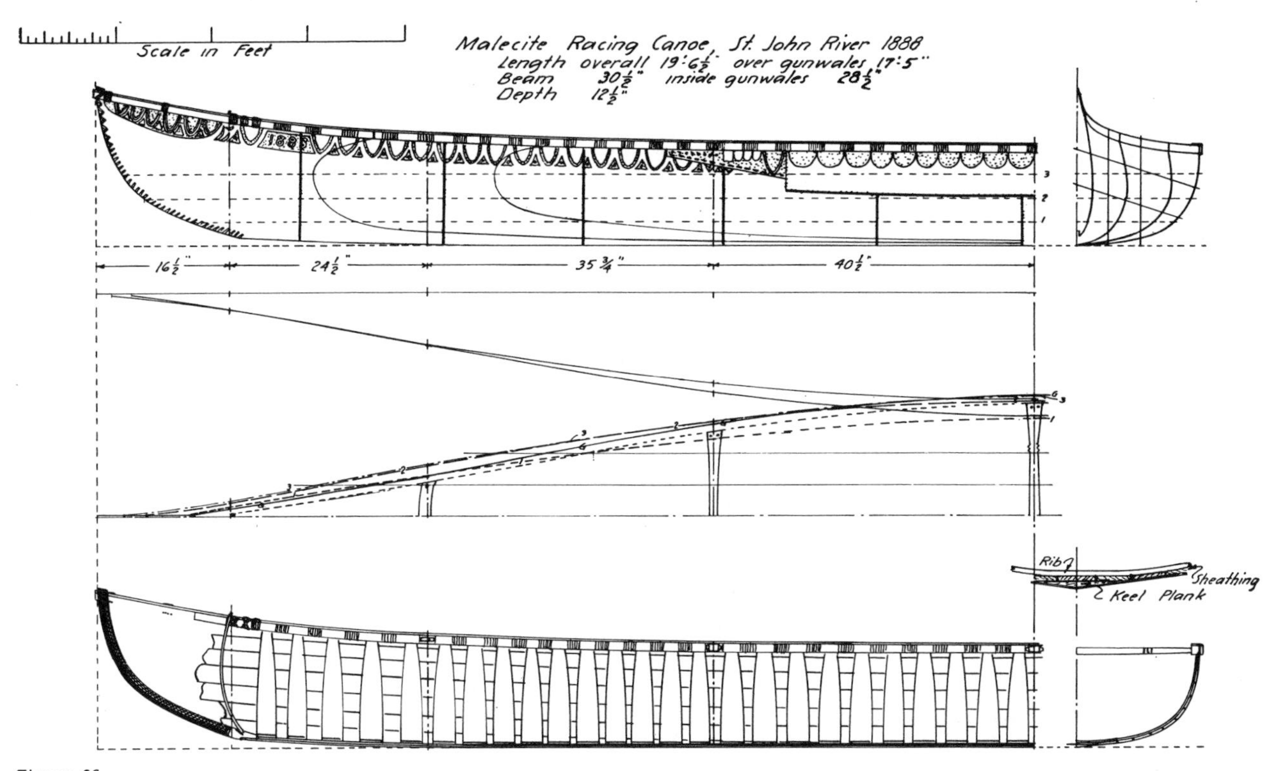


Figure 66

MALECITE RACING CANOE OF 1888, showing V-shaped keel piece placed between sheathing and bark to form deadrise.

½ inch thick. The sheathing was from ¼ to ⅜ inch thick and the rocker of the bottom, from 4 to 6 inches, took place within the last 4 or 5 feet of the ends. The midsection showed a well-rounded bottom, a slack bilge, and the high reverse to form the tumble-home seen in the old Penobscot canoe at Salem. These canoes were still being built well into the 1880's, if not later, and are to be seen in some old U.S. Fish Commission photographs of porpoise and seal hunting at Eastport, Maine. Seal- and porpoise-hunting canoes carried a sail, usually the spritsail of the dory. While this model probably was little changed in construction from early times, the surviving examples and models are of the period when nails were employed. The drawing on page 74 is of a small coastal hunting canoe of the same class, built in 1875.

The reasons for the gradual decline in the building of canoes of the old style are not known, and the transition from the high-peaked ends to the more modern low and rounded ends was not sudden. It apparently began in some inland areas, particularly on the St. Lawrence and the St. John Rivers, at least as early as 1849, and the new trend in appearance finally reached the coast about 25 years later. In the period of transition, the high-peaked model developed toward the St. Francis type, or that of the modern "Indian" canvas canoe, as well as toward the low-ended type.

One of the later developments took place on the St. John River, in New Brunswick, where two Indians, Jim Paul and Peter Polchies, both of St. Marys, in 1888 built for a Lt. Col. Herbert Dibble of Woodstock the racing canoe illustrated above (fig. 66). This canoe, 19 feet 6½ inches long overall and only 30½ inches extreme beam, was of a design perhaps not characteristic of any particular type of Malecite canoe, but it nevertheless shows two elements that may have appeared during the period of change in model. The sides amidships not only are without tumble-home, they flare outward slightly, but tumble-home is developed at the first thwart each side of the middle and continues to the headboards. The bottom shows a marked V-deadrise achieved by an unusual construction in a birch-bark canoe: the center strake of the sheathing is shaped in a shallow V in cross section, its width being about 2½ inches amidships and taper-

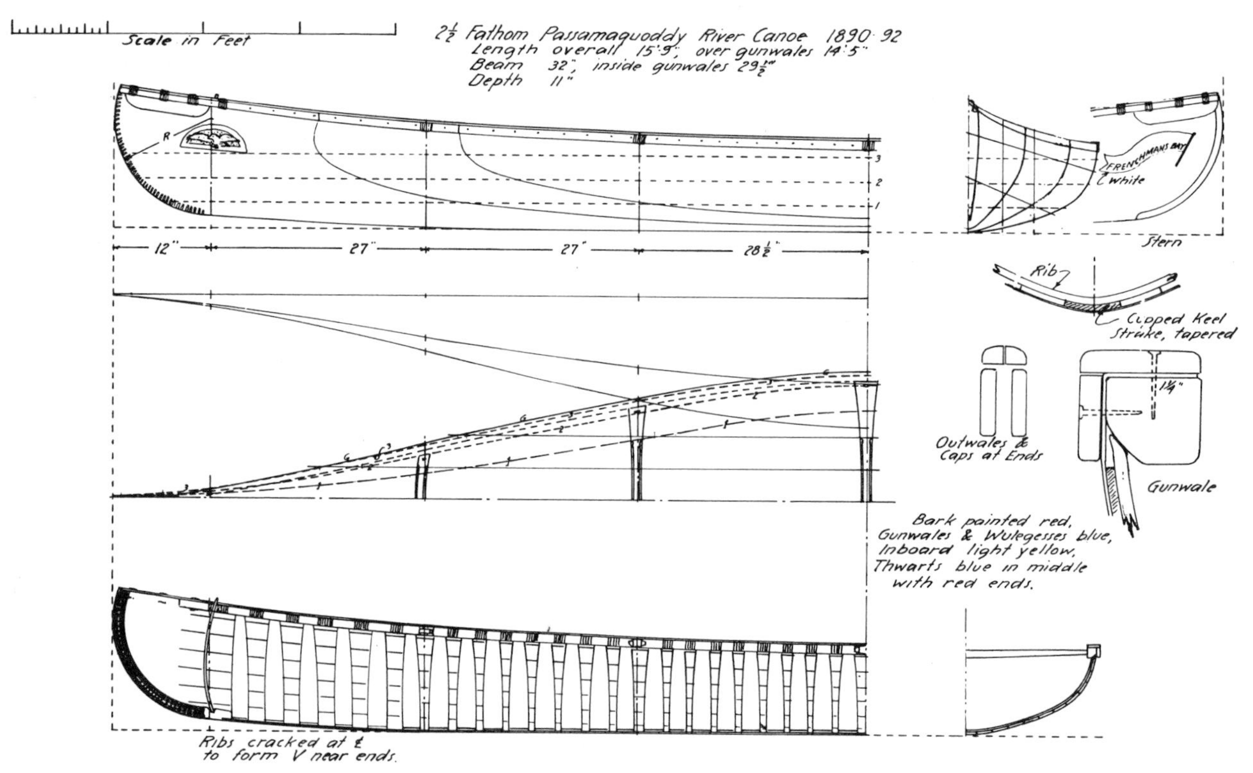


Figure 67

SHARP-ENDED 2½-FATHOM HUNTING CANOE for use on tidal river. Built by the Passamaquoddy Indian Peter Denis, it shows what may be the primitive construction method of obtaining a V-form in hull.

ing each way toward the ends, and its thickness along the longitudinal centerline being about ⅝ inch and tapering to about ¼ inch at the edges; the two lengths of the strake are butted, not lapped, amidships, though the rest of the sheathing is lapped at the butts in the usual way and is uniformly ¼ inch thick. In this manner a ridge that gives a V-deadrise is formed down the centerline of the bottom, though the frames are bent in a flattened curve from bilge to bilge. The bottom has very little rocker, the rise being only 1 inch, and this takes place in the last 2 feet inboard of the heel of the stem piece.

Another feature in this canoe is the end profile; the curved ends are strongly raked, the curve used being the same as that in the old Malecite type, but with the stem-pieces reversed, so that the quick turn is at the head, near the sheer, rather than at the heel. As a result, the gunwales come to the ends in a straight, rising line for the last 16½ inches rather than as a sudden lift near the ends. The stem-heads stand a little above the rail caps. The headboards belly toward the ends and are raked in the same direction.

The use of a V-shaped keel piece in the sheathing has been found in a St. Francis canoe from the St. Lawrence country; this may be a rather old practice. This racing canoe is very lightly built and much decorated, the date 1888 being worked into the hull near one end.

Another canoe having a marked V-deadrise was built sometime between 1890 and 1892 by Nicola (sometimes called Peter) Denis (sometimes spelled Dana), a Passamaquoddy, for his son Francis, who used it at Frenchman's Bay, Maine. The drawing above (fig. 67) shows a coastal-type hunting canoe, nailed along the gunwales but sewn elsewhere, and painted. The craft is 15 feet 9 inches overall and 14 feet 5 inches over the gunwales. The beam amidships is 32 inches over the gunwales, 29½ inches inside. The depth amidships is 11 inches, and at the headboards, 14½ inches. The ends are of the low rounded form; the profile shows a moderate tumble-home just below the sheer, which is a long fair curve without any quick lift toward the ends. The construction is of the usual Malecite type described in Chapter 3.

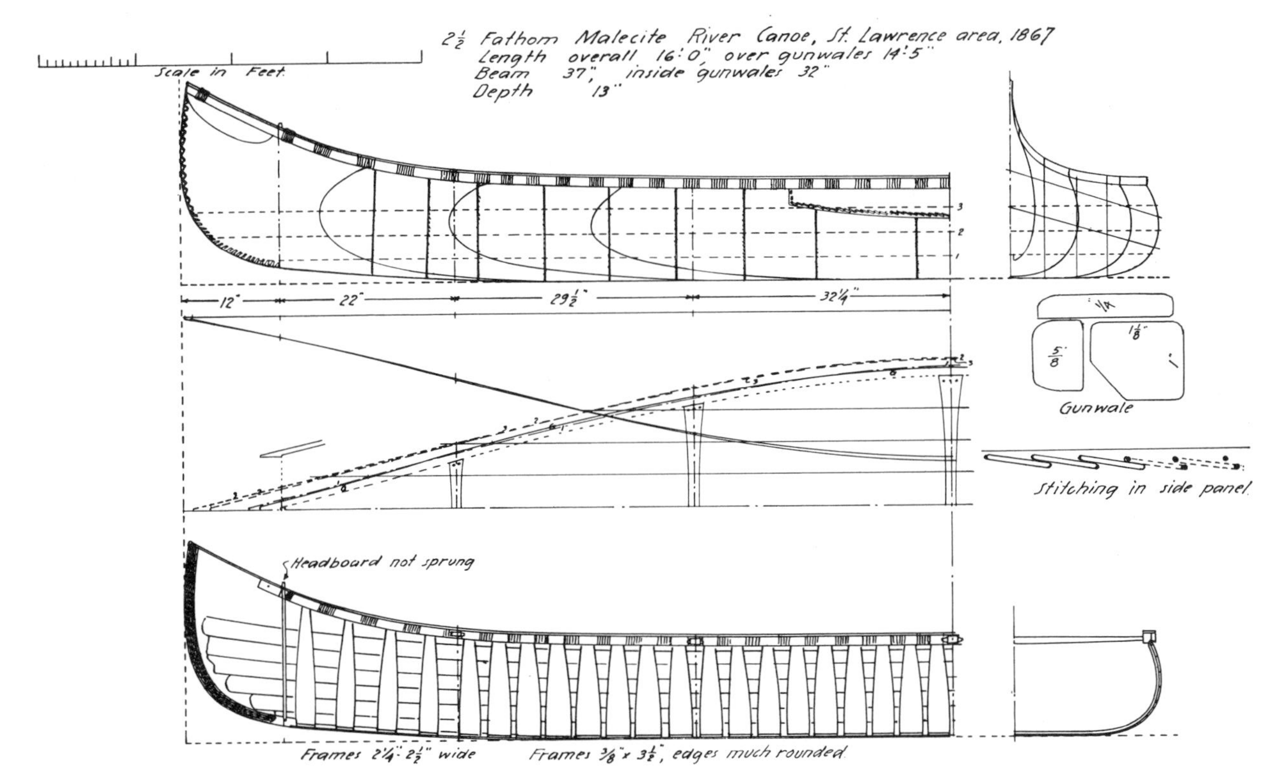


Figure 68

MALECITE 2½-FATHOM ST. LAWRENCE RIVER CANOE, probably a hybrid model. The high ends show a western influence.

The midsection shows a remarkable amount of V in the bottom without any tumble-home anywhere in the topsides. The V-bottom is rounded at the apex, where the keel would be; this is done by bending the ribs very sharply where they cross the centerline of the hull. A narrow strake of thin sheathing runs along the centerline of the canoe, and this is bent athwartwise to follow the bends in the ribs there. The canoe had 46 ribs, each 2½ inches wide and 5/16 inch thick, tapered slightly from the middle up to the gunwales. The gunwales, as previously noted, are nailed and the main gunwale members are of sawed spruce. The rest of the framework is cedar.

The outside of the canoe was painted red, the inside was a pale yellow, the gunwales and middle portions of the thwarts were cobalt blue, the ends of the thwarts were red. The *wulegessis* was blue, and the "canoe mark" was a painted representation of the spread eagle of the United States Seal, the border being in black and white and the eagle in black, yellow, and white, holding a brown branch with green leaves. The whole panel was outlined in red. On the side of the canoe, near the stern, was a white swallowtail pennant on which is lettered "Frenchmans Bay" in black capital letters. This canoe was used for fishing and also for porpoise and seal hunting.

The construction employed to form the V-bottom in a birch-bark canoe can be seen to have been done in two ways; that described on page 76 is undoubtedly the method used in prehistoric times, since laborious forming of a V keel-piece in the sheathing, using stone scrapers, would be avoided. The V-bottom, it should be noted, usually appears in canoes used in open waters, as this form tends to run straight under paddle, in spite of a side wind, and thus requires the minimum of steering to hold it on its course. It was this characteristic, too, that made the V-bottom suitable for the racing canoe on the St. John River, since stopping the stroke momentarily to steer diminishes the driving power of the stern paddler.

The various river canoes of the Malecite, built to the modern low, rounded-end profiles, or to the short-radii and straight-line forms, held rather closely to the same lines, that is, sharp ends with a rather flat bottom amidships and an easy bilge. Some of the canoes retained the characteristic tumble-home,

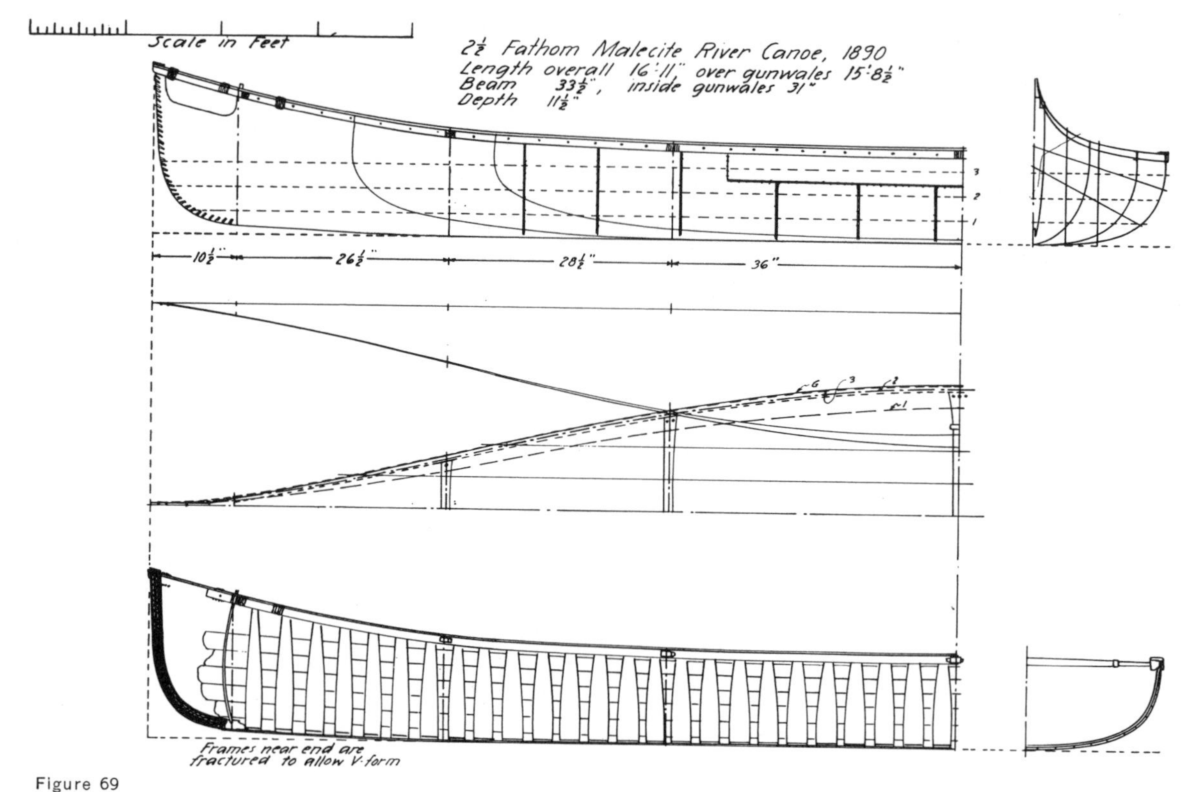


Figure 69

MALECITE 2½-FATHOM RIVER CANOE of 1890 from the Rivière du Loup region. Canoes in this area had straight stems and sharp lines from at least as early as 1857.

but others had nearly vertical sides or the curve of the bilge was carried so high that it ended at the gunwales.

On the St. Lawrence there was apparently a canoe having rather peaked ends as well as the rather straight-stemmed, low-ended type. A St. Lawrence River canoe found in the Chateau de Ramezay and built sometime before 1867 provides an example of the rather high-peaked ends. The canoe, as illustrated on page 77, has a well-rounded bilge working into a very round tumble-home above and into a rather flat bottom below, the tumble-home being carried into the extreme ends, so that the headboards are rather wide. The ends round up rather quickly and then continue up to the sheer in a very slight curve, having a very moderate tumble-home near the sheer. The latter follows somewhat the characteristic sheer of the old Malecite canoes, but the straight portion just inboard of the ends is much shorter, so that the quick upsweep of the sheer begins nearer the ends and thus appears somewhat more pronounced.

The construction is in the usual manner. The rocker of the bottom is 2 inches. The ribs are wider amidships than near the ends. The outwale is rounded on the outboard face so that the cap is slightly narrower than the thickness of inner gunwale and outwale combined. The headboard is rather unusual, however, as it is not bellied but stands straight and vertical. The lashing at the upper portion of the stems is the crossed stitch, below it is spiral. The gunwale groups are made up of six passes through the bark, and the spaces between groups are about 2½ inches. The side panels are sewn with the harness stitch. The canoe is 16 feet long overall and 14 feet 5 inches inside the gunwales; the extreme beam amidships is 37 inches and inside the gunwales 32 inches. The depth amidships is about 13 inches and the height of the ends 25 inches, with 2 inches of rocker at the headboards. This canoe, retaining the high ends, marks the transition from the old form to the new.

A later canoe built on the St. Lawrence about 1890, probably near Rivière de Loup, is shown above. It is 16 feet 11 inches long overall, the beam over the gunwales is 33½ inches and inside it is 31 inches, the curve of the bilge being carried up to the gunwales.

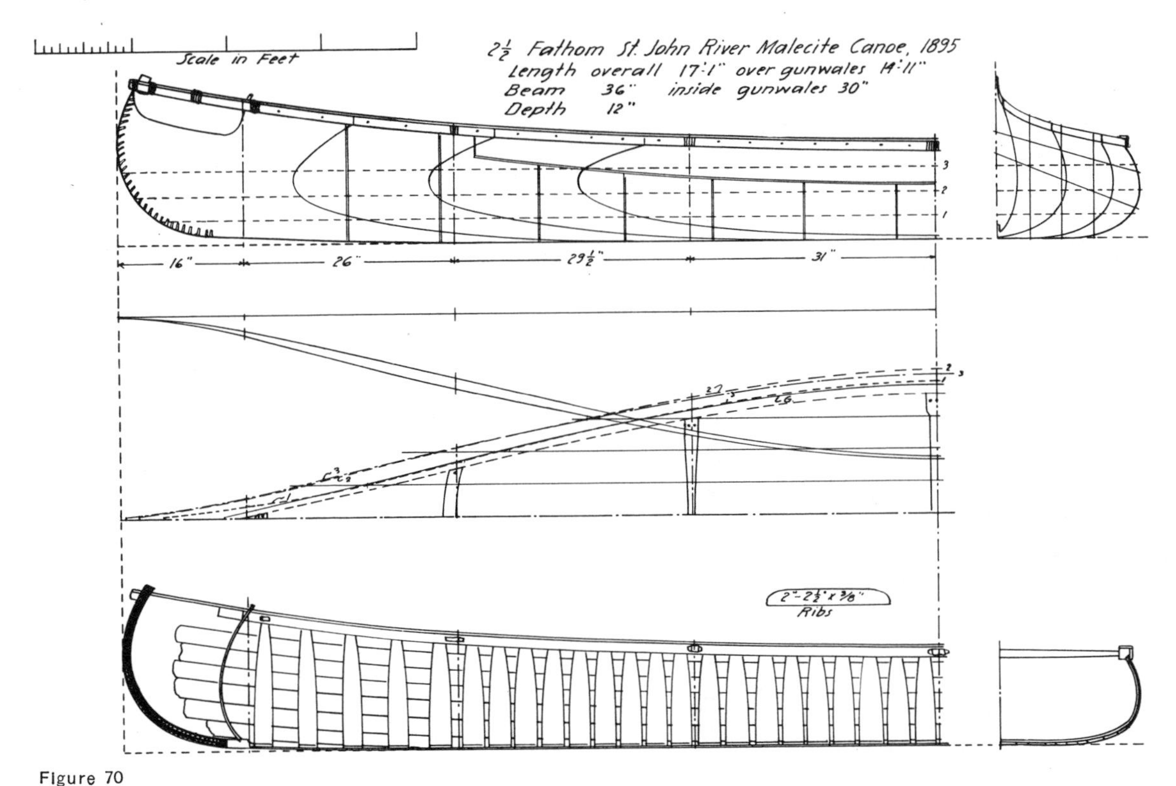


Figure 70

Modern (1895) Malecite 2½-Fathom St. John River Canoe, with low ends and moderate sheer, developed late in the 19th century.

The bottom is flat for only a short width. The depth amidships is 11½ inches and the height of the ends is 20 inches, with 1 inch of rocker in the last two feet of length. The sheer is a long fair sweep without any quick upward lift near the ends. The headboards are very narrow and belly only very slightly toward the ends. The end profile illustrates the short radii and straight line form that marked many of the last Malecite birch-bark canoes of the St. Lawrence Valley. It is possible that the end-form was copied from the white man's St. Lawrence skiff, which usually had ends that were straight and nearly vertical, with a sharp turn into the keel.

Since a Malecite canoe of the form having rounded low ends was the subject used to describe the construction of a birch-bark canoe in Chapter 3 (see p. 36), there is no need to discuss all the details here. There was some variety in the sewing and lashing used in Malecite canoes; the combination of cross and spiral stitches in the ends and the use of a batten and the over-and-over stitch in the side panels are, of course, very common in these canoes. The occasional use of other stitches in the side panels and even in the gores would probably be normal, since individual preferences in such details were not controlled by a narrow tribal practice.

The Malecite are known to have hauled their canoes overland in the early spring, before the snow was entirely gone, by mounting the canoe on two sleds or toboggans in tandem, binding the canoe to each. This was done as late as the 1890's for early spring muskrat hunts. The Malecite also fitted their river canoes with outside protection when much running of rapids or "quick water" work was done. This protection consisted of two sets of battens (see p. 80), each set being made up of five or six thin splints of cedar about ⅜ inch thick and 3 inches wide, tapering to 2 or 1½ inches at one end. These were held together by four strips of basket ash, bark cord, or rawhide. Each cord was passed through holes or slits made edgewise through each splint. The cords were located so that when the splints were placed on the bottom of the canoe, the cords could be tied at the thwarts. The tapered ends of the splints were at the ends of the canoe; the butts of the two sets being lapped amidships with the lap toward the stern. This formed a

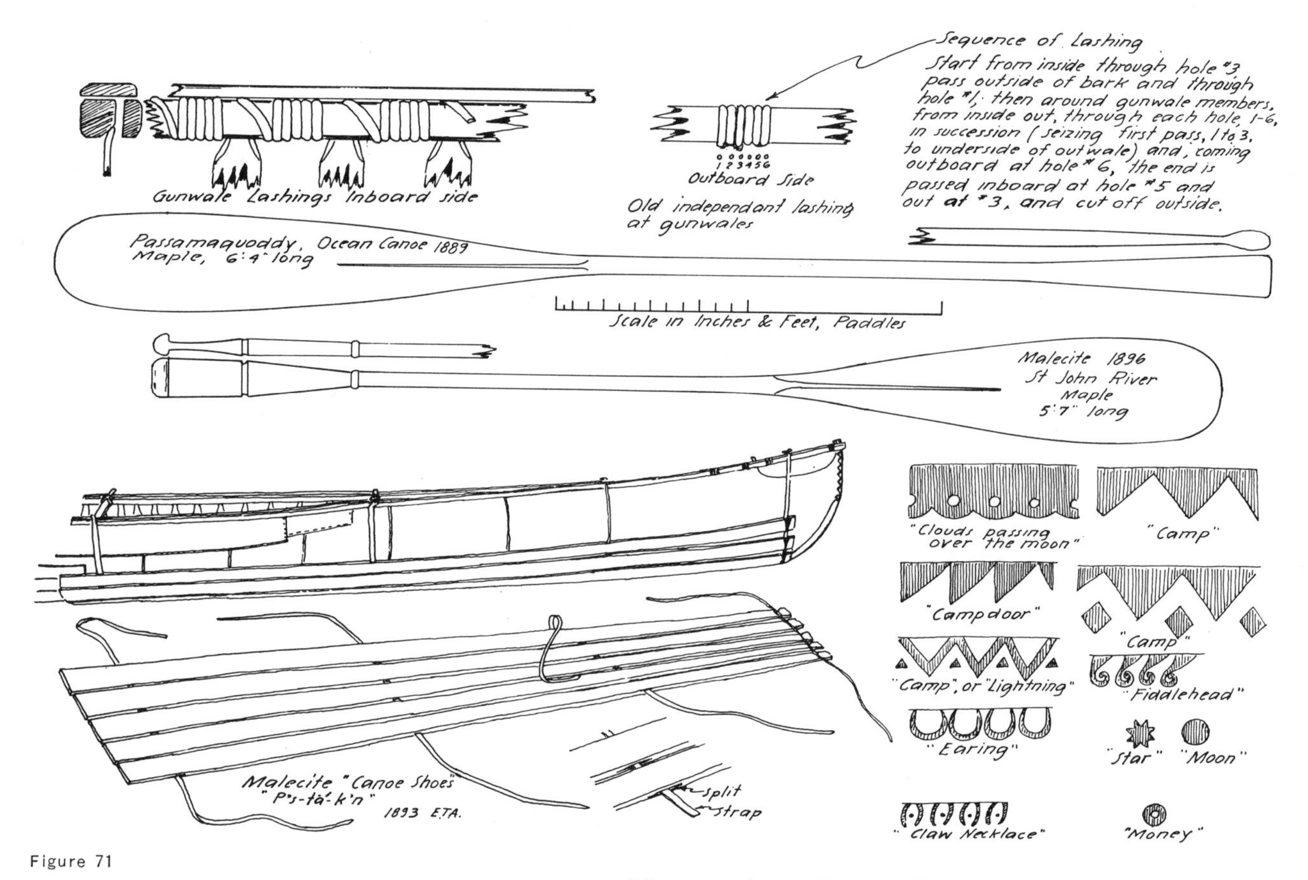


Figure 71

MALECITE CANOE DETAILS, GEAR, AND GUNWALE DECORATIONS.

wooden sheathing, outside the bottom, to protect the bark from rocks and snags or floating ice that might be met in rapids and small streams. The fitting was used also by the Micmac and Ojibway; it is not known whether this was an Indian or European invention. The French canoemen called it *barre d'abordage* and the Malecite, *P's-ta' k'n*; the English woodsmen called the fitting "canoe shoes."

The Malecite paddle was of various forms, as illustrated in figures 71 and 72, the predominant form being very similar to the paddle now used with canvas "Indian" canoes. The total length of the blade was usually about 28 to 30 inches; at 10 or 11 inches from the tip it was about 2½ inches wide. The handle was about 36 inches long. At just above the blade it was 1¼ inches wide and 1 inch thick. The handle was not parallel-sided. Near the top it widened gradually to about 2¼ inches at 2½ inches from the top; here the cross-grip was formed. The thickness of the handle reduced gradually from that given for just above the top of the blade to about ½ inch at about 5 inches below the cross-grip, and widened again to ⅝ inch at the point where the cross-grip was formed. The blade was ridged down its center. The lower end was rounded and the lower half of the blade was approximately half an ellipse in shape. The Passamaquoddy blade had its wide point within 7 inches of the lower tip, where it was about 6 inches wide. The handle was about 1⅛ inches in diameter just above the blade, and then tapered in thickness until it first became oval and then flat in cross section. The width remained nearly constant to a point within 12 to 16 inches of the cross-grip, then gradually widened to nearly 3 inches at the top. The blade was 33 to 36 inches long and the whole paddle somewhere between 73 and 76 inches long. The cross-grips were sometimes round, at other times they were merely worked off in an oval shape to fit the upper hand. The usual width of the cross-grip was just under 3 inches.

Formerly, the Malecite placed his personal mark, or *dupskodegun*, on the flat of the top of his paddle near the cross-grip. The mark was incised into the wood and the incised line was filled with red or black pig-

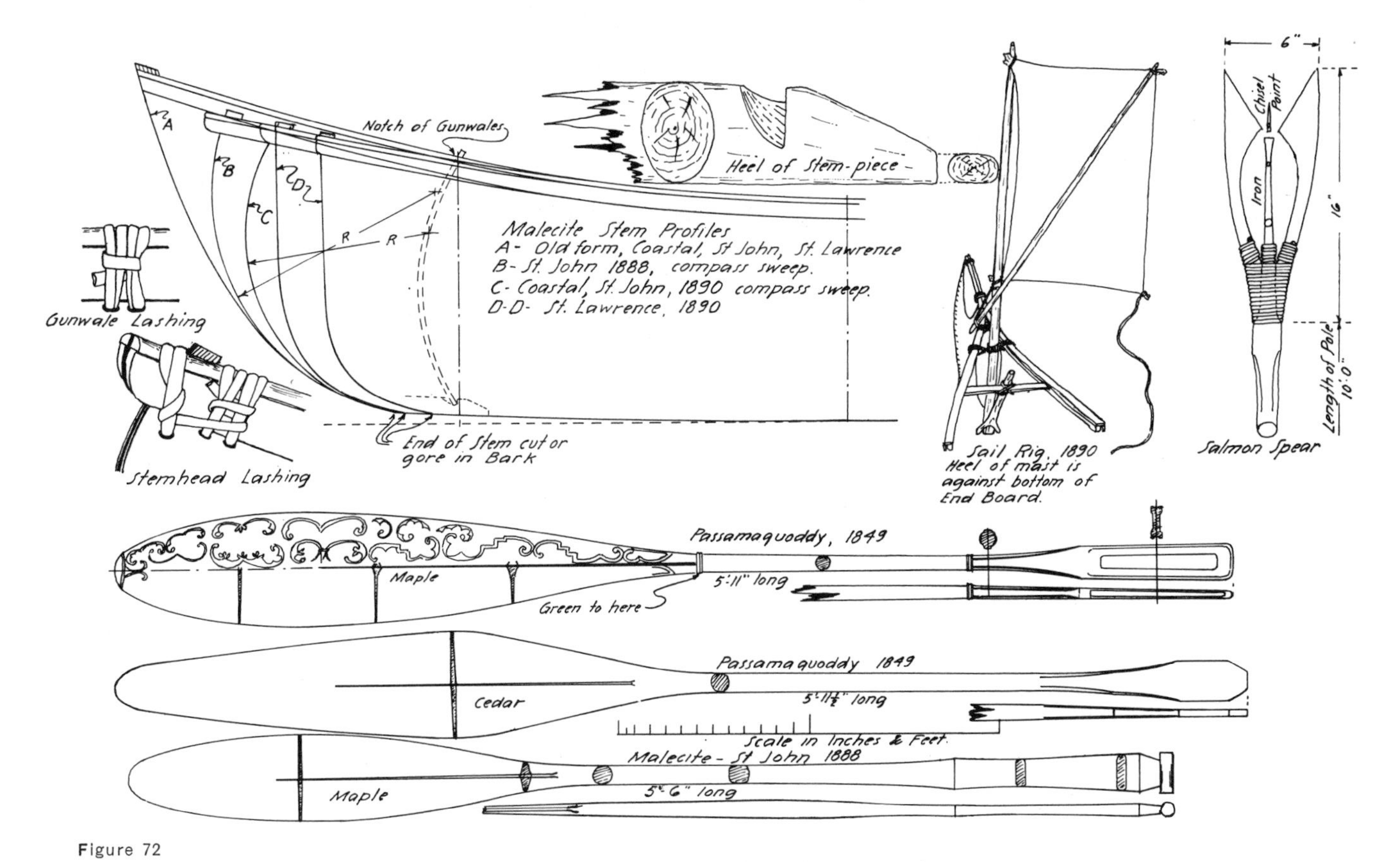


Figure 72

Malecite Canoe Details, Stem Profiles, Paddles, Sail Rig, and Salmon Spear.

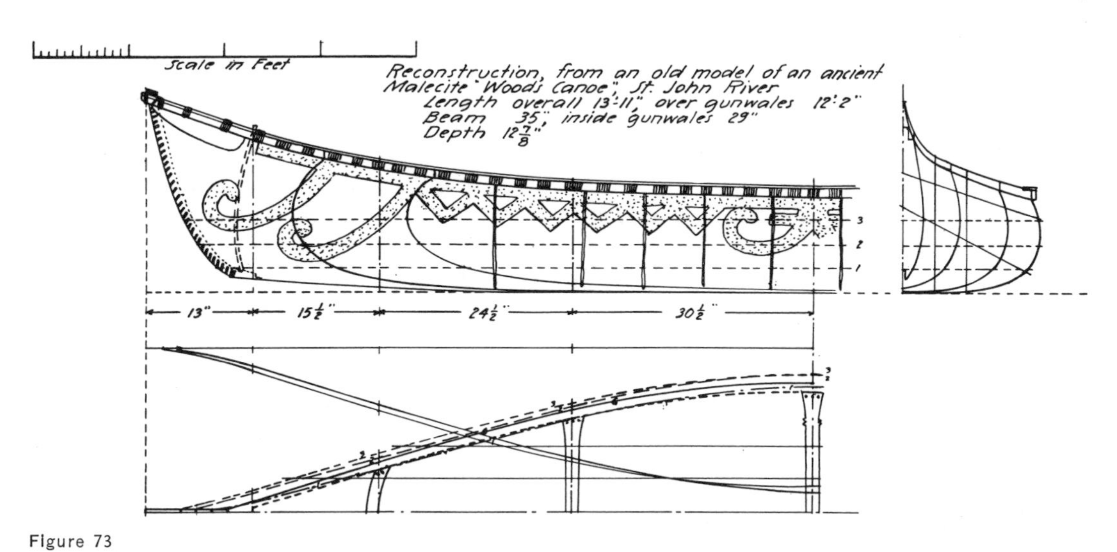


Figure 73

Lines and Decoration Reconstructed From a Very Old Model of an ancient woods, or pack, canoe, showing short ends and use of fiddlehead and firesteel form of decoration.

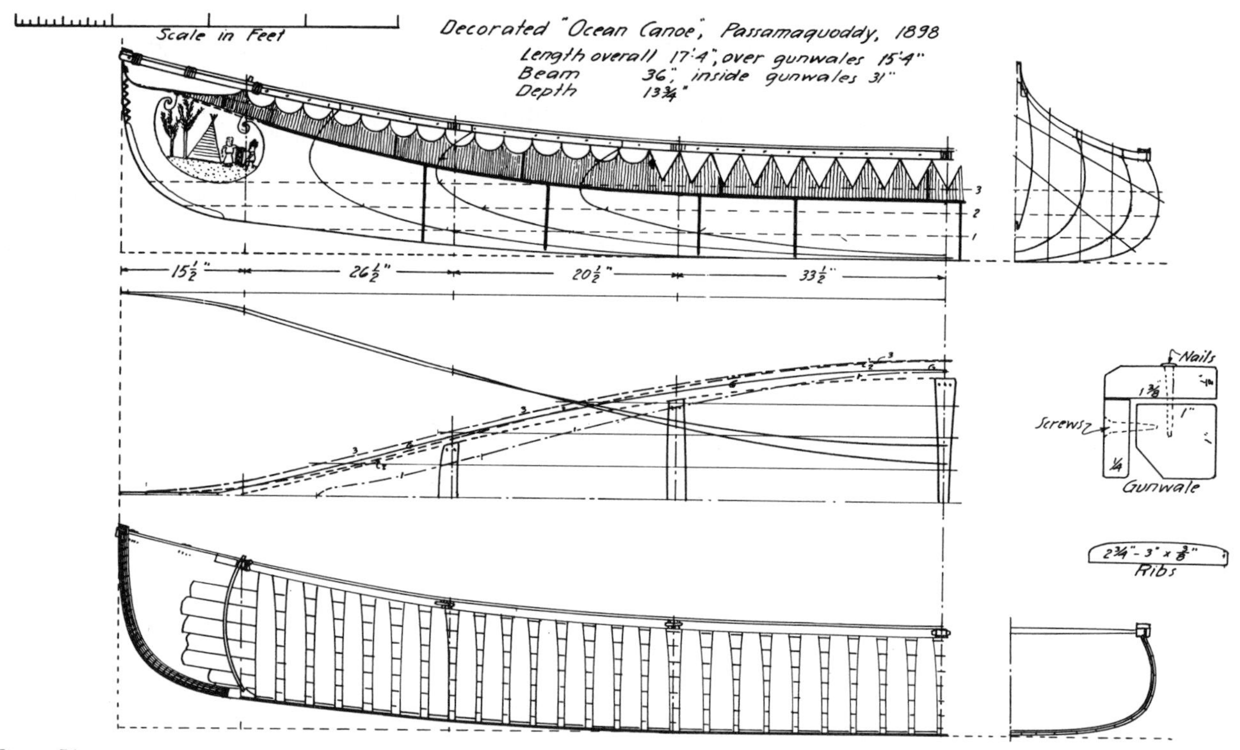


Figure 74

Last Known Passamaquoddy Decorated Ocean Canoe to be built. Constructed in 1898 by Tomah Joseph, Princeton, Maine, on the same model as a canvas porpoise-hunting canoe.

ment when available. Sometimes the whole paddle, including the blade, was covered with incised line ornamentation. This was usually a vine-and-leaf pattern, or a combination of small triangles and curved lines. The Passamaquoddy used designs suggesting the needlework once seen on fine linens. Sometimes other designs showing animals, camps, or canoes were used.

The Malecite, particularly the Passamaquoddy, were especially skillful in decorating bark canoes, as can be seen from the illustrations (pp. 81–87). Sometimes they used scraped winter bark decoration just along the gunwales; occasionally the whole canoe was decorated in this manner above the normal load waterline as described on page 87. Usually, however, the bark decoration was confined to a long panel just below the gunwales and to the ends of the canoe. The personal "mark" of the owner-builder would usually be on the flaps near the ends, the *wulegessis*, meaning the outside bark of a tree or a child's diaper, but in canoe nomenclature used to indicate the protective cover which it formed for the gunwale-end lashings. Sometimes the Malecite placed his mark in the gunwale decoration. Sometimes he placed a picture or a sign on each side of the ends below the *wulegessis*, in about the position used for insignia on the canvas "Indian" canoe.

The swastika was used by the Passamaquoddy in a war canoe in colonial times and has been used later. The Passamaquoddy mark for an exceptional canoe (such as a war canoe that won the race home) was often on the *wulegessis*, and on a relatively modern canoe this mark, or *gogetch*, was a picture of "a funny-looking kind of doll." A common form of decoration in Passamaquoddy canoes was the fiddlehead curve which resembles the top of young fern shoots. This appears in numerous combinations; often double and back to back, joined with a long bar, or "cross." This particular combination is known as the "fiddlehead and cross" or as the "fire steel"; the latter because of a fancied resemblance of the form to the shape of the old firemaking steels of colonial times. A zigzag line appears to represent lightning to most Indians. A series of half-circles along the gunwales,

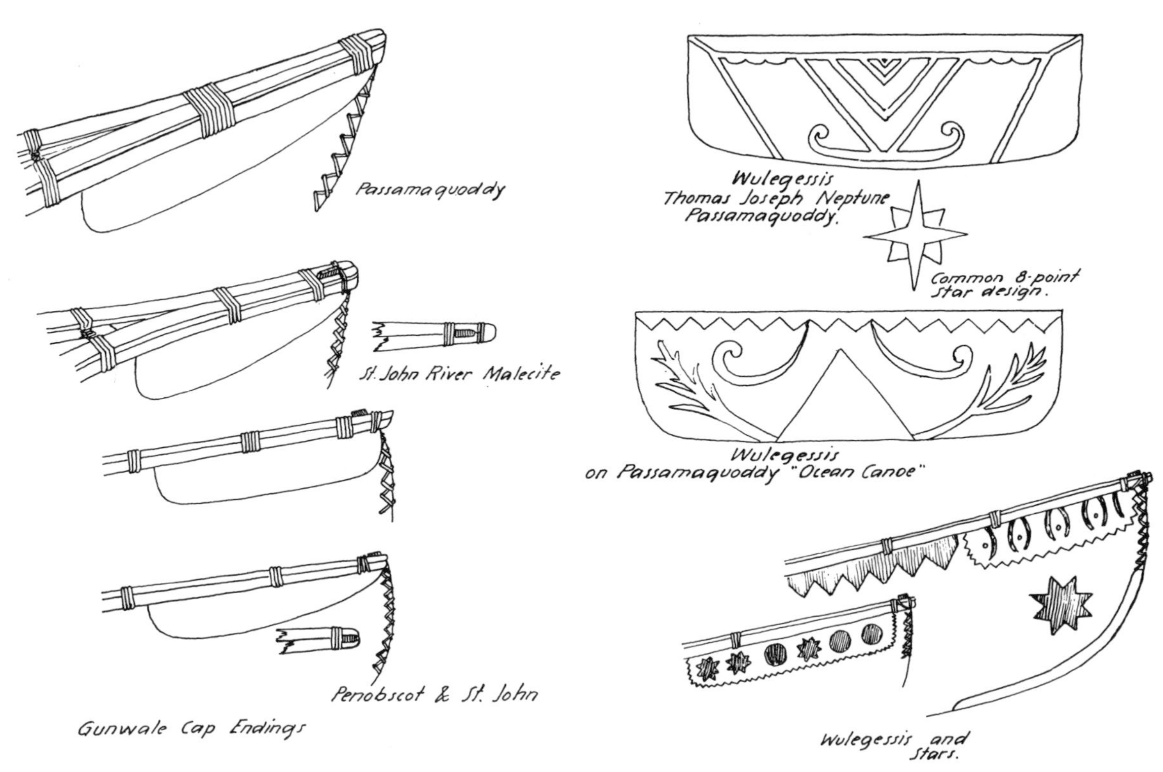


Figure 75

MALECITE CANOE DETAILS AND DECORATIONS.

with the rounded side down and just touching one another at the top, having a small circle in the center of each, represents "clouds passing over the moon." A similar series of half-circles without the center circles might mean the canoe was launched during a new moon; the number of half-circles shown would indicate the month.

Yet there is not full agreement among Indians about the meaning of decorative forms; the crooked or zigzag line might also mean camps or the crooked score stick used in a Malecite game. The circle could mean sun or moon or month. A half-moon form might also be "a woman's earring," or a new moon. A circle with a very small one inside might be a "brooch," as well as "money." Right triangles, in a closely spaced series along the gunwales, apparently meant "door cloth," or tent door ("what you lift with your hand"). Shown on pages 84 and 85 are some Indian marks on the *wulegessis*, based upon the statements of old Malecites or upon their sketches.

After the Malecite had become Roman Catholic, a fish on the middle panel of a canoe meant that it had been launched on Friday. Pictures on a canoe sometimes indicated a mythological story; a picture of a rabbit sitting and smoking a pipe on one side of the canoe and a lynx on the other would be such a case. In Malecite mythology the rabbit was the ancestor of the tribe. He was also a great magician. The lynx was the mortal enemy of the rabbit, but in the mythological tales he was always overcome and defeated by the rabbit's magic. Hence, the idea conveyed is that "though the lynx is near, the rabbit sits calmly smoking his pipe and as he knows he can overcome his enemy," or, in short, "self-confidence."

The Indian's mark on his canoe or weapons is not a signature to be read by anyone. The mark may, of course, be identified as to what it represents, but unless it is known as the mark used by a certain man it cannot be "read." Any mark could be used by an Indian, either because it had some connection with his activities or habits, or because he "likes it." The stone tobacco pipe used by Peter Polchies (see p. 85) as his mark had no known connection with this Indian's habits or activities. However, his son, of the same name and well known also as "Doctor Polchies," took the same mark, but in his case it had a personal meaning since he was noted locally for his skill in making stone pipes. Another case was

Figure 76

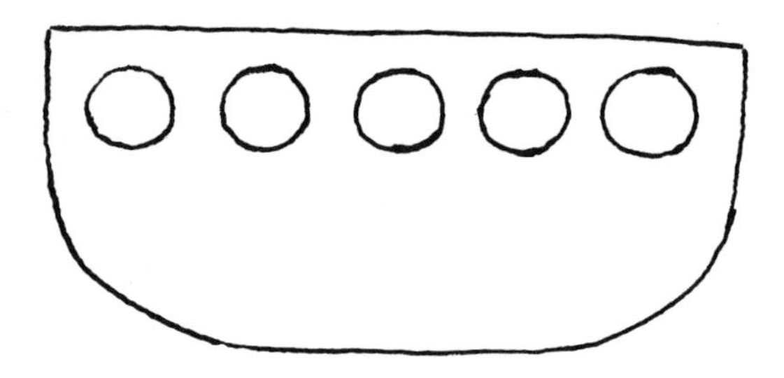

"mark of Mitchell Laporte"

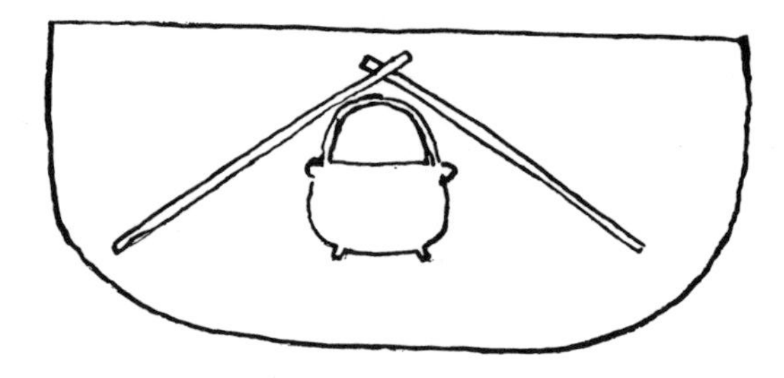

"that pot hanging was used by three or four generations—it was mark on John Lolar's canoe in 1872"

"I made marks like this on wulegessis and sometimes on middle" (Charlie Bear)

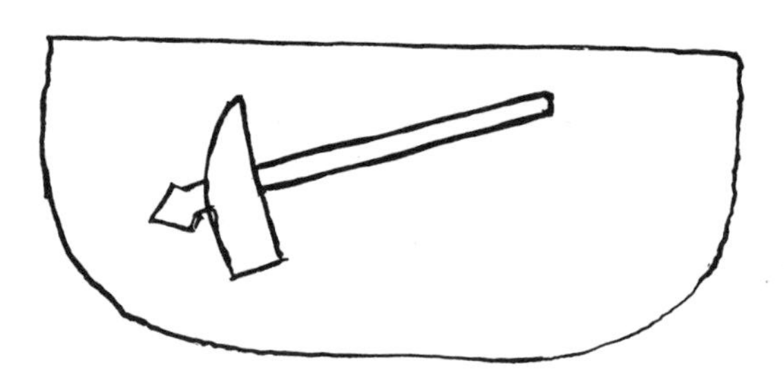

"mark of Noel John Sapier" (tomahawk)

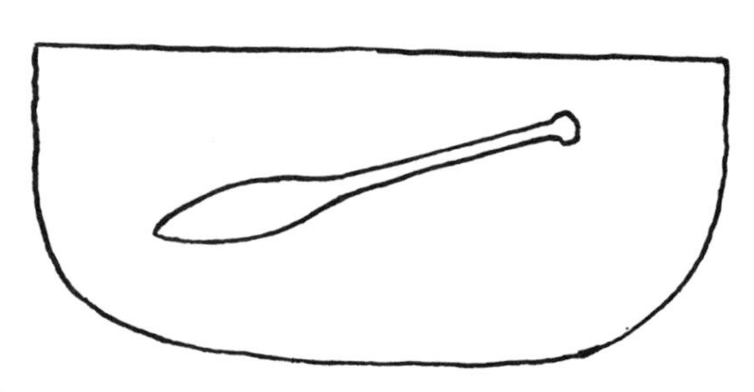

"mark of Noel Polchies" (paddle)

Wulegessis Decorations

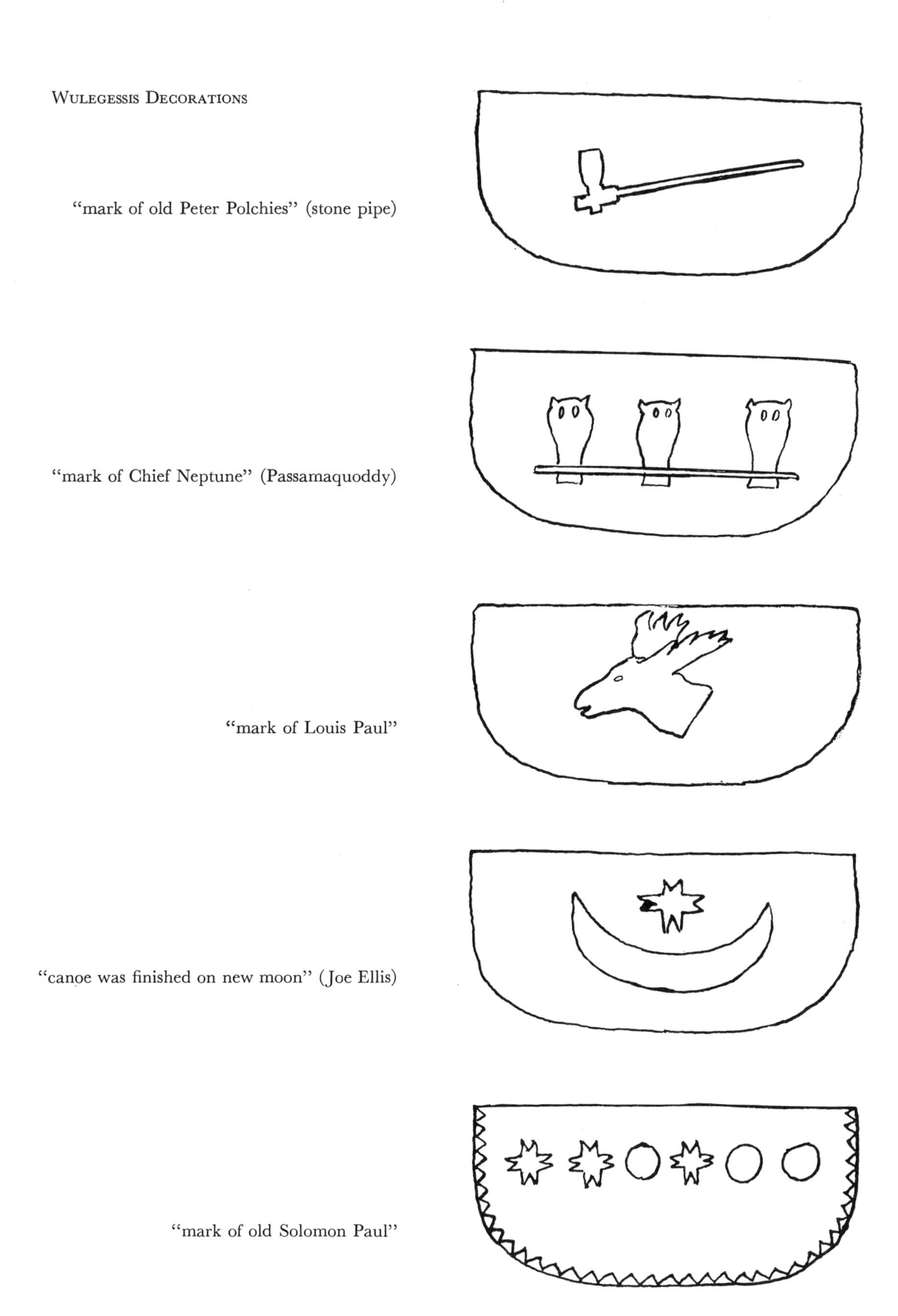

"mark of old Peter Polchies" (stone pipe)

"mark of Chief Neptune" (Passamaquoddy)

"mark of Louis Paul"

"canoe was finished on new moon" (Joe Ellis)

"mark of old Solomon Paul"

Figure 77

END DECORATIONS, PASSAMAQUODDY CANOE built by Tomah Joseph.

a Passamaquoddy who at every opportunity used to pole his canoe in preference to paddling. As a result he had become known as "Peter of the Pole" or "Peter Pole" and he then used as a canoe mark a representation of a setting pole. In submitting sketches of the marking on the *wulegessis* of canoes to old Indians it was seldom possible to learn the identity of the owner or builder, since the marks were usually not known to those questioned. In more recent times, the educated Malecite signed his name in English on his canoe and thus gave it more permanent identification.

In duplicating a design, the Malecite apparently used a pattern, or stencil, which was preserved to allow duplication over a long period of time. The stencil was usually cut from birch bark, apparently an old practice, although whether it was done in prehistoric times cannot be determined. The long contact of the Malecites with Europeans is a factor to be considered in such matters. This is sometimes shown in picture-writing on a canoe; one, for instance, showed a white man fishing with rod and line from a canoe with an Indian guide. On the opposite side was the representation of an Indian camp beside two trees, a kettle over the fire and the brave sitting cross-legged smoking his pipe, indicating, of course, "comfort and contentment."

Asking old Indians to identify or give the names of decorations, Adney recorded statements which indicate their thought in regard to such matters. There were used, for example, two forms of the half-moon or crescent; one was quite open at the points which plainly indicated a half-moon, but the other was more nearly closed: Mrs. Billy Ellis, widow of Frank Francis, a Malecite, said of them, "Old Indian earrings, that is only what I can call them. Also in nose. Wild Indian made them of silver or moosebone, I guess he thought he looked nice; it looked like the devil." Joe Ellis, an old canoe builder, also called this form "earrings" and when asked why an Indian would put these on a canoe, replied "He will think what he will put on here. He might have seen his wife at bow of canoe, and put it on [there]." Shown the right-triangle-in-series design, Mrs. Ellis said "I fergit it but I will remember; what you lift with your hand, we call it that—camp door" (referring to the cloth or hide hung over a camp door, and raised at one corner to enter, so that the opening is then divided diagonally).

In a later period, the Malecite usually confined decoration to the *wulegessis* and to the pieced-out bark amidships, the panel formed on each side. The *wulegessis* was of various forms; its bottom was sometimes shaped like a cupid's bow, sometimes it was

Figure 78

End Decorations, Passamaquoddy Canoe built by Tomah Joseph.

rectangular. A common form was one representing the profile of a canoe. Being of winter bark, it was red or brown, with the part where the design was scraped showing white or yellow. The center panel was also of winter bark, and the design on it showed a similar contrast in color. Even when the bark cover was not pieced out, the panel was formed by scraping all the cover except a panel amidships on each side. Old models indicate that the early Malecite canoes may have used decoration all over above the waterline (see p. 81) far more frequently than has been the recent custom. The decorations were a fiddlehead design in a complicated sequence so that it bore a faint resemblance to the hyanthus in a formal scroll, but the design apparently had no ceremonial significance; it was used for the same reason given Adney for so many forms of bark decoration, "it looked nice."

The drawings and plans on pages 71 to 87 will serve better than words to show these characteristic designs and decorations. It is doubtful that color, paint or pigment, was used in decorating the Malecite bark canoes before the coming of Europeans, but it was employed occasionally in the last half of the 19th century. The beauty of the Malecite canoe designs

Figure 79

Passamaquoddy Decorated Canoe built by Tomah Joseph.

lay not in the barbaric display of color characteristic of the large fur-traders' canoes, but in the tasteful distribution of the scraped winter bark decoration along the sides of the hull. The workmanship exhibited by the Malecite in the construction of their canoes was generally very fine; indeed, they were perhaps the most finished craftsmen among Indian canoe-builders.

## *St. Francis*

The tribal composition of the Abnaki Indians is somewhat uncertain. The group was certainly made up of a portion of the old Malecite group, the Kennebec and Penobscot, but later also included the whole or parts of the refugee Indians of other New England tribes who were forced to flee before the advancing white settlers. It is probable that among the refugees were the Cowassek (Coosuc), Pennacook, and the Ossipee. There were also some Maine tribes among these—the Sokoki, Androscoggin, (Arosaguntacook), Wewenoc, Taconnet, and Pequawket. It is probable that the tribal groups from southern and central New England were mere fragments and that the largest number to make up the Abnaki were Malecite. The latter in turn were driven out of their old homes on the lower Maine coast and drifted northwestward into the old hunting grounds of the Kennebec and Penobscot, northwestern Maine and eastern Quebec as far as the St. Lawrence. The chief settlement was finally on the St. Francis River in Quebec, hence the Abnaki were also known as the "St. Francis Indians." These tribesmen held a deep-seated grudge against the New Englanders and, by the middle of the 18th century, they had made themselves thoroughly hated in New England. Siding with the French, the St. Francis raided the Connecticut Valley and eastward, taking white children and women home with them after a successful raid, and as a result the later St. Francis had much white blood. They were generally enterprising and progressive.

Little is known about the canoes of these Abnaki during the period of their retreat northwestward. It is obvious that the Penobscot, at least, used the old form of the Malecite canoe. What the canoes of the other tribal groups were like cannot be stated. However, by the middle of the 19th century the St. Francis Indians had produced a very fine birch-bark canoe of distinctive design and excellent workmanship. These they began to sell to sportsmen, with the result that the type of canoe became a standard one for hunting and fishing in Quebec. When other tribal groups discovered the market for canoes, they were forced to copy the St. Francis model and appearance to a very marked degree in order to be assured of ready sales. It is obvious, from what is now known, that the St. Francis had adapted some ideas in canoe building from Indians west of the St. Lawrence, with whom they had come into close contact. However, they had also retained much of the building technique of their Malecite relatives. Hence, the St. Francis canoes usually represent a blend of building techniques as well as of models.

The St. Francis canoe of the last half of the 19th century had high-peaked ends, with a quick upsweep of the sheer at bow and stern. The end profile was almost vertical, with a short radius where it faired into the bottom. The rocker of the bottom took place in the last 18 or 24 inches of the ends, the remaining portion of the bottom being usually straight. The amount of rocker varied a good deal; apparently some canoes had only an inch or so while others had as much as four or five. A few canoes had a projecting "chin" end-profile; the top portion where it met the sheer was usually a straight line.

The midsection was slightly wall-sided, with a rather quick turn of the bilge. The bottom was nearly flat across, with very slight rounding until close to the bilges. The end sections were a U-shape that approached the V owing to the very quick turn at the centerline. The ends of the canoe were very sharp, coming in practically straight at the gunwale and at level lines below it. The gunwales were longer than the bottom and so the St. Francis canoes were commonly built with a building-frame which was nearly as wide amidships as the gunwales but shorter in length.

At least one St. Francis canoe, built on Lake Memphremagog, was constructed with a tumblehome amidships the same as that of some Malecite canoes. The rocker of the bottom at each end started at the first thwart on each side of the middle and gradually increased toward the ends, which faired into the bottom without any break in the curves. The end profiles projected with a chin that was full and round up to the peaked stem heads. The sheer swept up sharply near the ends to the stem heads.

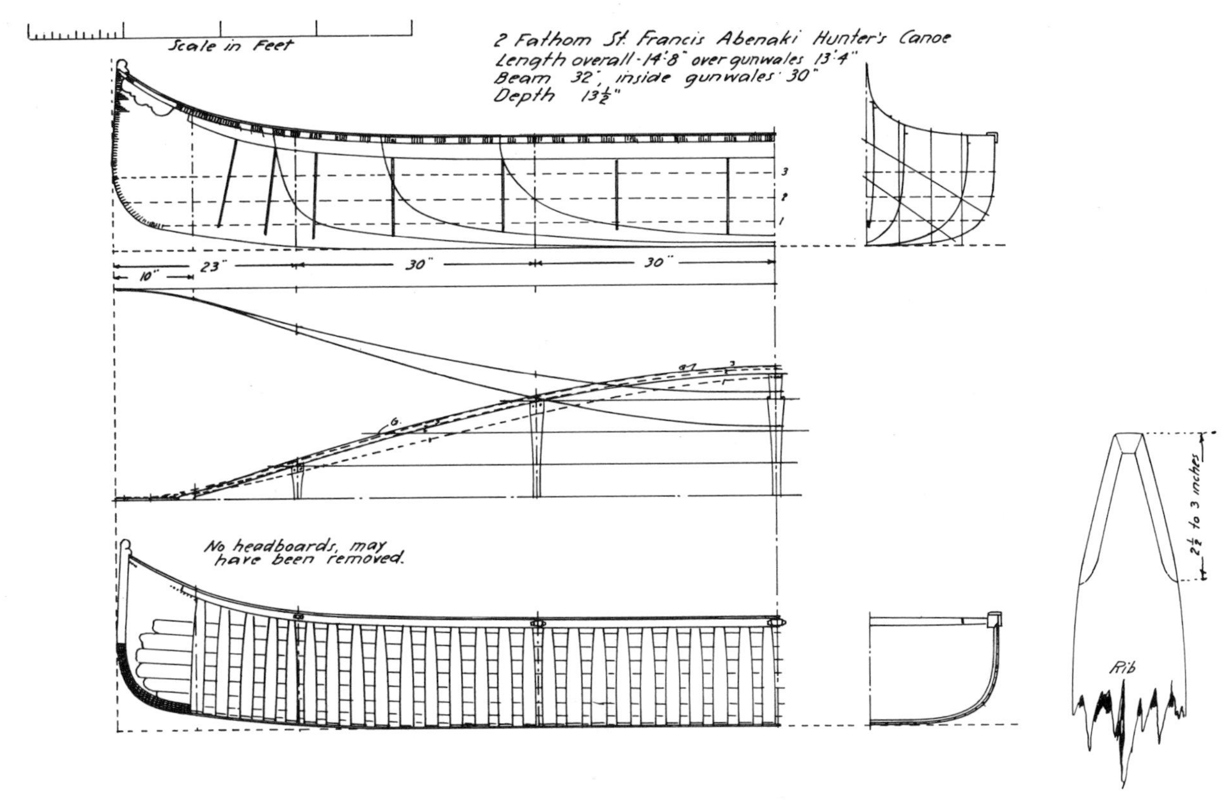


Figure 80

St. Francis 2-Fathom Canoe of About 1865, with upright stems. Built for forest travel, this form ranged in size from 12 feet 6 inches overall and 26½-inch beam, to 16 feet overall and 34-inch beam.

This particular canoe represented a hybrid design not developed for sale to sportsmen, and the sole example, a full-size canoe formerly in The American Museum of Natural History at New York and measured by Adney in 1890, is now missing and probably has been broken up.

The St. Francis canoes were usually small, being commonly between 12 and 16 feet overall; the 15-foot length usually was preferred by sportsmen. The width amidships was from 32 to 35 inches and the depth 12 to 14 inches. The 14-foot canoe usually had a beam of about 32 inches and was nearly 14 inches deep; if built for portaging the ends were somewhat lower than if the canoe was to be used in open waters. Canoes built for hunting might be as short as 10 or 11 feet and of only 26 to 28 inches beam; these were the true woods canoes of the St. Francis.

The gunwale structure of the St. Francis canoes followed Malecite design; it was often of slightly smaller cross section than that of a Malecite canoe of equal length, but both outwale and cap were of somewhat larger cross section. The stem-pieces were split and laminated in the same manner, but occasionally the lamination was at the bottom, due to the hard curve required where the stem faired into the bottom. Many such canoes had no headboards, the heavy outwales being carried to the sides of the stem pieces and secured there to support the main gunwales. If the headboard was used, it was quite narrow and was bellied toward the ends of the canoe. In some St. Francis canoes the bark cover in the rockered bottom near the ends showed a marked V. In the canoe examined by Adney at the American Museum of Natural History, the ribs inside toward the end showed no signs of being "broken," so it is evident that the V was formed either by use of a shaped keel-piece in the sheathing or by an additional batten shaped to give this V-form under the center strake. Since the V began where the rocker in the canoe started, in an almost angular break in the bottom, it is likely that a shaped batten had been

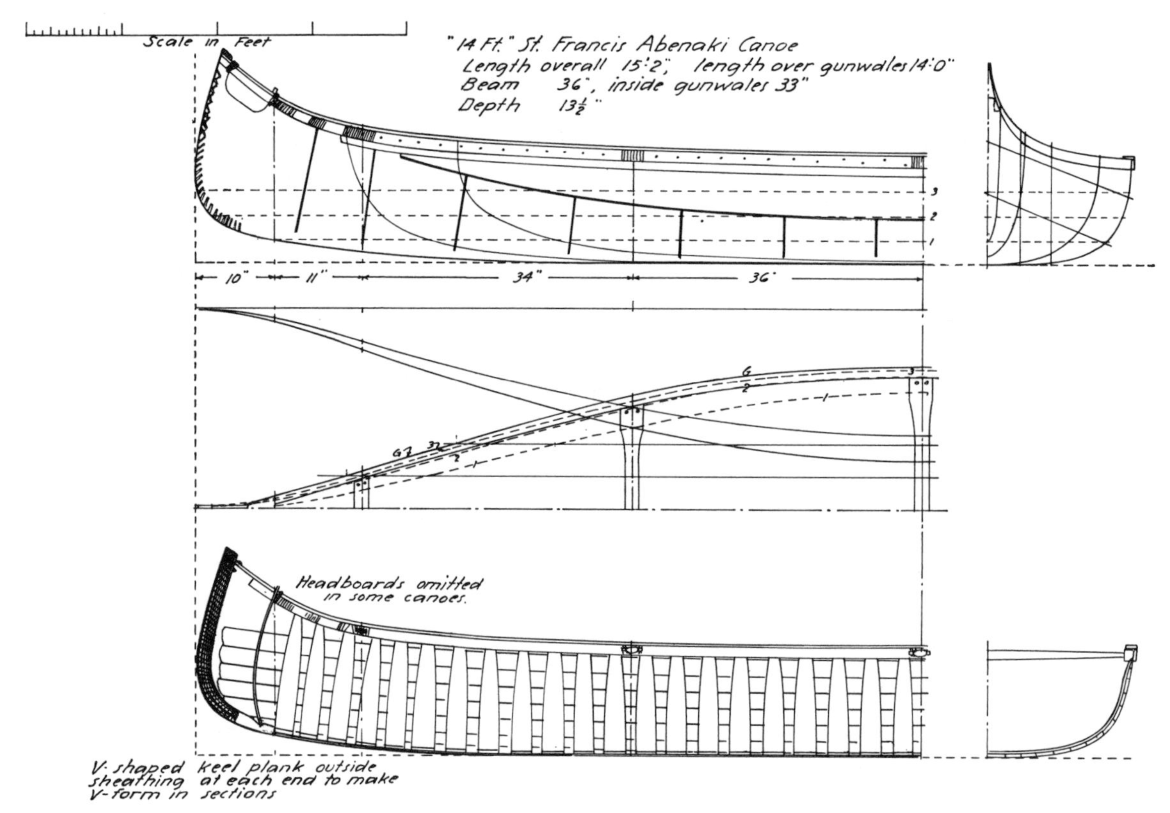


Figure 81

St. Francis Canoe of About 1910, with narrow, rockered bottom, a model popular with guides and sportsmen for forest travel.

used to form it. He could not verify this, however, as the area was covered by the frames and sheathing.

The sheathing was in short lengths with rounded ends which overlapped, and it was laid irregularly in the "thrown in" style found in many western birch-bark canoes. The ribs were commonly about 2 inches wide and nearly $\frac{3}{8}$ inch thick, the width tapering to roughly $1\frac{3}{4}$ inches under the gunwales. The ends of the ribs were then sharply reduced in width to a chisel point about 1 inch wide; the sides of the sharply reduced taper being beveled, as well as the end. A 15-foot canoe usually had 46 to 50 ribs.

The thwarts, unlike those of the Micmac and some Malecite canoes, in which the thwarts were unequally spaced, were equally spaced according to a builder's formula. The ends of the thwarts, or crossbars, were tenoned into the main gunwales and lashed in place through the three lashing holes in the ends of each thwart, except the end ones, which usually had but two. In some small canoes, however, two lashing holes were placed in all thwart ends. The design of the St. Francis thwart was as a rule very plain, gradually increasing in width from the center outwards to the tenon at the gunwale in plan and decreasing in thickness in elevation in the same direction. The ends of the main gunwales were of the half-arrowhead form, and were covered with a bark *wulegessis*, but the flaps below the outwales were sometimes cut off, or they might be formed in some graceful outline.

The bark cover was sometimes in one piece; when it was pieced out for width, the harness-stitch was used. In most canoes, the bark along the gunwale was doubled by adding a long narrow strip, often left hanging free below the gunwales and stopping just short of the *wulegessis*, which it resembled. It was sometimes decorated. A few St. Francis canoes with nailed gunwales omitted this doubling piece. When used, the doubling piece, as well as the end cover, were folded down on top of the gunwale before being sewn into place. The decoration of the St. Francis canoes seems to have been scant and wholly

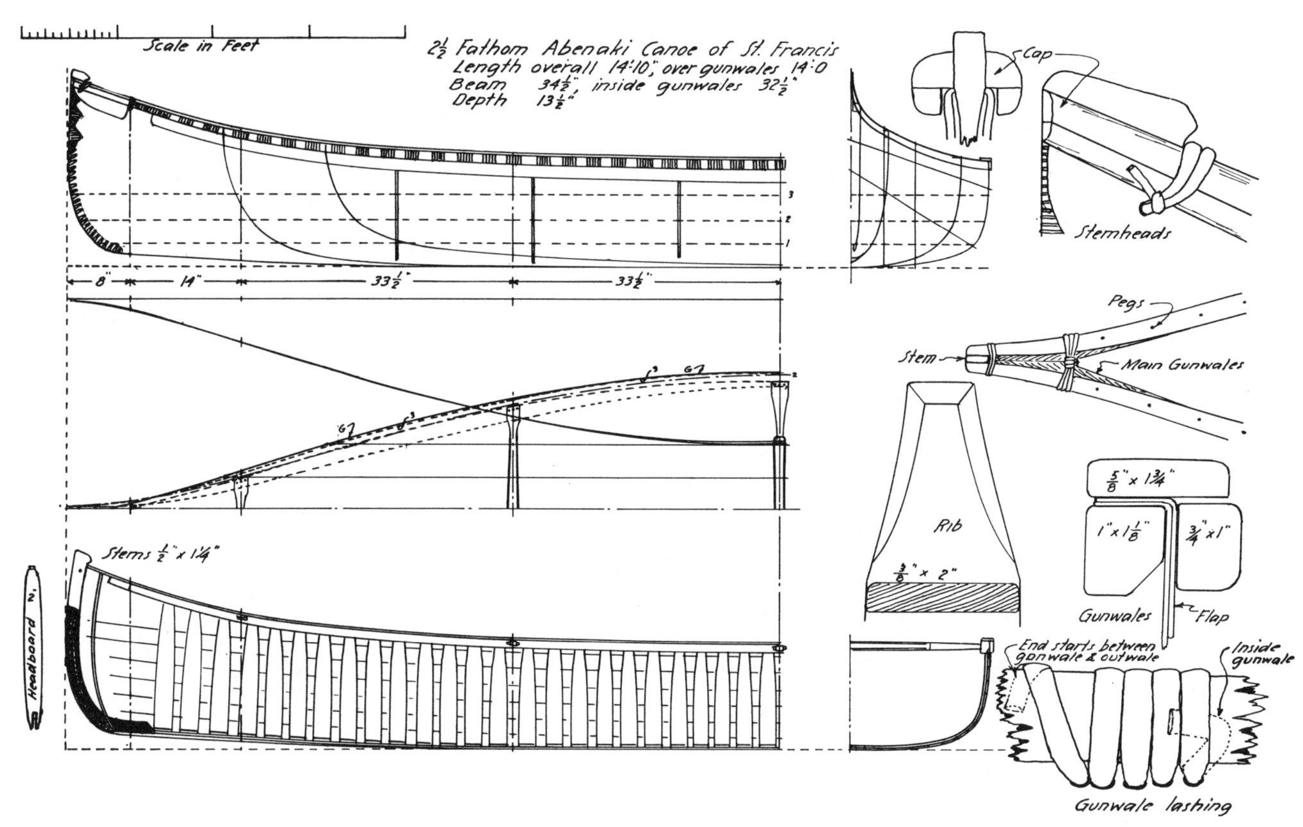


Figure 82

LOW-ENDED ST. FRANCIS CANOE with V-form end sections made with short, V-shaped keel battens outside the sheathing at each end. Note the unusual form of headboard, seen in some St. Francis canoes.

confined to a narrow band along the gunwale, or to the doubling pieces. The marking of the *wulegessis* had ceased long before Adney investigated this type of canoe and no living Indian knew of any old marks, if any ever had been used.

The ends were commonly lashed with a spiral or crossed stitch, but some builders used a series of short-to-long stitches that made groups generally triangular in appearance. The gunwale lashing was in groups about 2½ inches long, each having 5 to 7 turns through the bark. The groups were about 1½ to 1¼ inches apart near the ends and about 2 inches apart elsewhere. The groups were not independent but were made by bringing the last turn of each group over the top and inside the main gunwale in a long diagonal pass so as to come through the bark from the inside for the first pass of the new group. The caps were originally pegged, with a few lashings at the ends.

The ribs were bent green. After the bark cover had been sewn to the gunwales, the green ribs were fitted roughly inside the bark, with their ends standing above the gunwales, and were then forced into the desired shape and held there, usually by two wide battens pressed against them by 7 to 10 temporary cross struts. After being allowed to dry in place, the ribs were then removed, the sheathing was put into place, and the ribs, after a final fitting, were driven into their proper positions. Some builders put in the ribs by pairs in the shaping stage, one on top of the other, as this made easier the job of fitting the temporary battens. The forcing of the ribs to shape also served to shape the bark cover, and the canoe was placed on horses during the operation, so that the shape of the bottom could be observed while the bark was being moulded. Some builders used very thin longitudinal battens between the bark and the green ribs to avoid danger of bursting the bark.

The canoe was built on a level building bed, in most instances apparently, with the ends of the building frame blocked up about an inch. It seems pos-

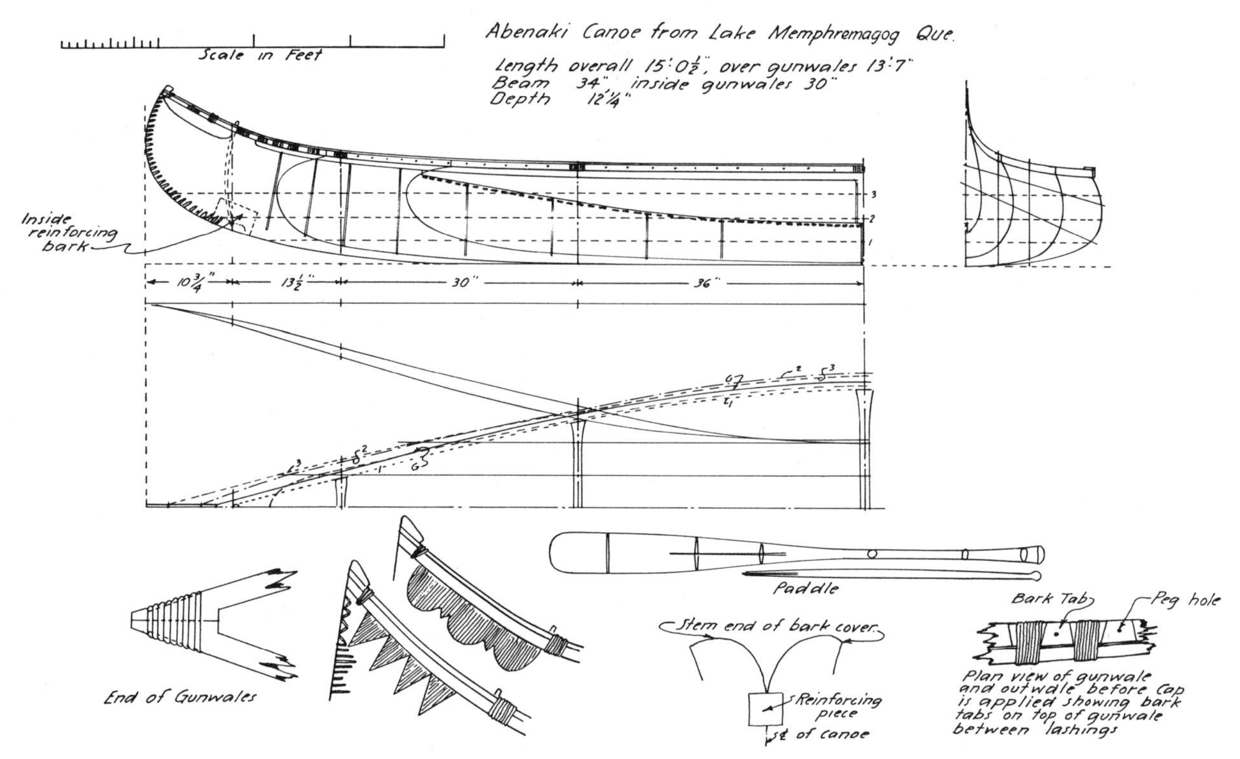


Figure 83

St. Francis-Abnaki Canoe for Open Water, a type that became extinct before 1890. From Adney's drawings of a canoe formerly in the Museum of Natural History, New York. Details of Abnaki canoes are also shown.

sible, however, that narrow bottom canoes may have been built with the bed raised 2 or 3 inches in the middle, rather than employing a narrow building frame. The construction of the building frame was the same as among the western Indians and as described in Chapter 3.

In preparing the ribs, a common practice was the following: Assume, for example, that there are 10 ribs from the center to the first thwart forward; these are laid out on the ground edge-to-edge with the rib under the center thwart to the left and the rib under the first thwart to the right. On the rib to the left the middle thwart is laid so that its center coincides with that of the rib, and the ends of the thwart are marked on the rib. The same is done to the rib on the far right, over which the first thwart is laid as the measure. On each side of the centerline the points marking the ends of the thwarts are then joined by a line across the ribs, as they lie together, to mark the approximate taper of the canoe toward the ends, at the turn of the bilge. Each rib is taken in turn from the panel and with it is placed another from the stock on hand to be set in a matching position on the other side of the middle thwart, toward the stern; the pair, placed flat sides together, are then bent over the knee at, or outside of, the marks or lines. The ribs in the next portion of the canoe's length are shaped in the same manner, using the lengths of the first and second thwarts as guides. Thus, the ribs are given a rough, preliminary bend before being fitted inside the bark cover and stayed into place to season. This method allowed the bilge of the canoe to be rather precisely determined and formed during the first stages of construction. At the ends, of course, the ribs are sharply bent only in the middle. Since the full thwart length makes a wide bottom, by setting the length of the rib perhaps a hand's width less than that of the whole thwart, the narrow bottom is formed.

The rough length of the ribs was twice the length of the thwarts nearest them. Hackmatack was used for thwarts by the St. Francis Indians, rock maple being considered next best. Cedar was first choice for ribs, then spruce, and then balsam fir. Longitudinals were cedar or spruce. All canoe measurements were

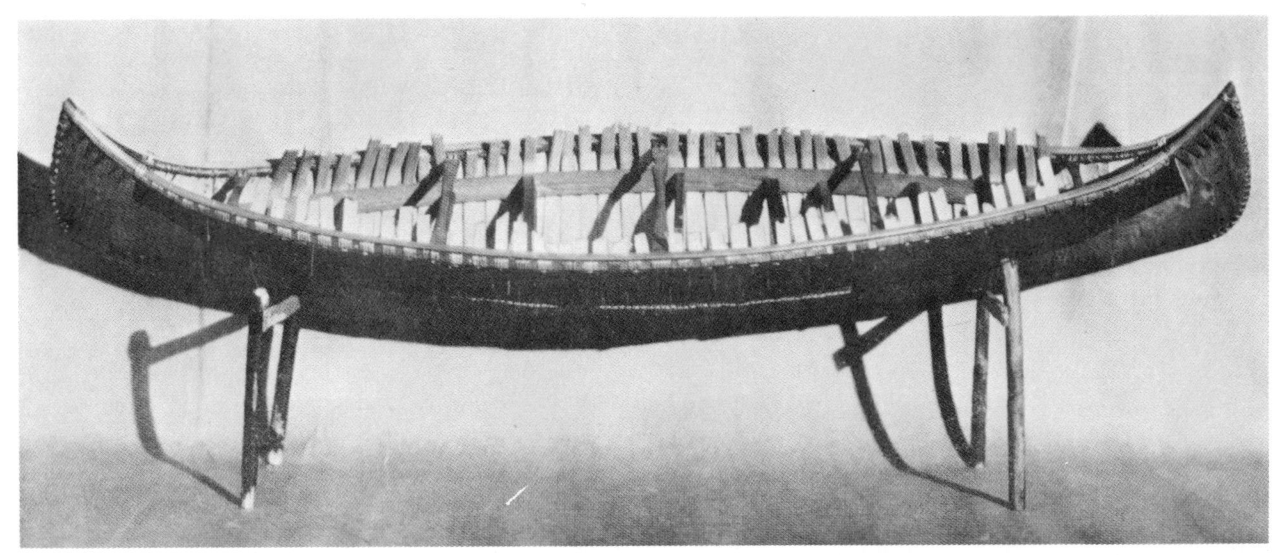

Figure 84

Model of a St. Francis-Abnaki Canoe Under Construction, showing method of moulding ribs inside the assembled bark cover.

made by hand, finger, and arm measurements. Basket ash strips were often used in transferring measurements.

From what has been said, it will be seen that the construction practice of the St. Francis did not follow in all details that of their Malecite relatives. The intrusion of western practices into this group probably took place some time after the group's final settlement at St. Francis. As they gradually came into more intimate relations with their western neighbors and drifted into western Quebec, beyond the St. Lawrence, their canoe building technique became influenced by what they saw to the westward. As would be expected, the St. Francis Abnaki began early to use nails in canoe building, but, being expert workmen, they retained the good features of the old sewn construction to a marked degree up to the very end of birch-bark canoe construction in southern Quebec, probably about 1915. It should perhaps be noted that what has been discovered about the St. Francis Abnaki canoes refers necessarily to only the last half of the 19th century, since no earlier canoe of this group has been discovered. The changes that took place between the decline of the Penobscot style of canoe and that of the later Abnaki remain a matter of speculation.

St. Francis-Abnaki Canoe

Figure 85

# *Beothuk*

The fourth group of Indians, classed here as belonging to the eastern maritime area, are the Beothuk of Newfoundland. Historically, perhaps, these Indians should have been discussed first, as they were probably the first of all North American Indians to come into contact with the white man. However, so little is known about their canoes that it has seemed better to place them last, since practically all that can be said is the result of reconstruction, speculation, and logic founded upon rather unsatisfactory evidence. The tribal origin of the Beothuk has long been a matter of argument; they are known to have used red pigment on their weapons, equipment, clothes, and persons. A prehistoric group that once inhabited Maine and the Maritime Provinces appears to have had a similar custom; these are known as the "Red Paint People," and it may be that the Beothuk were a survival of this earlier culture. But all that can be said with certainty is that the Beothuk inhabited Newfoundland and perhaps some of the Labrador coast when the white man began to frequent those parts. The Beothuk made a nuisance of themselves by stealing gear from the European fishermen, and by occasionally murdering individuals or small groups of white men. Late in the 17th century, the French imported some Micmac warriors and began a war of extermination against the Beothuk. By the middle of the 18th century the Newfoundland tribe was reduced to a few very small groups, and the Beothuk became extinct early in the 19th century, before careful investigation of their culture could be made.

Their canoes were made to a distinctive model quite different from that of the canoes of other North American Indians. The descriptions available are far from complete and, as a result, many important details are left to speculation. Some parts of the more complete descriptions are obscure and do not appear to agree with one another. In spite of these difficulties, however, some information on the canoes is rather specific; by using this, together with a knowledge of the requirements of birch-bark canoe construction, and by reference to some toy canoes found in 1869 in the grave of a Beothuk boy, a reasonably accurate reconstruction of a canoe is possible.

Captain Richard Whitbourne had come with Sir Humphrey Gilbert to Newfoundland in 1580 and revisited the island a number of times afterward. In 1612 he wrote that the Beothuk canoes were shaped "like the wherries of the River Thames," apparently referring to the humped sheer of both; in the wherry the sheer swept up sharply to the height of the oar tholes, in profile, and flared outward, in cross section.

John Gay, a member of the Company of Newfound-land Plantation, wrote in 1612 that Beothuk canoes were about 20 feet long and 4½ feet wide "in the middle and aloft," that the ribs were like laths, and that the birch-bark cover was sewn with roots. The canoes carried four persons and weighed less than a hundredweight. They had a short, light staff set in each end by which the canoes could be lifted ashore. "In the middle the canoa is higher a great deale, than at the bowe and quarter." He also says of their cross section: "They be all bearing from the keel to portlesse, not with any circular, but with a straight, line."

Joann de Laet, writing about 1633, speaks of the crescent shape of the canoes, of their "sharp keel" and need of ballast to keep them upright; he also states that the canoes were not over 20 feet long and could carry up to five persons.

The most complete description of the Beothuk canoe was in the manuscript of Lt. John Cartwright, R.N., who was on the coast of Newfoundland in 1767–1768 as Lieutenant of H.B.M. Ship *Guernsey*. However, some portions are either in error or the description was over-simplified. For example, Cartwright says that the gunwales were formed with a distinct angle made by joining two lengths of the main gunwale members at the elevated middle of the sheer. This hardly seems correct since such a connection would not produce the rigidity that such structural parts require, given the methods used by Indians to build bark canoes. The three grave models show that the sheer was actually curved along its elevated middle. It is possible that Cartwright saw a damaged canoe in which the lashings of the scarf of the gunwales had slackened so that the line of sheer "broke" there. Cartwright is perhaps misleading in his description of the rocker of the keel as being "nearly, if not exactly, the half of an ellipse, longitudinally divided." The models show the keel to have been straight along the length of the canoe and turned up sharply at the ends to form bow and stern. Cartwright also states the keel piece was "about the size of the handle of a common hatchet" amidships, or perhaps 1 inch thick and 1½ inches wide, and tapered toward the ends, which were about ¾ inch wide and about equally thick. The height of the sheer amidships was perhaps two-thirds the height of the ends.

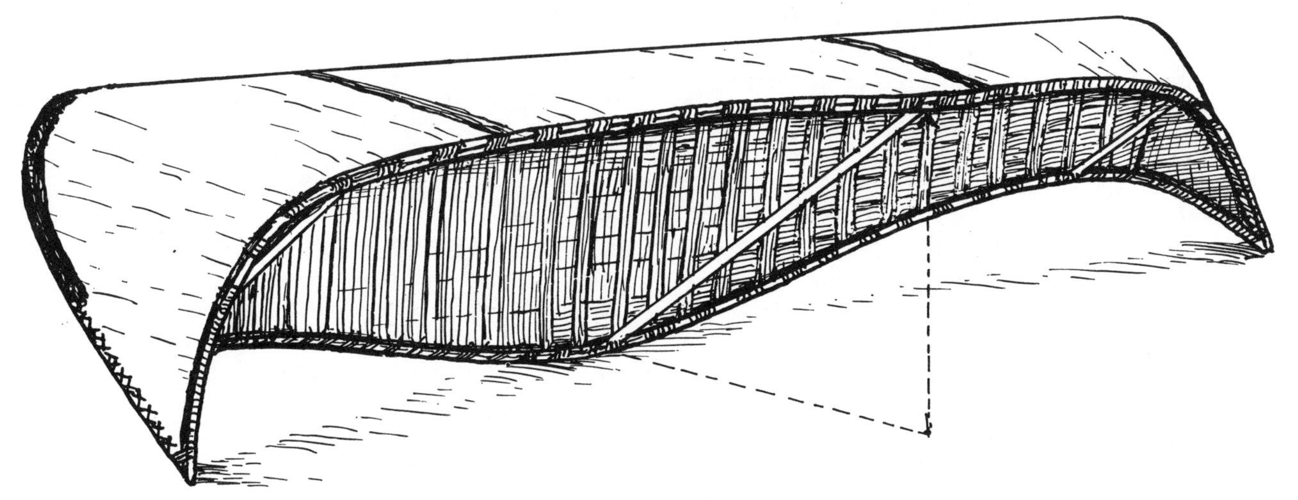

Figure 86

A 15-Foot Beothuk Canoe of Newfoundland with 42½-inch beam, inside measurement, turned on side for use as a camp. It gives headroom clearance of about 3 feet, double that of an 18-foot Malecite canoe with high ends. When the ends were not high enough to provide maximum clearance, small upright sticks were lashed to bow and stern. The shape of the gunwales would permit the canoe to be heeled to an angle (more than 35°) which would swamp a canoe of ordinary sheer and depth. (*Sketch by Adney.*)

Nearly all observers, Cartwright included, noted the almost perfect V-form cross section of these canoes, with the apexes rounded off slightly and the wings slightly curved. From an interpretation of Cartwright's statements, it appears that after the bark cover had been laced to the gunwales, the latter were forced apart to insert the thwarts, as in some western Indian canoe-building techniques. The three thwarts are described as being about two fingers in width and depth. It is stated that the gunwales were made up of an inner and outer member and all were scarfed in the middle to taper each way toward the ends, the outer member serving as an outwale or guard. Cartwright also states that the inside of the bark cover was "lined" with "sticks" 2 or 3 inches broad, cut flat and thin. He refers also to others of the same sort which served as "timbers" so he is describing both the sheathing and the ribs as being 2 or 3 inches wide. He does not say how the thwarts were fitted to the gunwales, how high the ends were, how the ends of the gunwales were formed, nor does he give any details of the sewing used. However, the grave models suggest the form of sewing probably used and the approximate proportions of sheer.

An old settler told James Howley that the Beothuk canoes could be "folded together like a purse." Considering the construction required in birch-bark canoes, this is manifestly impossible; perhaps what the settler had seen was a canoe in construction with the bark secured to shaped gunwales, ready for the latter to be sprung apart by thwarts, as in opening a purse. Howley also obtained from a man who had seen Beothuk canoes a sketch which shows a straight keel and peaked ends, confirmed in all respects by the grave models or toys.

The toy canoes so often referred to here were found by Samuel Coffin in an Indian burial cave on a small island in Pilley's Tickle, Notre Dame Bay (on the east coast of Newfoundland), in 1869. Among the graves in the cave, one of a child, evidently a boy, was found to contain a wooden image of a boy, toy bows and arrows, two toy canoes and a fragment of a third, packages of food, and some red ochre. With one of the canoes was a fragment of a miniature paddle. One of the canoes was 32 inches long, height of ends 8 inches, height of side amidships 6 inches, straight portion of keel 26 inches and beam 7 inches, as shown by Howley.

In Newfoundland there was very fine birch but no cedar. There was, however, excellent spruce which would take the place of cedar. It seems certain, then, that all the framework of the Beothuk canoes was of spruce. It seems likely that they were never built of a single sheet of birch but were covered with a number of sheets sewn together, as in other early Indian birch-bark canoes. The canoe birch of Newfoundland grew to a diameter of 2 to 2½ feet at the butt, which would produce a sheet of birch of 6 to 7 feet width; the length would be decided by how far up the tree the Indian could climb to make the upper cut. As has been stated, the prehistoric Indians seemingly made little attempt to build birch-bark canoes of long lengths of bark, preferring to use only the bark obtainable near the ground and above the height of the winter snows.

The form of the Beothuk canoes, particularly the lack of bilge and the marked V-form, has caused much speculation. One writer assumed that the form was particularly suited for running rapids. Actually, the Beothuk appeared to have used canoes for river travel very rarely, as few rivers in their country were suited for navigation. Instead, they seem to have been coast dwellers and to have used canoes for coastal travel and for voyages from island to island.

Their canoes were undoubtedly designed for open-water navigation, and the V-form was particularly suitable for this. The draft aided in keeping the canoe on its course with either broadside or quartering winds, and if the Beothuks knew sail, the hull-form would have served them well. It is quite evident that the Beothuk canoes used ballast in the form of stones or heavy cargo. Stones would have been placed along the keel piece and covered with moss and skins. The strongly hogged sheer was useful in protecting cargo amidships from spray and, in picking up a seal or porpoise, the canoe could be sharply heeled without taking in water. The V sections fore and aft were suitable for rough-water navigation; because of its form and the weight of ballast, the canoe would pass partly over and through the wave-top without pounding. If a wave of such height as to overtop the gunwales just abaft the stem were met, the strongly flaring sides would give reserve buoyancy, causing the canoe to lift quickly as the wave reached up the sides.

The small sticks in the ends, mentioned by John Gay, served not only for lifting the canoe but also as braces to support the canoe at a given angle when turned over ashore to serve as a shelter. The Beothuk canoe, because of its form, was not well suited for portaging, and it must be concluded that little of this was done. In coastal voyages, the canoe would be unloaded and brought ashore each night to serve as a shelter.

It is believed that the gunwale lashing of these canoes was in groups, as in the Malecite. Howley questioned an old Micmac who had seen the Beothuk lashing; he likened it to the continuous lashing used by his own people, indicating some form of group wrapping, at least. It is probable that the group lashings were let into the gunwales by shallow notching at each group, a common Indian practice when no rail cap was used, to prevent abrasion from the paddle or from loading and unloading the canoe. The lacing of the ends appears to have been in the common spiral stitch, judging by the grave models. These, however, show a continuous wrapping at the gunwales, a common simplification found in Indian canoe models, representing either group or continuously wrapped gunwales indiscriminately.

The paddle of the Beothuks had a long, narrow blade, probably with a pointed tip and a ridged surface. The shape is nearly spatulate. The handle is missing from the grave model but was perhaps of the usual "hoe-handled" form without a top cross-grip.

From these descriptions and on the basis of common Indian techniques in birch-bark canoe construction, the form and methods of building the Beothuk canoe can be reconstructed. The drawing on page 97 shows the probable shape and appearance of the finished canoe. It seems likely that a level building bed was first prepared. The keel, probably rectangular in cross section, was then formed of two poles placed butt-to-butt, worked to shape, and scarfed. The fastening of the scarf was probably two or more lashings let into the surface of the wood. These lashings are assumed to have been of split-root material but may have been sinew. Possibly to strengthen the scarfs, pegs were also used, a technique consistent with the state of Beothuk culture. The keel probably had its ends split into laminae to allow the sharp bend required to form the bow and stern pieces; and it was probably treated with hot water and staked out to the desired profile. The main gunwales were similarly made and worked to the predetermined sheer which, in staking out, was hogged to a greater degree than was required in the finished canoe. The ends of the gunwales were apparently split into laminae to allow the shaping of the sharp upsweep of the sheer close to bow and stern. The outwales were probably formed in the same manner, after which the three thwarts were made and the material for ribs and sheathing prepared. The ribs were apparently bent to the desired

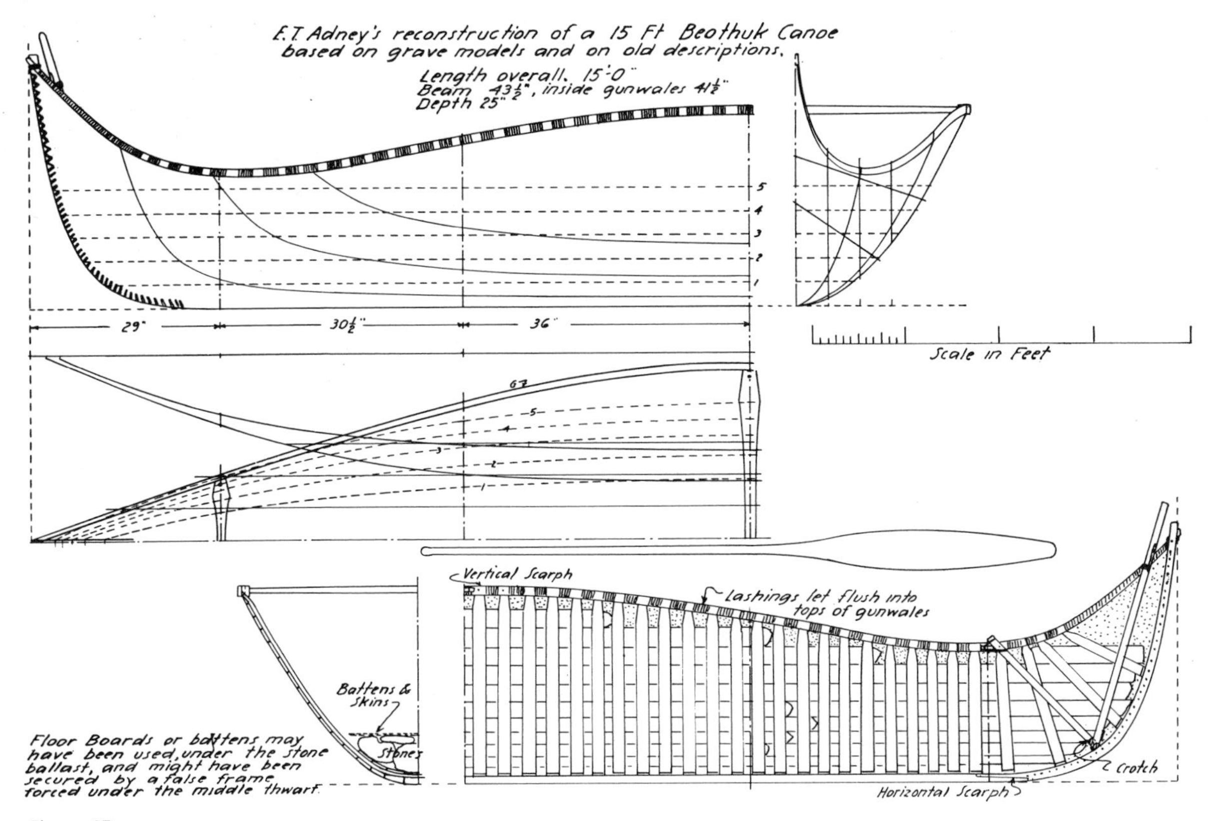


Figure 87

BEOTHUK CANOE, APPROXIMATE FORM AND CONSTRUCTION

shape, using hot water, and were either staked out or tied to hold them in form until needed.

The keel was then laid on the bed and a series of stakes, perhaps 4½ feet long, were driven into the bed on each side of the piece in opposing pairs at intervals of perhaps 2 or 3 feet. The stakes and keel piece were then removed and the bark cover laid over the bed. This may have been in two or three lengths, with the edges overlapped so that the outside edge of the lap faced away from what was to be the stern. The keel was then placed on the bark and weighted down with a few stones or lashed at the stem heads to the end stakes; then the bark was folded up on each side of the keel, and the stakes slipped back into their holes in the bed and driven solidly into place, perhaps with the tops angled slightly outward. The heads were then tied together across the work and battens placed along the stakes and the outside of the bark to form a "trough" against which the cover could be held with horizontal inside battens. These were secured by "inside stakes" lashed to each outside stake in the manner used in building eastern Indian canoes (see p. 45). The bark cover now stood on the bed in a sharp V form, with the keel supported on the bed, the ends of the bark supported by the end stakes, and both held down by stones along the length of the keel. An alternative would have been to fix heavy stakes at the extreme bow and stern of the keel and to lash the stem-heads firmly to these in order to hold the keel down on the bark.

Next the main gunwales, pre-bent to the required form, were brought to the building bed and their ends temporarily lashed to stem and stern. The bark was brought up to these, trimmed, folded over their tops, and secured by a few temporary lashings. Then the outwales were placed outside the bark with their ends temporarily secured, and a few pegs were driven through outwale, bark, and main gunwales, or a few permanent lashings were passed. The bark cover was next securely lashed to the gunwales and outwales combined, all along the sheer to a point near the ends. The excess bark was then trimmed away at bow and stern and the cover was laced to the end pieces to form bow and stern. This lacing must have passed through the laminations of the stem and stern pieces in the usual manner, avoiding the spiral lashing that held the laminae together. The ends of the gunwales and outwales were next permanently lashed together with root or other material and to the stem and stern pieces. This done, the gunwales were spread apart amidships, pressing the stakes outward still more at the tops. At this point the tenons may have then been cut in the main gunwales and the thwarts inserted. This method, incidentally, was used in building some western Indian bark canoes.

The usual steps of completing a birch-bark canoe would then follow—the insertion of sheathing, held in place by temporary ribs, and then the driving home of the prebent ribs under the main gunwales, with their heads in the spaces between the group lashings along the gunwales and against the lower outboard corner of the main gunwale member, which was probably beveled as in the Malecite canoe. The sheathing may have been in two or three lengths, except close to the gunwale amidships where one length would serve. On each side of the keel piece a sheathing strake was placed which was thick on the edge against the keel but thin along the outboard edge, in order to fair the sheathing into the keel piece.

At some point in this process, the bark cover was pieced out to make the required width, and gores were cut in the usual manner. In spreading the gunwales, the bow and stern would have to be freed from any stakes, as these would tend to pull inboard slightly as the gunwales were spread in the process of shaping the hull. The ribs could have been put in while green and shaped in the bark cover by use of battens and cross braces inside, as were those of the St. Francis canoes.

The sewing of the bark cover at panels and gores would take place before the sheathing and ribs were placed, of course. A 15-foot canoe when completed would have a girth amidships of about 65 to 68 inches if the beam at the gunwales were 48 inches, and a bark cover of this width could be taken from a tree of roughly 20 inches in diameter. Hence, there may have been little piecing out of the bark for width. In the form of the Beothuk canoe as reconstructed there is nothing that departs from what is possible by the common Indian canoe-building techniques. The finished canoe would, in all respects, agree with most of the descriptions that have been found and would be a practical craft in all the conditions under which it would be employed.

These were the only birch-bark canoes supposed to have made long runs in the open sea clear of the land. In them the Beothuk are supposed to have made voyages to the outlying islands, in which runs in open water of upward of 60 miles would be necessary, and they probably crossed from Newfoundland to Labrador.

The V-form used by the Beothuk canoe was the most extreme of all birch-bark canoe models in North America, although, as has been mentioned, less extreme V-bottoms were used elsewhere. The Beothuk canoe may have been a development of some more ancient form of bark sea canoe also related to the V-bottom canoes of the Passamaquoddy. The most marked structural characteristic of the Beothuk canoe was the keel; the only other canoe in which a true keel was employed was the temporary moosehide canoes of the Malecite.

The Beothuk keel piece may have sometimes been nearly round in section like the keel of the Malecite moosehide canoe (p. 214). The two garboard strakes of the sheathing may have been shaped in cross section to fair the bark cover from the thin sheathing above to the thick keel and at the same time allow the ribs to hold the garboards in place. They could, in fact, be easily made, since a radial split of a small tree would produce clapboard-like cross sections. This construction would perhaps comply better with Cartwright's description of the keel than that shown in the plan on page 97.

The sheer of the Beothuk canoe is an exaggerated form of the gunwale shape of the Micmac roughwater canoe but this, of course, is no real indication of any relationship between the two. Indeed, the probable scarfing of the gunwales of the Beothuk canoe might be taken as evidence against such a theory. On the other hand, the elm-bark and other temporary canoes of the Malecite and Iroquois had crudely scarfed gunwale members, as did some northwestern bark canoes.

Most of the building techniques employed by Indians throughout North America are illustrated by these eastern bark canoes, yet marked variation in construction details existed to the westward, as will be seen.

# *Chapter Five*

# CENTRAL CANADA

THE INDIANS INHABITING central Canada were expert builders of birch-bark canoes and produced many distinctive types. The area includes not only what are now the Provinces of Quebec (including Labrador), Ontario, Manitoba, and the eastern part of Saskatchewan, but also the neighboring northern portions of Michigan, Wisconsin and Minnesota in the United States. The migrations of tribal groups within this large area in historical times, as well as the influence of a long-established fur trade, have produced many hybrid forms of bark canoes and, in at least a few instances, the transfer of a canoe model from one tribal group to another. It is this that makes it necessary to examine this area as a single geographic unit, although a wide variation of tribal forms of bark canoes existed within its confines.

The larger portion of the Indians inhabiting this area were of the great Algonkian family. In the east during the 18th and 19th centuries, however, some members of the Iroquois Confederacy were also found, and in the west, from at least as early as the beginning of the French fur trade, groups of Sioux, Dakota, Teton, and Assiniboin. From the fur trade as well as from normal migratory movements there was much intermingling of the various tribes, and it was long the practice in the fur trade, particularly in the days of the Hudson's Bay Company, to employ eastern Indians as canoemen and as canoe builders in the western areas. These apparently introduced canoe models into sections where they were formerly unknown; as a result, the tribal classification of bark canoes within the area under examination cannot be very precise and the range of each form cannot be stated accurately. It was in this area, too, that the historical *canot du maître* (also written *maître canot*), or great canoe, of the fur trade was developed.

Most of central Canada, except toward the extreme north in Quebec and toward the south below the Great Lakes, is in the area where the canoe birch was plentiful and of large size. There the numerous inland waterways, the Great Lakes, and the coastal waters of James and Hudson Bays make water travel convenient, and natural conditions require a variety of canoe models. Hence, when Europeans first appeared in this area they found already in existence a highly developed method of canoe transportation. This they immediately adopted as their own, and in the long period lasting until very recent times, during which the development of the northern portion of this area was slow, the canoe remained the most important means of forest travel.

In the northeastern portion of the area, including the Province of Quebec (with Labrador) from a line drawn from the head of James Bay eastwardly through Lake St. John and the Saguenay River Valley to the St. Lawrence and thence northward to the treeline in the sub-Arctic, dwelt the eastern branch of the far-ranging Cree tribe. Those living on the shores of Hudson and James Bays, along the west side of the Labrador Peninsula, were known as the Eastern, Swamp, or Muskeg Cree. To the north, at the Head of Ungava Bay, around Fort Chimo, and to the immediate southward, were the Nascapee, or Nascopie, supposedly related to the Eastern Cree. In southern Labrador and in Quebec along the north shore of the Gulf of St. Lawrence and for some distance inland, dwelt another related tribal group now known as the Montagnais.

Although the most recent canoe forms employed by these three Indian groups were very much the same, this may not have been the case earlier. A common canoe model in this area was the so-called "crooked canoe," in which there was a very marked fore-and-aft rocker to the bottom without a corresponding amount of sheer; as a result the canoe was much deeper amidships than near the ends. Another common model had a rather straight bottom fore

Figure 88

MONTAGNAIS CROOKED CANOE. (*Canadian Geological Survey photo.*)

and aft, with some lift near the ends and a corresponding amount of sheer. Between these was a hybrid which had some fore-and-aft rocker in the bottom and a very moderate sheer. Not until the 1870's was any detailed examination made of the canoes in this area; then it appeared that the crooked canoe might be the tribal model of the eastern Cree only, while the Nascapee employed a straight-bottom model, but it is possible that the examination was limited and that Nascapee use of the crooked canoe was simply not observed. By 1900, however, the crooked model was in use not only by the eastern Cree and the Nascapee but also by the Montagnais.

In the area around Fort Chimo and at the northern ranges of the eastern Cree and of the Montagnais the lack of good birch bark made it necessary to make up the bark cover out of many small pieces. This not only was laborious but made a rough and rather unsightly cover. Hence, some of the northern builders, particularly the Nascapee, substituted spruce bark, which was available in quite large sheets. The use of the spruce bark, however, did not cause any of these people to depart markedly from the model or the method of constructing birch-bark canoes, as it did for the Indians in the maritime area.

At the time (1908) when Adney was carefully observing the canoes in this area he found that both crooked and straight-bottom canoes were being used by all three tribal groups, but with a variation in midsection form among individual builders. Both types were built with a midsection that had a wide bottom and vertical sides, or, as an alternative, a narrow bottom and flaring sides. The end profile of all these canoes showed chin. In some crooked canoes the profile was apparently an arc of a circle, but in most canoes the form was an irregular curve. The stem met the gunwale in a marked peak rounded very slightly at the head, as the result of the method by which the stem was constructed, but in the hybrid model used by the Nascapee the ends were low and not much peaked and the quick upward rise of the sheer near the ends was lacking. In cross section all these canoes became V-shaped close to the ends, regardless of the midsection form. For the straight-bottom canoe and in the hybrid form this resulted in very sharp level lines, but the very great rocker of the crooked canoe brought the ends well above the normal line of flotation, so that this type was quite full-ended at the level line in spite of the V-section.

It is apparent upon examining the crooked canoe that there was actually less variation in its form, in spite of differences in midsection shape, than in that of the straight-bottom canoe, owing to its very great depth amidships in proportion to its width. This proportion made necessary a very moderate flare in in the narrow-bottom midsection and resulted in a rather wall-sided appearance, even in this model. The hybrid form, which fell between the extremes of the crooked canoe and the straight-bottom canoe, had a narrow-bottomed flaring-sided midsection, and its relatively moderate depth made obvious the flare in the topsides and thus created a distinctive model.

Figure 89

BIRCH-BARK CROOKED CANOE, UNGAVA CREE. (*Smithsonian Institution photo.*)

# *Eastern Cree*

The construction of canoes of the eastern Cree and related tribes seems generally like that of the Micmac craft. Instead of the gunwale method employed in the Maritime area, a building frame was used, and as a result the gunwales were longer than the bottom. In constructing the crooked canoe, the building frame must be heavily sheered, and there is evidence that the building bed was depressed amidships, rather than raised as was usual in the east. The great amount of rocker in the bottom in this form of Cree canoe made it necessary to block up the ends of the building frame to a very great height, and there was no need to raise the building bed at mid-length, since the rocker extended the full length of the bottom. The bark cover had to be gored at closely spaced intervals to allow the rocker to be formed, and even in the straight-bottom model, the quick rise of the bottom near the ends required closely spaced gores there. In the straight-bottom model, however, the building bed was raised at midlength, as in eastern canoe-building, and the building frame was ballasted to a cupid's-bow profile, when on the bed, so as to acheive the combination of straight bottom amidships with sharply rising ends.

The gunwales were formed of the main gunwale member and a light gunwale cap, no outwale being employed. They were joined at the ends and, after hot water had been applied, were staked out with posts under the ends to obtain the required sheer. The thwarts were then tenoned into the main gunwales, though occasionally a canoe was built with "broken" gunwales, that is, the thwart-ends were let flush into the top and covered by the caps. Some builders did not spread the gunwales and place the thwarts until after the bark cover was lashed at the sheer; others used the eastern methods of assembling the gunwale structure prior to securing the bark cover at sheer. The bark cover was attached to the main gunwales with a continuous lashing, as in the Micmac canoes, but the bark was not always brought over the top of the gunwales. As a result, some canoes had a batten placed under the lashing, near the edge of the cover, to prevent the lashing from tearing away. Due to the lack of good root material, the lashing was often of rawhide. For all horizontal seams in the side panels of the bark cover, rawhide sewing over a root batten was used. The ends of the gunwales were supported by sprung headboards; in some canoes these were bellied toward the ends to such a degree that they almost paralleled the end profiles.

The ends were formed by means of the same technique used for Micmac canoes; no inside stem-piece was employed and the bark cover was stiffened by outside battens covered by the lashing. In the Cree canoes, however, the stem battens were "broken" sharply at the sheer to form a slightly rounded peak where the end met the gunwale caps. The "break" in the battens was made by bending them very sharply, so that they were almost fractured. The Cree practice also differed from that of the Micmac, although not universally, by passing the lower end of

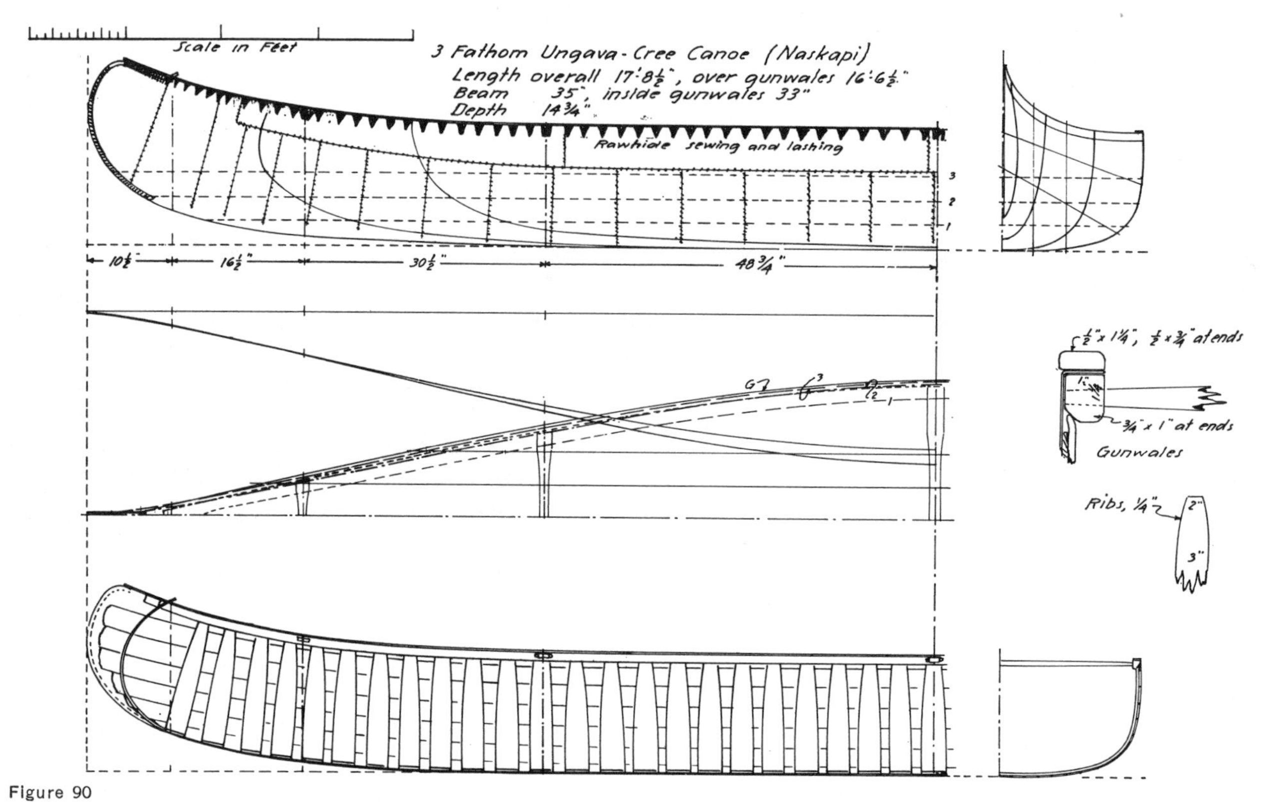


Figure 90

NASCAPEE 3-FATHOM CANOE, EASTERN LABRADOR. Similar canoes, with slight variations in model and dimensions, were used by all Ungava Indians: the Montagnais and the Eastern, or Swamp, Crees.

MONTAGNAIS 2-FATHOM CANOE OF SOUTHERN LABRADOR AND QUEBEC, showing old decoration forms. Drawing based on small model of a narrow-bottom canoe built for fast paddling.

Figure 91

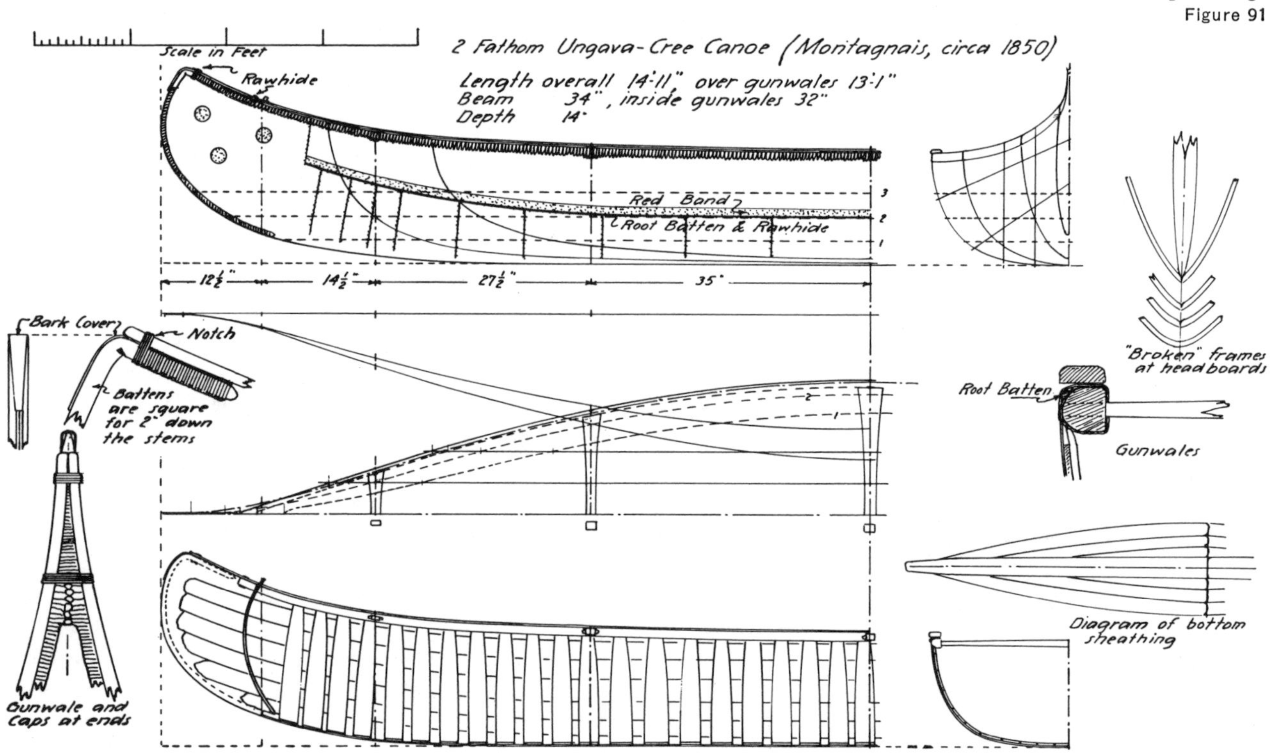

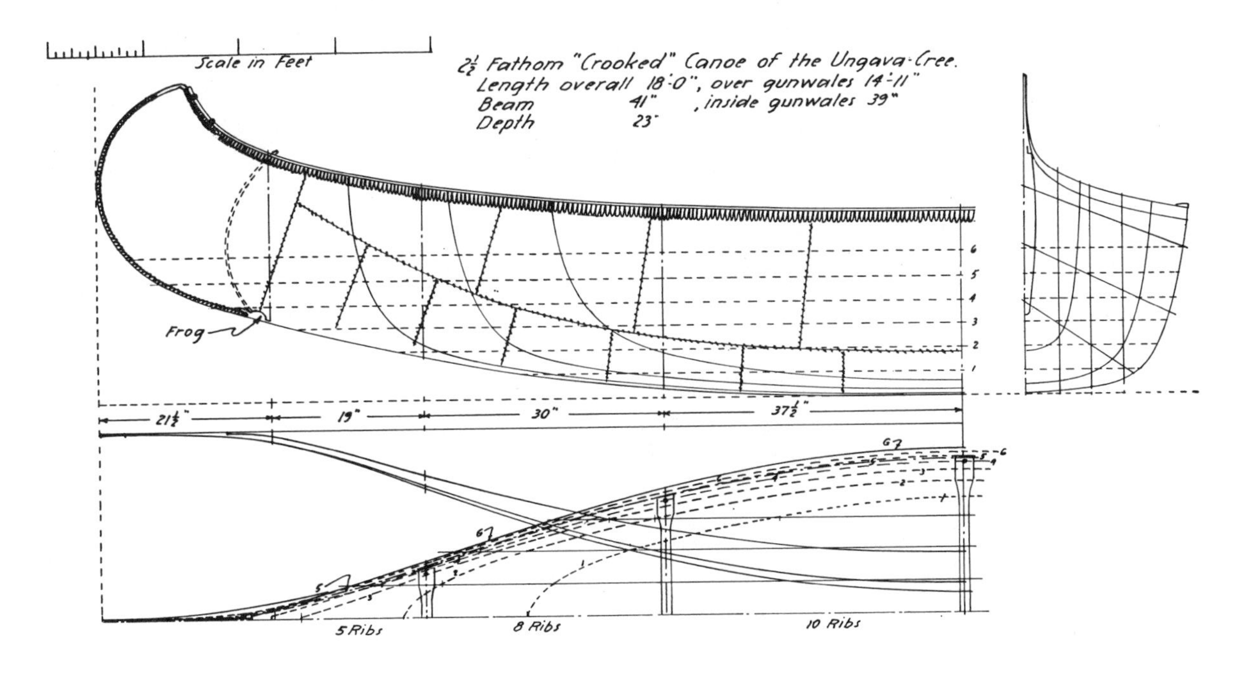


Figure 92

Crooked Canoe, 2½-Fathom, of the Ungava Peninsula, used by the Ungava-Cree, Montagnais, and Nascapee. Also built with a wide bottom and a slight tumble-home in the topsides.

Hybrid Model of the Nascapee-Cree Canoe, 2-Fathom, built or spruce or birch bark, with details of canoes and paddles. Figure 93

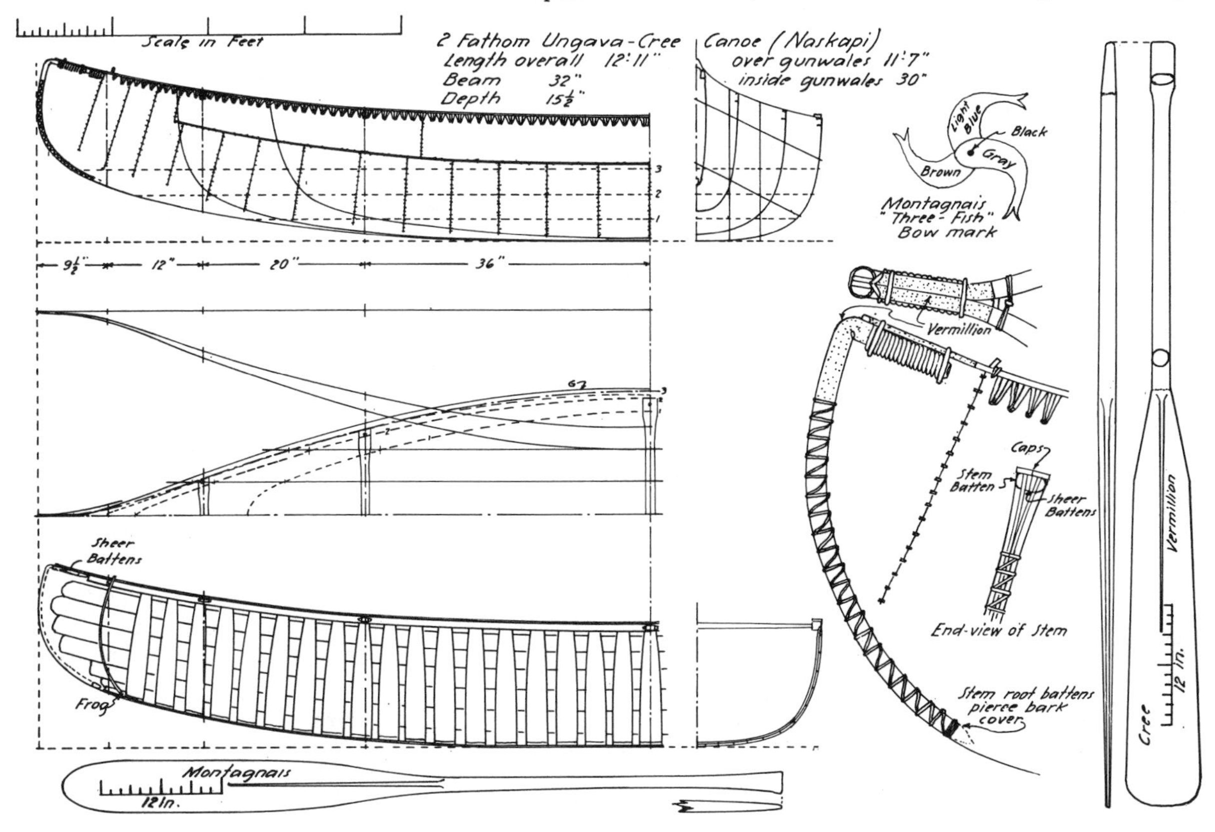

Figure 94

EASTERN CREE CROOKED CANOE of rather moderate sheer and rocker. (*Canadian Pacific Railway Company photo.*)

the stem batten through the bark cover at the point where the stem met the bottom. The slit thus made was sealed with gum or, more recently, covered with cloth impregnated with gum. The stems were lashed in various ways; the most common was a spiral form up to the sheer. Near the gunwale caps crossed stitches or small, closely spaced wrappings were also employed. The tops of the battens, forming the peak of the stem, were brought along under the rail caps, in line with the gunwale lashings inboard, and secured with a continuous lashing for about 6 inches. In the northern parts of the area under discussion the stem lashing was often of rawhide.

Gunwale caps were wider than the gunwales and thus gave some protection to the lashing there. The ends of the gunwale caps were heavily tapered to allow the sharp bends necessary to carry them out on the stems. They were pegged or nailed to the gunwales, but at the ends were lashed; usually with two or three small group lashings over and under the stem battens, below the caps.

The most recent canoes had canvas covers instead of bark. Nails, tacks, and twine for sewing were used; otherwise they were built as the Indians built birch- and spruce-bark craft, and not as white men built canvas canoes and boats.

The framework of the canoes was usually spruce or larch. Toward the south and along the St. Lawrence some white cedar was used, and in the south maple was sometimes used for thwarts. The ribs of the canoes inspected by Adney were usually about 3 inches wide, and a short taper brought them to about 2 inches at the ends, where they were cut square across. They were spaced about 1 inch apart edge-to-edge amidships and somewhat further apart toward the ends of the canoe. The canoes usually had an odd number of ribs, as the first was placed under the thwart amidships. The last three ribs at the ends were "broken" at the centerline to allow them to take the necessary V-section there; but the fourth rib from each end was only sharply bent. In some canoes the heel of the very narrow headboard was stepped on

Figure 95

Straight and Crooked Canoes, Eastern Cree.

the sheathing against the endmost rib, in others it was stepped, as in the Micmac canoes, on a frog which rested against the endmost rib.

In more recent times the sheathing was laid in one of two ways, according to the preference of the builder, but the existence of the two styles suggests that each was once a tribal-group method. One method of shaping the bottom sheathing was to employ a center, or keelson, piece in two lengths, the butts being overlapped amidships, parallel-sided except toward the stems, where it was tapered to fit the V-sections rather closely. The next strake outboard was short and was in the form of a shallow triangle with its base along the middle portion of the first strakes and about one-third the length of the bottom. Its apex was under the middle thwart. The next strake outboard was in two lengths lapped amidships, parallel sided along the arms of the triangular strake, and snied off at the ends to fit along the sides of the first strake. Another strake outboard of this was similar in form and position, but longer. Thus seven strake widths would complete the bottom sheathing. The side sheathing was narrow and slightly tapered; each strake in two lengths overlapped slightly amidships. The ends of the topside sheathing ran well into the ends, in most canoes, where they apparently served as stiffening. The second method of sheathing employed parallel-sided strakes throughout, laid side by side on the bottom, with the ends snied off to fit the form of the bark bottom. The existence of a model canoe made about 1850 (see p. 91) supports the theory that the first method was originally the Montagnais tribal construction and that the more primitive second method was probably Cree or Nascapee.

The ribs were preformed and fitted to the canoe after drying out. They were bent to the desired shape in pairs and tied with a thong across the ends to hold their shape while drying. Some builders inserted a strut inside the bent ribs, parallel to the thong, protecting the surface of the inner rib by a pad of bark placed under each end of the strut. The pair of ribs might also be wrapped with a bark cord to help hold them together. To aid in handling, one pair of ribs might be nested inside another. As in eastern canoes

Figure 96

MONTAGNAIS CANVAS-COVERED CROOKED CANOE under construction. (*Canadian Geological Survey photo.*)

the ribs under the gunwales were driven into place. At the ends they were canted toward the center, so that in the straight-bottom models they stood nearly perpendicular to the rocker of the bottom there; in the crooked canoe the ribs were all somewhat canted in this manner.

The paddles used in this area were made with parallel-sided blades, the end of the blade being almost circular. The handle might be fitted with a wide grip at the head or it might be pole-ended. It is impossible to say how early sails were used to propel canoes, but is is probable they were introduced by the fur traders. Square sails were being used on the coastal canoes at the time the earliest reference was made to these canoes, in the 1870's.

Little is known about the decorations employed by the eastern Cree. The Montagnais birch-bark model canoe of about 1850 (see p. 91) has three small circles placed in a triangular position on the bow and a band along the bottom of the side panels. The circles and the bands are in red paint, but may have been intended to represent the dark inner rind left after scraping the winter bark cover. The use of decoration in this area after 1850 has not been noted in any available reference.

As a rule, the straight-bottom canoes were small, commonly between 12 and 18 feet overall, and the most popular size was 14 to 16 feet overall. A canoe of this size was usually employed as a hunters' canoe for forest travel, though it might be used occasionally along the coasts. These canoes were light and, in this respect, resembled the Micmac models shown in Chapter 4.

The original purpose of the crooked canoe is in question. Those travelers who saw this canoe in use on the Hudson Bay side of the Labrador Peninsula believed that it was designed for use in rough, exposed water. While it would be a desirable form for beach work in surf, the high ends would make paddling against strong winds very difficult. On the other hand the Montagnais used the crooked canoe for river navigation, particularly where rapids were to be run, and for this work it appears to have been well adapted. The crooked canoe was commonly built larger than the straight-bottom model, between 16 and 20 feet in length overall, and was a vessel of burden rather than a hunting canoe. Canoes up to 28 feet in length have been mentioned by travelers in this area but investigation indicates strongly that these were not the tribal form but the *canot du nord*, or north canoe of the Hudson's Bay Company traders.

Along the southern borders of their territory and to the westward the eastern Cree often built and used canoes modeled on those of their neighbors, the Têtes de Boule and the Ojibway. Hence the tribal classification does not hold good in these localities. Also, the eastern Cree were employed by the Hudson's Bay Company as builders of forms of the *maître canot* and *canot du nord* that are unlike their typical tribal model.

# *Têtes de Boule*

The Têtes de Boule, particularly the western bands, were skilled canoe builders and had long been employed by the Hudson's Bay Company in the construction of large fur-trade canoes. Apparently made up of bands of Indians inhabiting lower Quebec, in the basin of the St. Maurice River and on the Height of Land, these bands had come down to the lower Ottawa River to trade with the local Algonkin tribe there in early times. They were known to the Algonkins, who had had some contact with civilization, as "wild Indians." They also came into close trading relations with the French colonists, as the Ottawa River was the early French canoe route between Montreal and Lake Superior. Because they cut their hair short, unlike the other Indians, these northern bands were nicknamed "Bull Heads," or "Round Heads," by the French traders, and the tribesmen soon came to accept this rather than their own designation of "White Fish People" as the tribal name. In more recent times, the name has been applied to groups of Indians living in western Quebec Province, near Lake Barrière and Grand Lake Victoria, but these do not consider themselves related to the St. Maurice bands.

It seems apparent that the canoe models of all these groups had been altered as a result of long contact with other tribal groups. Although the St. Maurice and the western bands were apparently not of the same tribal stock, their relations with the Algonkin may have brought about the use of a standard model by all.

The Têtes de Boule lived in an area where very superior materials for birch-bark canoe construction were plentiful. This, with the need for canoes imposed by the numerous waterways and the demand for canoes from white traders, made many of the tribesmen expert builders. Their small canoes, ranging from the 8- to 12-foot hunter's canoes to the 14- to 16-foot family canoes, were very similar in profile to the canoes of the St. Francis Abnaki. The Têtes de Boule canoes, however, were commonly narrower on the bottom, and in their construction a building frame was always used. The Têtes de Boule model was straight along the bottom for better than half the length and then rose rather quickly toward the ends. Similarly, the sheer was moderate amidships and increased toward the ends. The stems showed a chin and were much peaked at the gunwale ends. Most commonly the midsection had a flat bottom athwartships and a well-rounded bilge, giving the topsides, near the gunwale, a very slight outward flare. Some Têtes de Boule canoes had rather V-section ends in which the endmost rib was "broken" at the centerline. As a result the lines were sharp and the canoes paddled very easily.

Figure 97

FIDDLEHEAD OF SCRAPED BARK on bow and stern of a Montagnais birch-bark canoe at Seven Islands, Que., 1915.

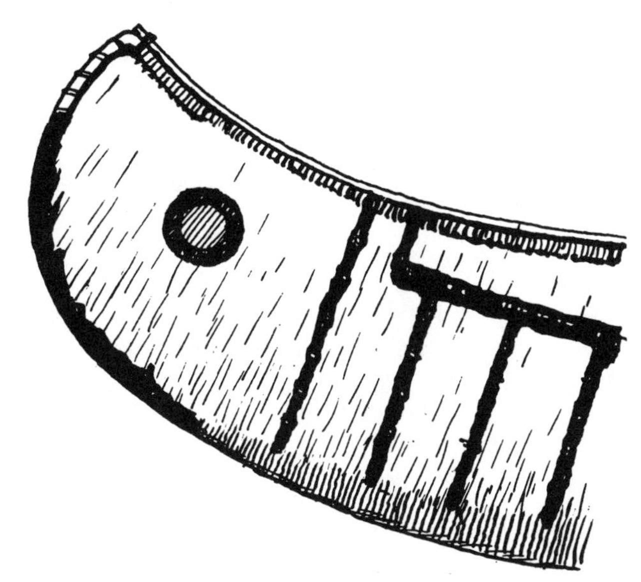

Figure 98

DISK OF COLORED PORCUPINE QUILLS decorating canoe found at Namaquagon, Que., 1898. Within the 4-inch disk may have been an 8-pointed star.

Figure 99

A Fleet of 51 Birch-Bark Canoes of the Têtes de Boule Indians, assembled at the Hudson's Bay Company post, Grand Lake Victoria, Procession Sunday, August 1895. (*Photo, Post-Factor L. A. Christopherson.*)

For construction of the Têtes de Boule canoe, which was marked by good structural design and neat workmanship, the building bed was slightly raised at mid-length, as was the general practice of the St. Francis builders. The building frame was usually about 6 inches less in width amidships, inside to inside, than were the gunwales, and from 15 to 18 inches shorter. The building frame was made quite sharp toward the ends so that, viewed from above, it rather approached a diamond form; this produced the very sharp lines that are to be seen in many examples of the Têtes de Boule canoes. The building frame was of course removed from the canoe as soon as the gunwales were in place and the bark cover lashed to them.

The gunwale structure, comprised of main gunwale members, caps, and outwales, was the same as in the Malecite canoes. The main gunwales were rectangular in cross-section, some being almost square, with the lower outboard corner bevelled off. Compared to those of eastern canoes of equal length, the main gunwales were unusually light; their depth and width rarely exceeded 1 inch, and in very small hunter's canoes these were often only about ¾ inch. Toward the ends, they tapered to ½ inch, or even slightly less. The ends of the main gunwales, usually of the common half-arrowhead form, were held together by rawhide or root thongs passed back and forth through horizontal holes in the members. After being thus lashed together, they were securely wrapped with thongs which usually went over gunwales and outwales and through the bark cover.

The gunwale caps, also light, were usually between ¼ and ½ inch thick and from 1 to 1½ inches wide. At the ends they were tapered in width and thickness, often to 3/16 by ½ inch, so as to follow the quickly rising sheer there. The ends of the gunwales, caps, and outwales required hot-water treatment to obtain the required curve of the sheer. The caps were pegged to the gunwales and were secured at each end with two or three groups of lashings which passed around the outwales as well, and through the bark cover.

The outwales were likewise light battens between ¼ and ½ inch thick and from ¾ to 1¼ inches deep, the depth near the ends being tapered to ⅜ to ¾ inch so as to sheer correctly.

The bark cover had four or five vertical gores on each side of the middle thwart, the gore nearest each stem being commonly well inboard of the end thwarts. The side panels were usually deep amidships and narrowed toward the ends. A root batten was used under the stitching of the longitudinal seams of the side panels, which were sewn with a harness-maker's

Figure 100

Têtes de Boule Canoe.

stitch. The top edge of the bark cover was brought over the top of the main gunwales, as in the Malecite canoes, and was secured by group wrappings passing over the gunwales and outwales, under the caps. These groups were not independent, the root thong being carried from group to group outside the bark in a long pass under the outwales. The groups of seven to nine turns were roughly an inch apart in many small canoes, and perhaps 1½ inches in the large craft. In the last birch-bark canoes in which no nails or tacks were used, wrappings of root thongs began with a stop knot, but this does not appear to have been the earlier practice.

The Têtes de Boule canoes had inside stem-pieces split, according to the size of the canoe, in four to six laminations and lashed with a bark or root thong in an open spiral in some canoes but close-wrapped in others. The stem-piece was as in the Malecite canoes, except that it ended under the rail cap, and did not pass through it as in the Eastern canoes; the heel was notched to receive the heel of the headboard. The bark was usually lashed through the stem, as in the Malecite construction. However, in some Têtes de Boule canoes, the stem close to the heel was not laminated and the bark was lashed to the solid part by an in-and-out stitch passing through closely spaced holes drilled in the stem piece. Above this, the lashing was the usual spiral which, in at least a few instances, was passed through the bark just inboard of the stem piece. Near the top of the stem the lashings sometimes were rather widely spaced and passed inboard of the stem-pieces; at other times, however, these lashings were more closely spaced and passed through the stem.

Ordinarily, at the ends of the canoe no *wulegessis*, or covers of bark, were used under the gunwale caps, although in one example examined a small cover had been inserted over the gunwale ends and under the caps, it did not extend below the outwales to form a *wulegessis*. In some canoes the bark cover was pieced up at the peak of the stems by a panel whose bottom faired into the bottom of the side panels.

A variety of methods was used to fit the gunwale caps at the ends of the canoe. Some builders carried the cap out beyond the gunwale ends, flat, over the edges of the bark cover and the top face of the outwale, but others tilted the cap outboard and downward. The ends of the caps came flush with the face of the stems. In an apparently late variation, the gunwales, instead of ending in the half-arrowhead, were snied off the inside and a triangular block was inserted between the ends. The gunwales were then pegged or nailed to the block and the whole secured with a root wrapping around them, before the outwales were in place. The first turn began by passing the root through a hole in the block near its inboard end, with a stop knot in the root.

The ends of the gunwales were supported by a narrow headboard sharply bellied toward the end of the canoe. The top of the headboard was notched to stand under the main gunwales; the center portion often was carried high and ended with a cylindrical

Figure 101

TÊTES DE BOULE CANOES.

top that was slightly swelled like the handle of a gouge or chisel. The heel was sometimes held in the stem-piece notch with a root lashing.

The thwarts, spaced equal distances apart, were tenoned into the gunwales as in the old Malecite canoes, and were secured with a peg and lashing through the two holes in the thwart ends. The middle thwart was usually formed with a shoulder, viewed in plan, that started 6 or 7 inches inboard of the inside face of the main gunwale. In form, this thwart usually swelled outward in a straight line from the tenon shoulder, then reduced in a curved line to about the width of the tenon tongue and, finally, increased again in a right-angle cut to the greatest width. From here it was reduced again in a long curve to the canoe's center line. The other thwarts usually had simple ends, wide at the tenon shoulder and reduced in a long curve to a narrow center. In elevation, all the thwarts were thin outboard and thick at the centerline of the canoe. The cross section of the center thwart at the centerline was square or nearly so, the first thwart on each side was rectangular in cross section at the center, and the end thwarts were similar, but very thin.

The sheathing of the Têtes de Boule canoes was thin, particularly at the ends of the strakes. The bottom was laid with a parallel-sided center strake going in first. This strake was in two lengths in a small canoe and three lengths in a large, the butts overlapping slightly. The rest of the strakes in the bottom were tapered toward the ends of the canoe. At the extremities of the canoe, the narrow ends of the strakes were very thin and overlapped along their edges, the bottom sheathing, when in place, thus following the diamond form of the building frame. The topside sheathing was laid up in short lengths with overlapping butts and edges in an irregular plan, those strakes along the bilges being longer than above. Toward the ends of the canoe these strakes were slightly tapered and the edges were very thin. The sheathing ended irregularly, outboard of the headboards, in narrow butts as in most eastern canoes.

The ribs, like the rest of the structure, were very light, usually ¼ to ⅜ inch thick and from about 1¼ to 1¾ inches wide, depending upon the size of the canoe. A few examples had ribs 2 inches wide, and still fewer had ribs up to 2½ inches wide. The spacing was usually close, somewhat more than an inch edge to edge amidships and a little more between the end thwarts and the headboards. The spacing amidships would average perhaps 3¼ inches, center to center. The ends of the ribs, in the last 2 or 3 inches, were reduced in width very sharply in a hollow, curved taper to ½ to ¾ inch wide, and were usually beveled on the inside edge. The thickness was also reduced by a cut on the inside, so that the ends were chisel-pointed with a short bevel on the inboard side. The rib ends were forced between the main gunwales

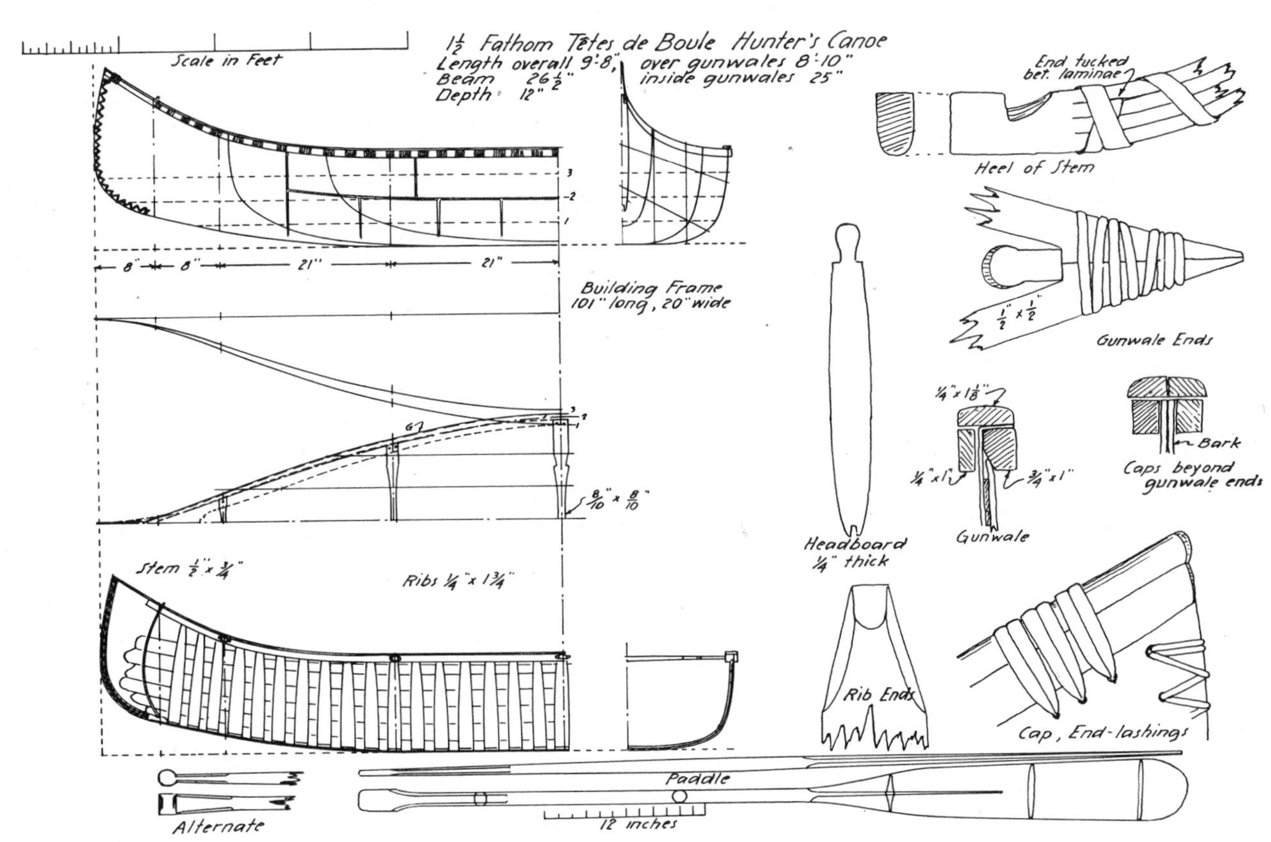


Figure 102

TÊTES DE BOULE HUNTING CANOE, 1½-FATHOM, with typical construction details and a paddle.

TÊTES DE BOULE CANOE, 2½-FATHOM, with some construction details.

Figure 103

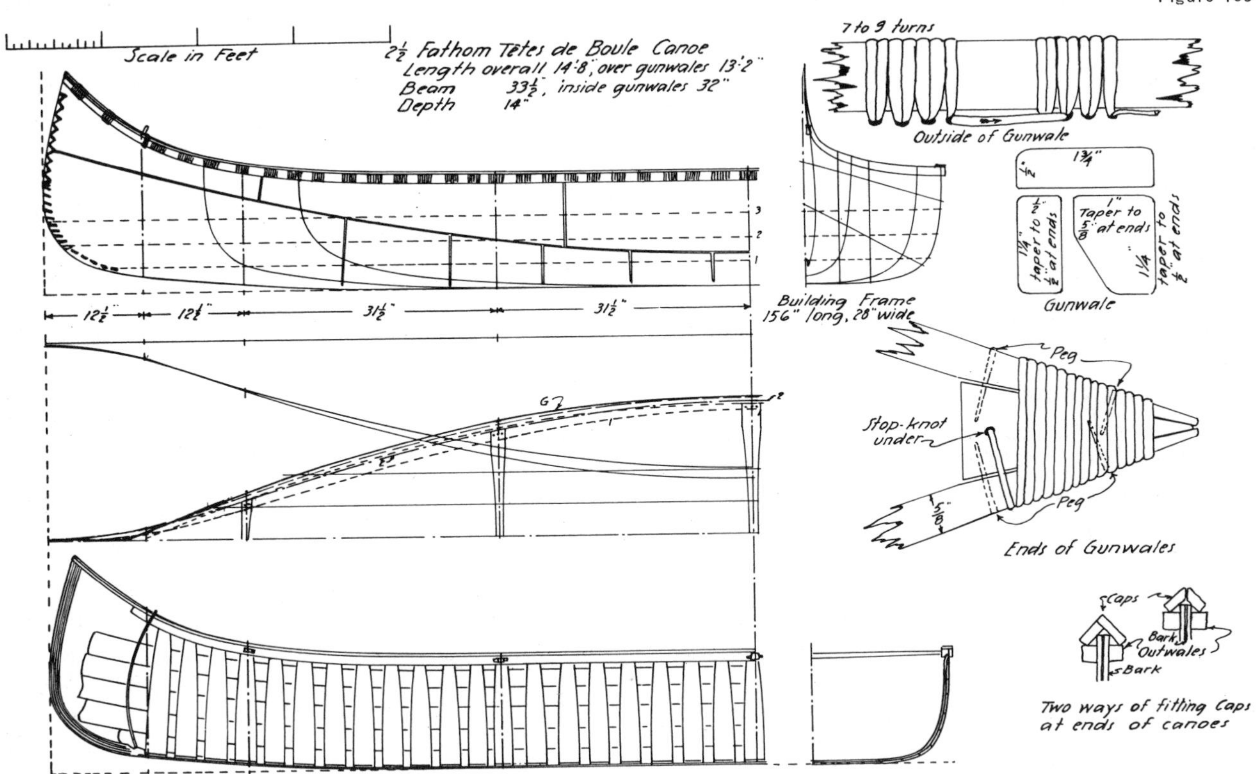

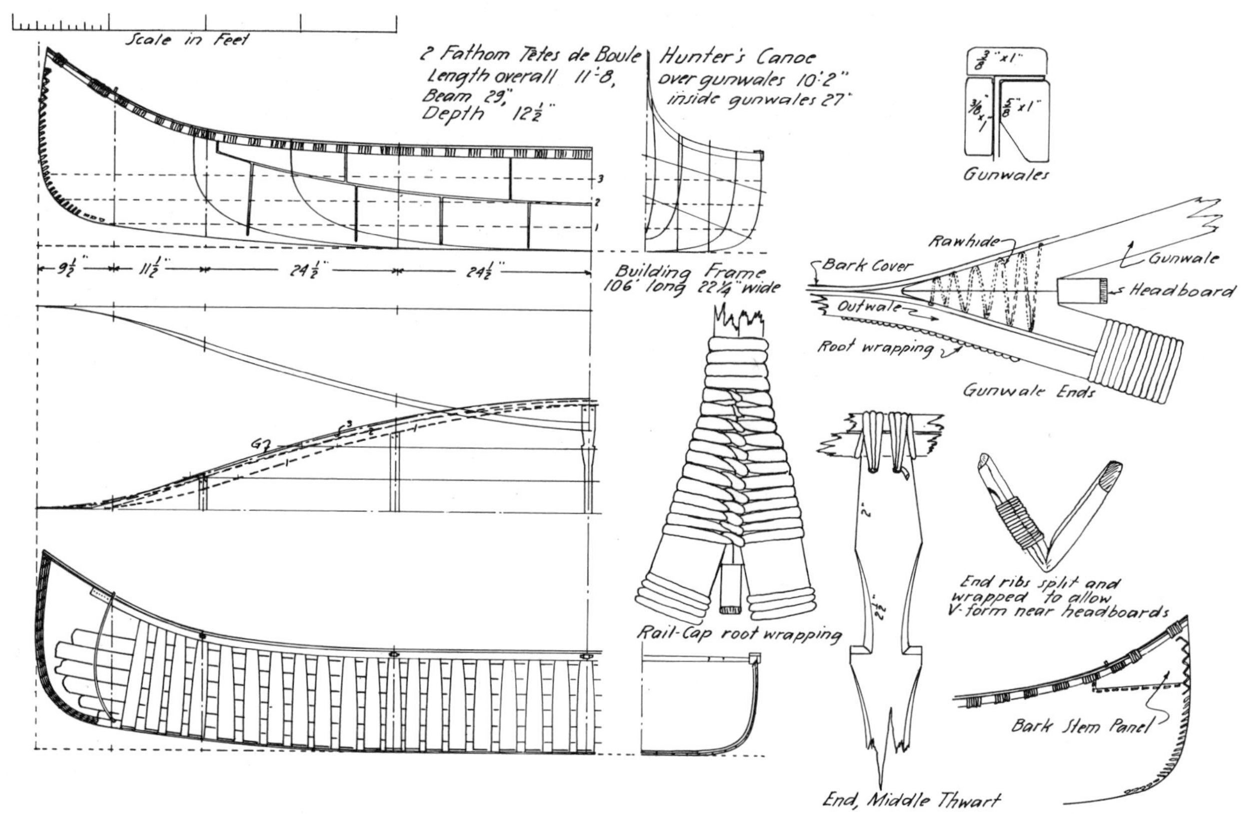


Figure 104

TÊTES DE BOULE HUNTING CANOE, 2-FATHOM, with wide bottom, showing structural details.

and the bark cover, coming home in the bevel of the lower outboard edge of the main gunwales between the group lashings of the bark cover as in the Malecite canoes. The ribs were not prebent but were placed in the canoe when green, treated with hot water, and then allowed to dry into place. In preparing the rib, it was first bent over the knee. It was the custom of some builders to place under the building frame the ribs that were to go near the ends of the canoe, and to mark the point where they would be bent. Sometimes the endmost ribs that were to be "broken" at the centerline to form the V-section were split edgewise. A piece of the inner lamina was then cut out to one side of the center so that the inner laminae would lie flat against each other, and to prevent the inner half from buckling the rib was wrapped with a thong to one side of the "break."

It does not appear to have been the common practice of the Têtes de Boule to decorate their small canoes, though when building for white men they would decorate if the buyer requested it.

The paddles used by the Têtes de Boule were somewhat like those of the eastern Cree but the blade was slightly wider near the tip than near the handle. The top grip was formed wide and thin, the taper from the lower grip to the upper one often being very long. The paddles were usually of white birch, but maple was used in a few of the examples examined.

The gunwales, outwales, and caps of the Têtes de Boule canoes were usually of spruce; the ribs and stem pieces, white cedar; the thwarts, white birch; the headboards, white cedar in all but one of the canoes inspected (in this, birch had been used). Jack pine was used also for thwarts, and cedar was sometimes used for the gunwale members; as would be expected, the builders used the materials that were at hand near the building sites.

Têtes de Boule fur-trade canoes, like those of the eastern Cree, appear to have had no relationship to the smaller tribal types, since they were constructed under supervision of white men. They will be discussed as a group on page 135.

## *Algonkin*

The Algonkins were a tribe residing on the Ottawa River and its tributaries, in what are now the provinces of Quebec and Ontario, when the French first met them. They appear to have been a large and powerful tribe and were apparently competent builders and users of birch-bark canoes. They were not the same tribe as the Ottawa, who controlled the Lake Huron end of the canoe route between Montreal and Lake Superior, by way of the Ottawa River. These Ottawa were related to the Ojibway tribe and received their name from the French, who gave the name *Outaouais*, or "Ottaway," to all Indians, except the Hurons, who came from the west by way of the Ottawa. The Algonkins, because of their location, were much influenced by the French fur trade. Early in the 18th century they intermingled with certain Iroquois whom they allowed to settle with them, near Montreal, at the Lake of Two Mountains, later Oka. Thence they gradually spread out and lost tribal unity, until only small groups were left. These lived on the Golden Lake Algonkin Reserve, Bonshere River, Ontario; at Oka, Quebec; and elsewhere in western Quebec and eastern Ontario. It is possible that they were the first to build fur-trade canoes for the French, but evidence to support such a claim with any certainty is lacking.

Due to intermixing with other tribal groups and to the influence of the fur trade, in which they were long employed as canoe men and builders, the Algonkins no longer used a single tribal model of canoe. However, one of their models, which had high ends resembling those of the large fur-trade canoe, may have been the tribal type from which the fur-trade canoe was developed, as will be seen.

The high-ended model, the oldest form known to have been used by this tribe, was narrow-bottomed, with flaring sides. The canoes seen were built with careful workmanship and in the old manner, without iron fastenings. They were light and easily paddled, yet would carry a heavy load. The ends were sharp at the line of flotation. The bottom was straight to a point near the ends, where it lifted somewhat. The sheer was rather straight over the middle portion of the canoe, then lifted slightly until close to the stem, where it rose sharply, becoming almost perpendicular at the ends of the rail caps. The midsection was slightly rounded across the bottom, with a well-rounded bilge and a gently flaring topside. The cross-section became V-shaped close to the headboards. The most marked feature in the appearance of this canoe was the profile of the ends. The stem line, beginning with a slight angle where it joined the bottom, bent outward in a gentle curve, reaching the perpendicular at a point a little more than half the height of the end, and from there it tumbled home slightly. In most of the canoes examined the top of the stem then rounded inboard in a quick, hard curve, usually almost half a circle, so that the stem was turned downward as it joined the outwale and gun-

Figure 105

OLD ALGONKIN CANOE.

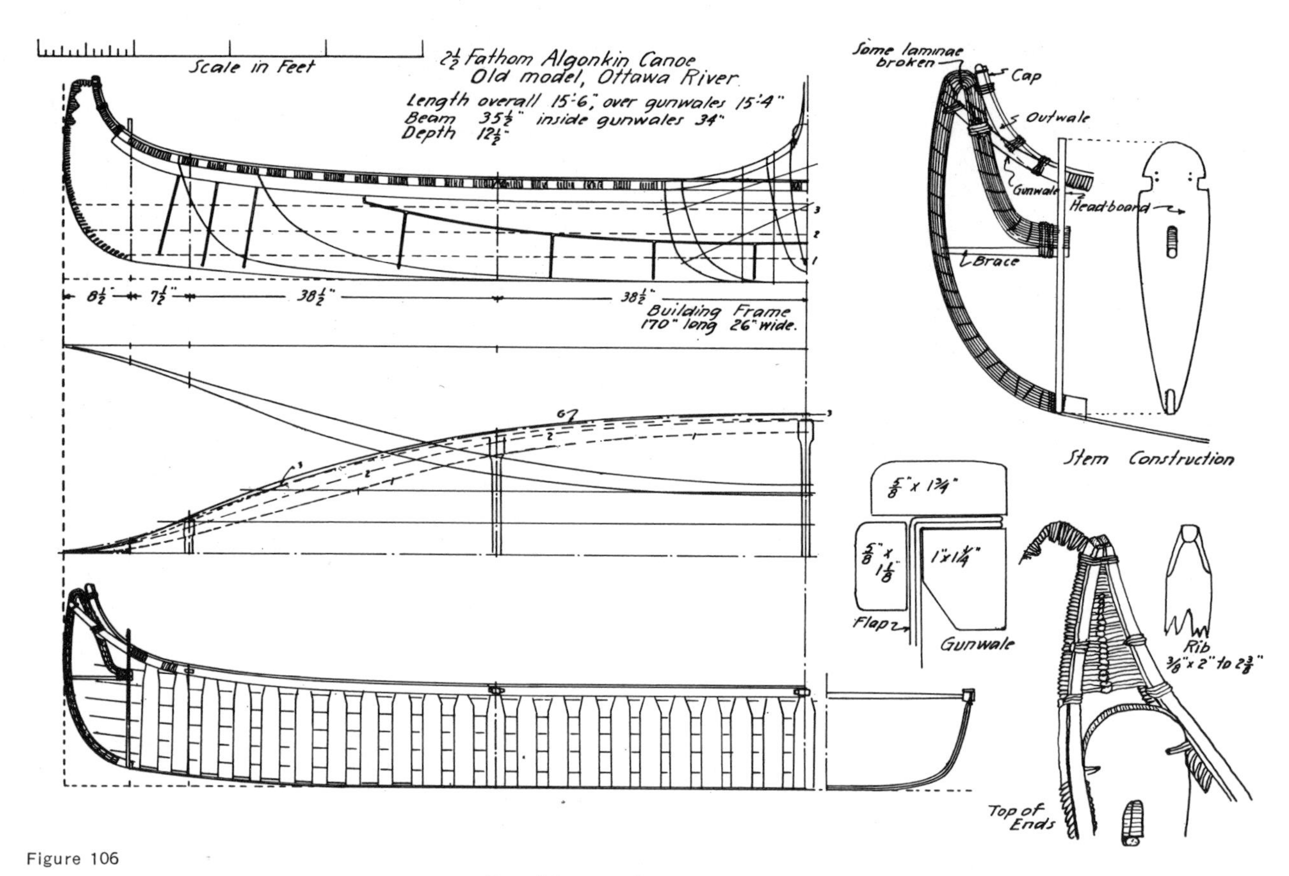


Figure 106

Old Model, Ottawa River, Algonkin Canoe, combining capacity with easy paddling qualities.

wale cap. In a variation of this stem form, the top of the stem was cut off almost square, forming a straight line that ran parallel to the rise of the bottom below the stems to the point where it would meet the upturned outwale and cap. The ends of the outwales and caps were thus 3 or 4 inches inboard of the extremities. This form of stem, particularly when to top was rounded in a half-circle, approached the basic form of the ends of the fur-trade canoe.

All the examples of this form of canoe that were examined were small, from 14 to a little over 16 feet in length overall, but this is not proof that larger canoes of this type had not existed earlier.

The later and more common form of Algonkin canoe was the *wabinaki chiman.* A corruption of Abnaki, *wabinaki* to the later Algonkin meant the Malecite as well as the St. Francis Indians. The *wabinaki chiman* was built in lengths from 12 to 18 feet.

Iroquois living in the Algonkin terriotry during the period built this form of canoe as well as the older, high-ended form. The *wabinaki chiman* was very much like the St. Francis and Malecite canoes in appearance, but it was not an exact copy. The Algonkin version was commonly a narrow-bottom canoe with flaring topsides. There was some variation in the end profiles; most had the rather high, peaked ends of the St. Francis canoe. The sheer was rather straight until near the end, where it rose rapidly to the stem. The stem was rounded and was faired into the bottom. The top of the stem was often rather straight and tumbled home slightly, but on some it raked outward, much as did the stem of some Malecite canoes.

Another form of Algonkin canoe had a low sheer with only a slight lift toward the ends. In this canoe the stem might have a short, hard curve at the heel and an upper portion that was quite straight and slightly tumbled home; or the full height might be well rounded, with a slight tumble-home near the stem head.

In appearance these canoes were very like the straight-stem Malecite models. The *wabinaki chiman*

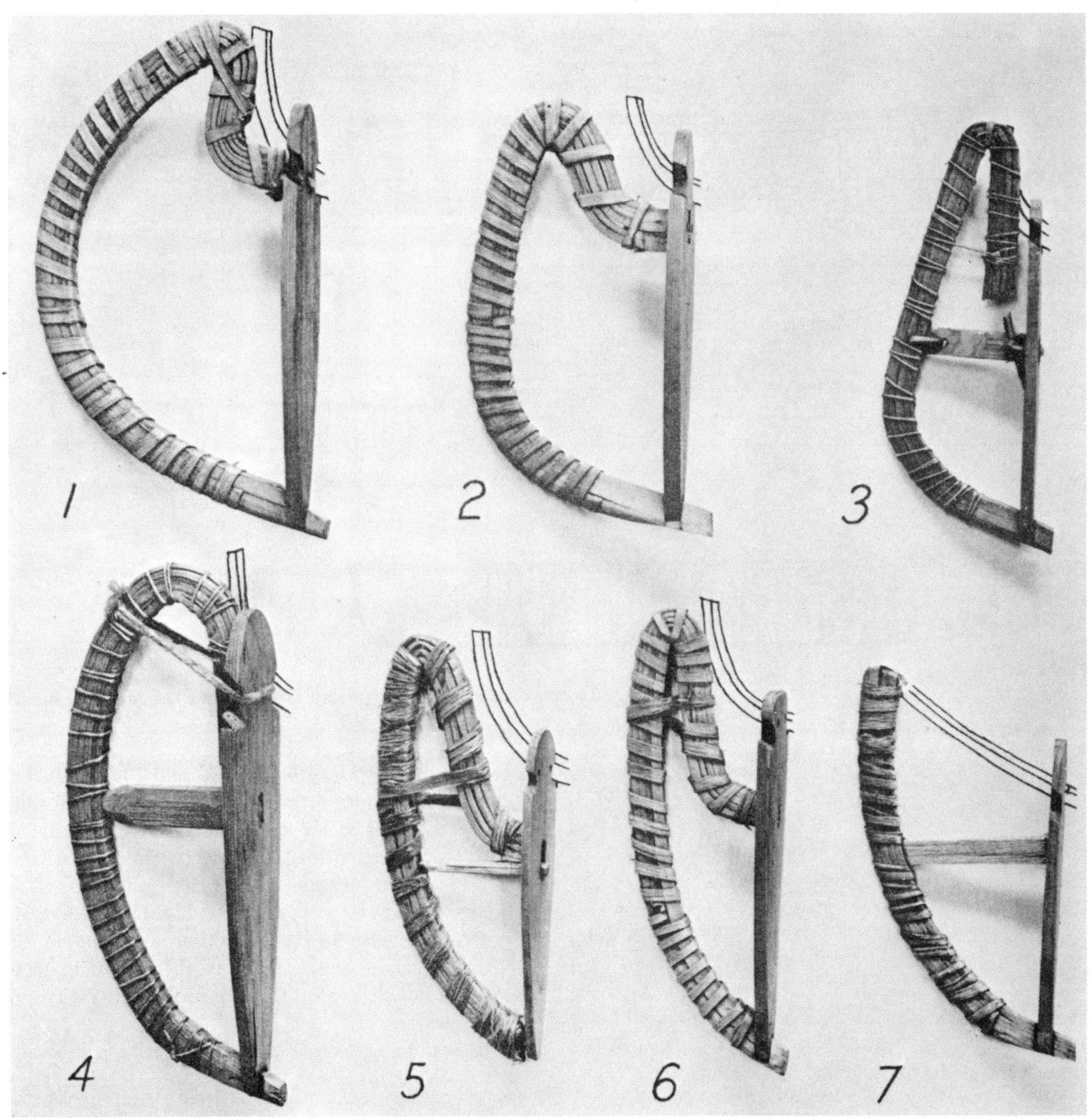


Figure 107

ALGONKIN AND OJIBWAY STEM-PIECES, models of old forms made by Adney: 1, 2, 3, Ojibway; 4, 5, 6, 7, Algonkin.

was unquestionably copied from the eastern canoes that came into popularity among the Algonkin late in the 19th century, when white sportsmen were demanding canoes of the St. Francis and Malecite models. However, the Algonkin canoes differed somewhat from the eastern canoes not only in model but also in methods of construction.

Algonkins used the same construction methods in both their canoe models, though the framework was not alike in all respects. The building frame was

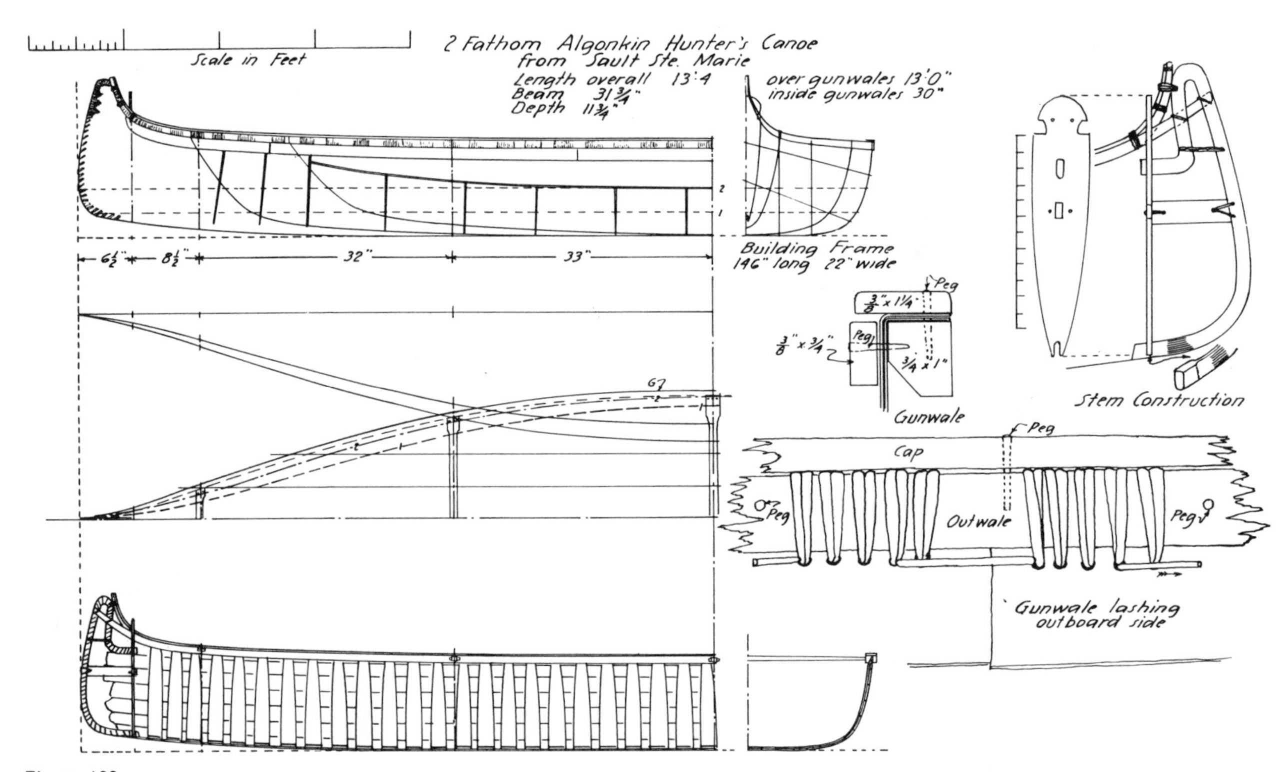


Figure 108

LIGHT, FAST 2-FATHOM HUNTING CANOE of the old Algonkin model.

always used. For a 2- or 2½-fathom canoe this was made of two strips of cedar, 1½ inches wide and ¾ inch deep, that were bent edgewise, notched, and tied together at the ends with thongs of the inner bark of the basswood. These strips were held apart in the required shape by cedar crosspieces 1 inch wide and 1¾ inches deep, with the ends notched ¾ inch deep (the depth of the longitudinals) and the tops well rounded. The crosspieces, five in all, were fastened to the longitudinals with thongs passing through holes in the ends. The middle one was about 19½ inches between the inside faces of the longitudinals, those on each side of it were about 15½ inches long by similar measure, and the end ones were nearly 6 inches long and were located a foot or so from the extremities of the longitudinals. The outside width of the building frame amidships would thus be about 22½ or 23 inches.

The building bed was level, with a 6-inch-wide board, some 6 to 8 feet in length, sunk into the earth flush with the surface to insure a true line for the bottom. The outside stakes were of the usual sort described in building the Malecite canoe (pp. 40–41). The wedge-shaped inside stakes, or clamp pieces, were 1½ inches wide, 1 inch thick, and 20 to 25 inches long. The posts for setting the height of the gunwales at the ends and at the crosspieces were not cut off square at the top as for the Malecite canoe, but were notched on the outside to take the gunwales. The heights of the posts were graduated, of course, to form the required sheer in the gunwales. Like the canoes of the Têtes de Boule, these of the Algonkin were generally less deep amidships than the general run of eastern canoes.

Building procedure was as follows: The gunwales were made, bent, and the ends fastened, but instead of being mortised and fitted with thwarts, they were spread by temporary crosspieces, or "spalls," made of a splint, or plank-on-edge, with the lower edge notched in two places to take the gunwale members. Sometimes the spalls were lashed, pegged or nailed to the gunwales as well. The stakes were set along the building frame and these were generally driven sloping, so that their heads stood outboard of the points. They were then pulled and laid aside, the building frame was removed, and the bark cover placed on the building bed. After the building frame has been reset in its original position and the

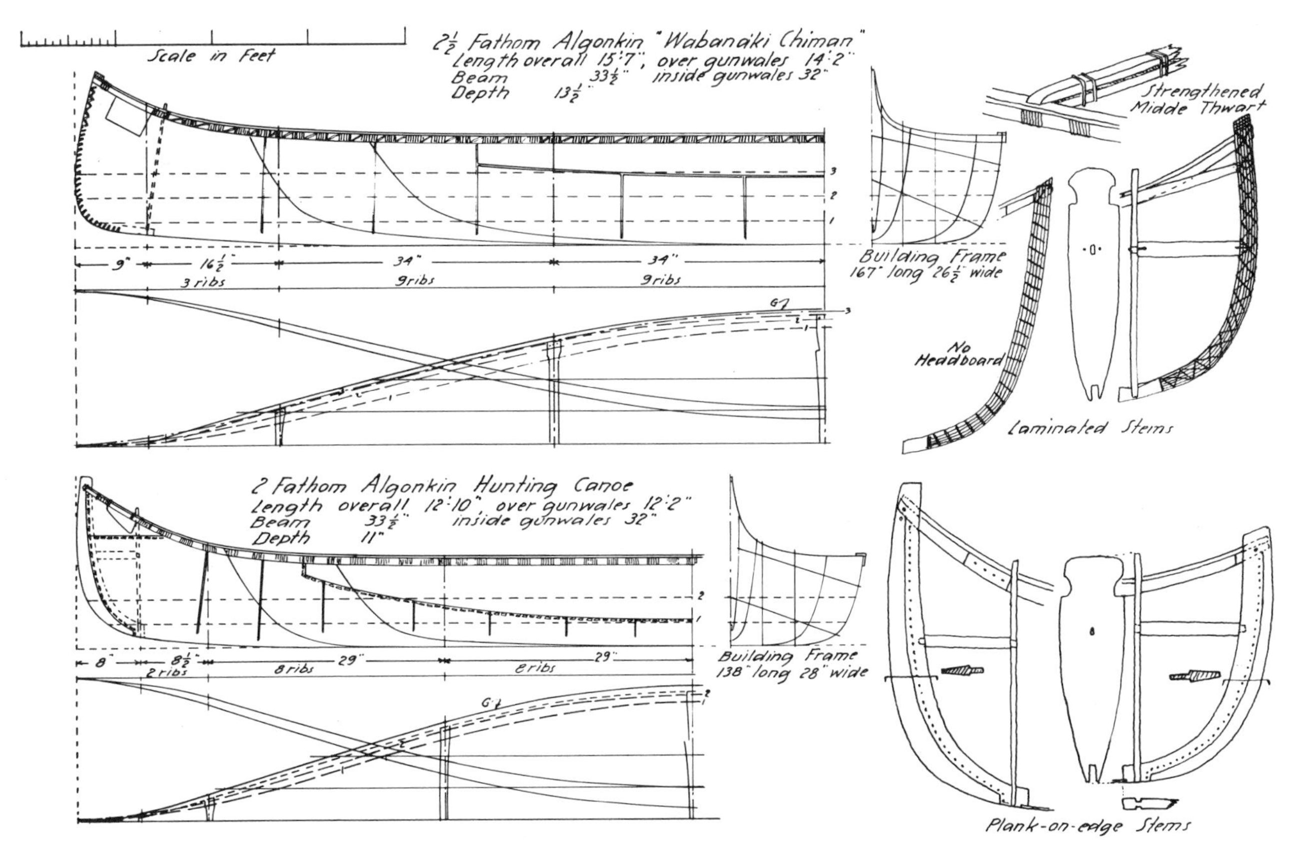


Figure 109

Hybrid Algonkin Canoes: Eastern 2½-fathom (above) and northeastern 2-fathom adaptation, with sketches of stems used in each.

bark cover turned up along the sides, the stakes were again driven in their holes. The cover was then pieced out with side panels as necessary and gored, and longitudinal strips of wood were set in place by means of the clamp pieces, about as in Malecite construction. The gunwales were then placed on the posts, which had been set to the required sheer, and the bark trimmed and fitted to them. The old method was to lash the bark to the main gunwale members and to peg on the outwales at intervals of about a foot. In earlier times most builders inserted along the gunwales an extra reinforcing strip of bark extending a little below the outwales, as in the St. Francis canoes, but in the nailed-and-tacked bark canoes built during the decadent period this was sometimes omitted.

Mortises for the thwarts were next cut and the middle thwart was forced into place, after the spall there had been removed. This required that the gunwales be spread slightly, thus increasing the amount of sheer somewhat. Much judgment was needed to do this correctly. The increase in the sheer lifted the ends slightly and put some rocker in the bottom toward the ends. The building frame was lifted out before the rest of the thwarts were placed; usually it was taken apart in the process. In forming the ends of the bark cover, the two sides were held together by a clothespin-like device made of two short, flat sticks lashed together.

Increasing the beam at the gunwales by fitting the thwarts after the bark cover had been secured to the gunwales not only increased the sheer but decreased the depth of the canoe amidships as established by the posts placed under the gunwales in setting up. In order to retain the required sheer and the desired depth of side, the gunwales had been sheered up at the ends while being shaped, and had also been treated with hot water and hogged upward amidships by being staked out to dry into shape. The spreading of the gunwales tended to lift the ends of the bottom line, a condition that was controlled in two ways: the usual one apparently was to employ, in combination

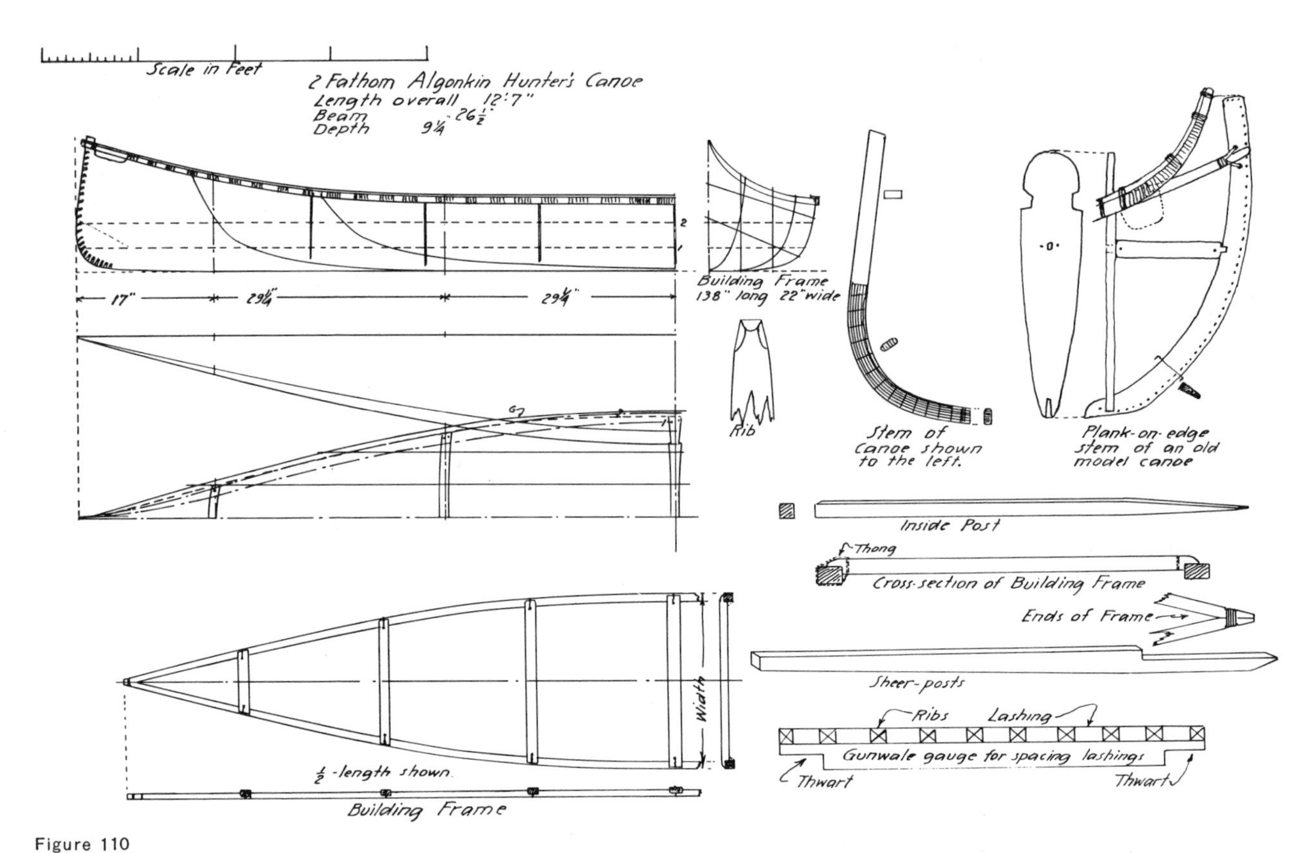


Figure 110

ALGONKIN, 2-FATHOM HUNTER'S CANOE, without headboards. Details of building frame, stakes or posts, gauge, and stem.

with a level bed, a building frame slightly wider than was desired for the finished bottom; the second way was to follow Malecite procedure and elevate slightly the middle of the building bed while employing a building frame the width of the finished bottom. The Algonkin procedure of spreading the gunwales during construction was that employed in the northwest and in the building of the fur-trade canoes, as will be seen. The amount of spread to be given the gunwales also affected the angle, or slope, at which the side stakes were driven on the building bed. Even so, some builders who spread the gunwales a good deal would set the stakes almost vertically, instead of at a slant, as this made sewing the side panels easier, particularly in large canoes and in canoes whose covers were made up of a large number of small pieces of bark.

The gunwales of the Algonkin canoes were made up of three members—main gunwales, outwales, and caps. The main gunwales, usually of cedar, were rectangular in cross section and bent on the flat. The lower outboard corner was bevelled off to take the rib ends, as in the Malecite canoes. The gunwales were rather light ranging in the examples found from about 1 inch square to 1 by 1⅝ inches, the ends being tapered to a lesser size. The outwales were light battens, rectangular in cross-section, about as deep as the main gunwales and about two-thirds their thickness or less; they tapered in depth toward the ends to ⅜ or ½ inch in order to follow the sheer, while the thickness might be constant or only slightly reduced. The caps, which were pegged to the gunwales, were also light and were about equal to the combined width of the main gunwales and outwales and had a depth of about ⅜ to ½ inch amidships. At the ends they were tapered in both width and depth, becoming ½ inch wide and ⅜ inch deep. The amount of taper in the ends of the gunwale members depended upon the form of sheer; the Algonkin practice in the old form of canoe was to sheer the outwales and caps to the top of the stem, while the gunwales sheered less and met the sides of the stem piece at a lower point,

Figure 111

ALGONKIN CANOE, OLD TYPE.

as in the drawing (p. 116). In the *wabinaki chiman*, however, the gunwales and other members, as a rule, all followed the sheer of the ends of the canoe.

The Algonkins used inside stem-pieces in both models, but the stem-piece of the old high-ended canoe was quite different from that of the *wabinaki chiman*, for it was built to give a profile in which the top of the high stem ended in a line straight across to the sheer. The piece consisted of a crooked stick, without lamination, worked out of a thin board, ⅜ to ½ inch thick. It was shaped to the desired profile inside and out, and was slightly sharpened, or sometimes rabbeted and sharpened, toward the outboard face. The headboard was mounted on this stem-piece by means of the usual notch but was not bellied; instead it stood approximately vertical and a short strut was tenoned into both the headboard and the inside face of the stem at a point about half the height of the stem. Sometimes two struts were used, side by side, with the outboard ends lashed at the sides of the stem. Thus the stem-pieces and headboards were placed as a single unit, not independently as in eastern canoes. The gunwale ends were lashed to the sides of the stem-piece, between the strut and the stem-head, at a height determined by the sheering of the main gunwale members. The outwales and caps did not touch the stem-piece, ending with a nearly vertical upward sweep, a few inches inboard. The ends of the outwales and caps were always higher than the top of the stem-piece so that, when the canoe was turned upside down, the bark cover over the stem-head was kept off the ground and thus preserved from damage. The top of the stem-piece was held rigid not only by the strut to the headboard but also by the ends of the main gunwale members lashed to it a little higher up. The headboard was in the form of a rounded V that was widest at midheight, at the gunwales, which were let into its sides.

When the stem-head was rounded in the style of the fur-trade canoe, the stem-piece except near the heel was split into very thin laminations about $\frac{1}{16}$ inch, or a little more, thick. The carefully selected cedar of which these were made was treated with boiling water, then bent to profile; the head was sharply bent over and down, inside the stem, then sharply up again so the end stood at about right angles to the face of the stem at midheight. The headboard was mounted as previously described, except that the end of the stem-piece was inserted into a hole in the headboard just above the strut. The laminations of the stem-piece were wrapped in the normal manner and the lashing was often brought around the strut as well, up against the outboard face of the headboard. The whole structure was thus made rigid and very strong. As in the other form, the main gunwale members did not follow the sheer near the ends of canoe but were secured at a point lower down on the sides of the stem-piece. In the round-head form, however, the outwale and cap ends were fastened on the after face of the stem-head where the laminations were curved downward as illustrated in the drawing (p. 116).

The headboards for both models were thicker than those in the eastern canoes; this aided in holding the stem line in form. Tension on the bark cover was obtained by making the cover V-formed toward the ends and then spreading the sides of the V with the headboard, thus bringing pressure on the strakes of the sheathing and forcing the sides outward in a slight curve.

The stem-pieces of the *wabinaki chiman* were either cut out of a thin board or laminated. In the straight-stem form, only the forefoot part was laminated, and no headboard was used. Ordinarily, however, the rigid headboard with a single strut was used. The head of the stem-piece was carried through the rail

Figure 112

ALGONKIN "WABINAKI CHIMAN."

caps and showed above them; the ends of the caps and main gunwales were notched to permit this, but neither these nor the cap extended outboard of the face of the stem.

The bark cover was lashed to the gunwales with group lashings in which the thong was carried from group to group by a long stitch outside the cover, under the outwale. The turns in each group were passed through five or six holes in the cover and reinforcing piece, two turns of the thong going through each hole. The connecting stitch between groups, which were usually about 1½ inches apart, usually passed from the last hole in a group to the second hole in the next. Some builders laid a wooden measuring stick along the gunwales to space the lashings; this was perhaps the practice of many tribal groups.

The lashing of the ends of the cover was passed through the stem pieces; when the latter were not laminated, holes through the soft, thin cedar were made by a sharp awl and an in-and-out or harness stitch was quite commonly used. On laminated stem pieces the form of lashing varied; in the *wabinaki chiman* it was commonly some combination of spiral and crossed turns; in the old form of high-ended canoe multiple turns through a single hole (usually at the top of the stemhead) were also used in combination with closely spaced long-and-short turns in triangular groups near the top of the stem profile. Below, in the forefoot, spiral or crossed stitches were used. The ends of the outwales were lashed together with a close wrapping of turns in contact where they turned upward sharply, and the caps were secured there by two or more group lashings. The head of the headboard was lashed to each gunwale by passing the thong through holes each side of the headboard; these lashings were in a long group and were passed around gunwale and outwale before the caps were in place. With plank stem-pieces the ends of the bark cover were slightly inboard of the cutwater line, sometimes protected by a rabbet.

The side panels were sewn on with in-and-out stitches, back stitches, or a double line of either. The gores were sewn spirally in the usual manner or were stitched with a closely spaced lacing.

Some of the old Algonkin canoes examined had what appeared to be a *wulegessis* just outboard of the headboards. No marking was found on these and they were too far aft to protect the ends of the gunwales. The bark was carried across the gunwales, under the caps, and hung down a little below the outwales. On top, it reached from the headboard out to the lashings of the outwales, forming between the headboards and the lashings a short deck that may have been intended to keep dirt and water out of the ends of the

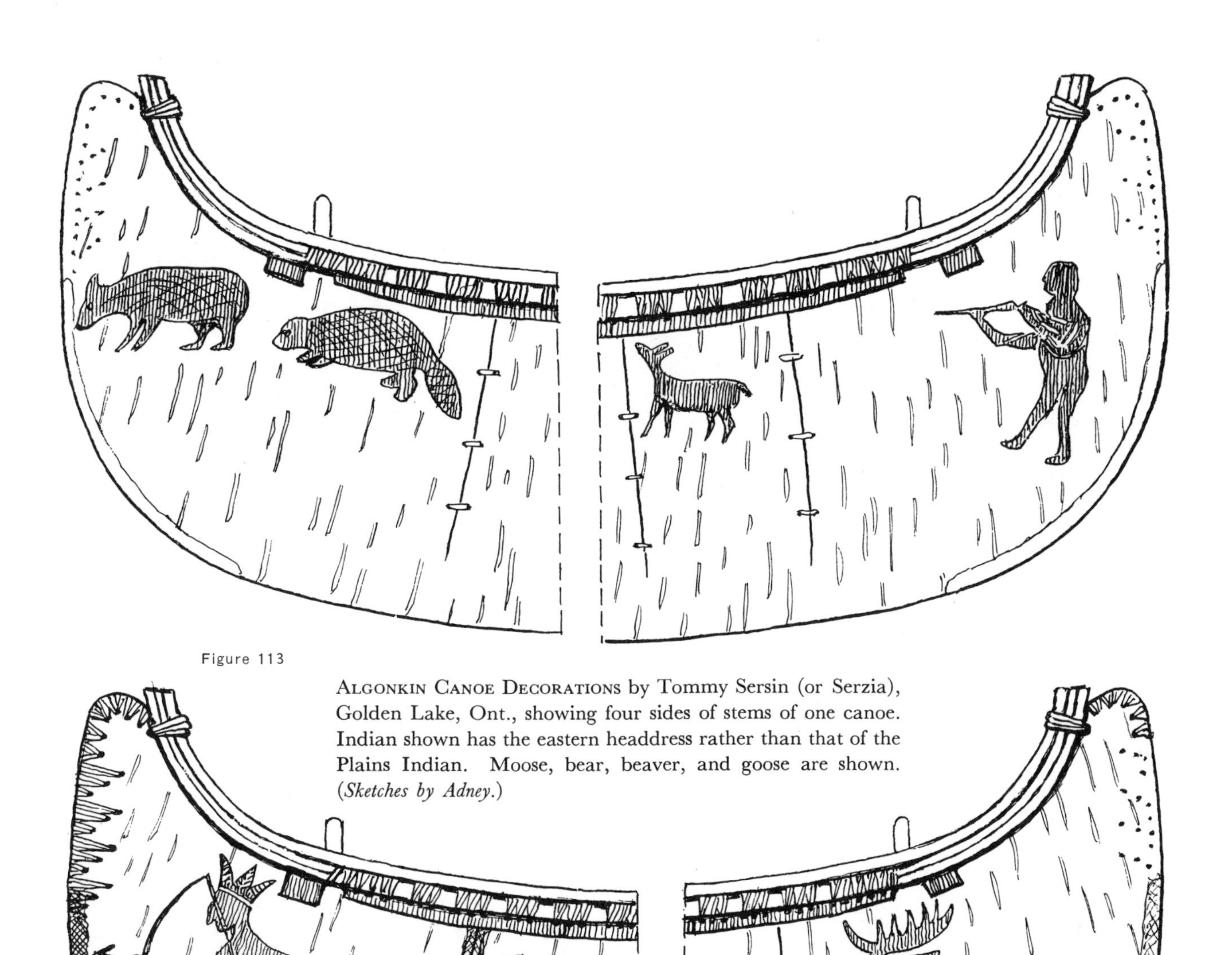

Figure 113

ALGONKIN CANOE DECORATIONS by Tommy Sersin (or Serzia), Golden Lake, Ont., showing four sides of stems of one canoe. Indian shown has the eastern headdress rather than that of the Plains Indian. Moose, bear, beaver, and goose are shown. (*Sketches by Adney.*)

canoe. Sometimes a modern *wabinaki chiman* has a *wulegessis*, copying the Eastern practice but without markings.

The thwarts were of various designs; a common one had parallel sides in plan. The old canoes had thwarts much like those of the Têtes de Boule. The end lashings of these were usually passed through three holes in the thwart ends, but some had only two holes.

Sheathing was laid somewhat as in the Têtes-de-Boule canoe, with overlapping edges and butts. The end sheathing was short and was laid first; the centerline strake was parallel-sided to a point near the sharp end of the canoe. The strakes on each side of it were tapered and were laid with their wide ends toward the middle of the canoe and with the sides and narrow end lapped. In the middle of the canoe the strakes were parallel-sided and their butts were on top of those of the strakes in the end of the canoe. The sheathing was carried up to within about three inches of the gunwales. The edges were not thinned or feathered as much as were those in the Têtes de Boule canoe.

Ribs were of cedar from 2 to 3 inches wide, closely spaced and, as usual, without taper until near the ends, which were formed with a narrow chisel edge as in the Têtes-de-Boule canoe. The ribs were first roughly bent, using the building frame as a general guide for length, in order to obtain a somewhat dish-shaped cross section; by this means the width of the bottom could be established to the builder's satsifaction.

The foregoing description of building methods and construction is based largely upon what is known of the old canoes. In later times the Algonkin copied the eastern canoes and their procedure altered. Not only did they copy extensively the appearance of the St. Francis and Malecite canoes, but they built some canoes much like those of the Têtes de Boule and Ojibway. As a result, it has become difficult to determine what their tribal practices were.

Their paddles were of the same design as those of the Têtes de Boule, round-pointed and with the blade parallel-sided for most of its length. In portaging, the Algonkin, like many forest Indians, placed a pair of paddles a foot or so apart fore-and-aft over the middle thwart and those on each side of it. These were lashed in place with the ends of a band of hide or the inner bark of a tree like the basswood or elm. This band had been first passed around the ends of the middle thwart, outside the shoulders, and hitched with ends long enough to secure the paddles in place. The shoulder on the middle thwart, a few inches inside the gunwales, was placed there for just this purpose, not as a mere decoration, so that the line could not slide in along the thwart. The canoe was then lifted and turned over by raising one end, or by lifting the whole canoe, and was placed on the carrier's shoulders, so that the paddle handles were on his shoulders. This brought the middle thwart to just behind the carrier's head. The loop of the bark or hide cord was then placed around the forehead of the carrier in order to keep the canoe from slipping backward. In this fashion one man could carry a canoe for miles if the canoe were small—and all woods, or portage, canoes were small and light. The headband was known to white men as a "tump line." The Indians used it to carry not only canoes but other heavy or awkward loads (see p. 25).

There is no certainty about the decorations of Algonkin canoes. Some of the older Indians claimed that the old form of canoe was often decorated with figures formed by scraping the winter bark; usually these depicted the game the owner hunted. Five-pointed stars, fish, and circular forms are known to have been used on the *wabinaki chiman*, but it is not known whether these were really Algonkin decorations or merely something that had been copied "because it looked good."

The Algonkin called the large fur canoes *nabiska*, a name which the Têtes de Boule rendered as *rabeska*. The word may be a corruption of the Cree word for "strong." At any rate, the name *rabeska* (sometimes pronounced ra-bas-ha), rather than the French *maître canot*, was long applied by white men in the fur trade to the large canoes built in the Ottawa River Valley for their business. In late years the *rabeska* was a "large" 2½-fathom high-ended birchbark canoe, but originally it meant a fur-trade canoe, with the characteristic ends, of from 3 fathoms upward in length.

## *Ojibway*

The Indian bands that were called "*Outaouais*" by the early French do not appear to have been an independent tribe, as has been mentioned, but were largely made up of Ojibway from the Great Lakes region. Perhaps some Têtes de Boule were among these bands before these people were given their nickname. The Ojibway were a powerful tribal group, made up of far-ranging bands, located all around Lake Superior and to the northwest as far as Lake Winnipeg. They had been in the process of taking over the western end of Lake Superior when the earliest French explorers reached that area; they pushed the Sioux from these forest lands into the plains area, joining with the western Cree in this movement. In the process they seem to have absorbed both some Sioux and some Cree bands. Within the Ojibway tribal group, later called Chippewa or Chippeway by the English and Americans, the bands had local names, or were given nicknames, such as the Menominee, Saltreaux, Pillagers, etc. All the important bands within the tribal group were expert canoemen and builders. As far as can be discovered now, the Ojibway added to their own tribal types the models of canoes they encountered in their expansion westward. It has long been true that the Ojibway canoe can be one of at least three forms, depending upon which area of their territory is being discussed.

What is believed to be their old tribal form was a high-ended canoe in all respects very much like the high-ended Algonkin type. This was the model used by the Lake Nipigon Ojibway, north of Lake Superior in Ontario, and by those of the same tribe that once lived near Saginaw, Michigan, as well as by the Menominee of Wisconsin. At the late period, from the middle of the 19th century onward, for which information was available or in which investigation was possible, it appears that the Ojibway canoes of this high-ended model were built in larger sizes than contemporary Algonkin canoes of like design. The Ojibway canoes had the same end structure as these; the early examples found had "chin" in the end profiles and the tumble-home of the stem was straight, or nearly so, between the large curve of the forefoot and the very short hard curve at the stem head. The Ojibway used the same inner stem-piece, laminated and brought downward abaft the stem-head and then inboard so that the end fitted into a slot in the headboard a little above its midheight, at which point was fitted a strut from the headboard to the back of the stem-piece. The midsection of the Ojibway canoe was very much like that of the Algonkin; it had a narrow bottom somewhat rounded athwartships, a well-rounded bilge, and flaring topsides.

A small Ojibway portage canoe built in the middle of the 19th century had an end profile somewhat different from that described above; the ends were well rounded and had a heavy chin, the stem was carried into the tumble-home with a full rounded curve all the way to the stem-head, where the stem piece was bent in and downward very sharply and then inboard sharply again, so that the end pierced the vertical headboard at sheer height. The S-curve was so located that the main gunwales could be lashed to the stem piece at the point where they paralleled it well below the stem head. In these canoes the Ojibway followed Algonkin practice in ending the gunwales; there was, therefore, no strut. Where this canoe was built is uncertain.

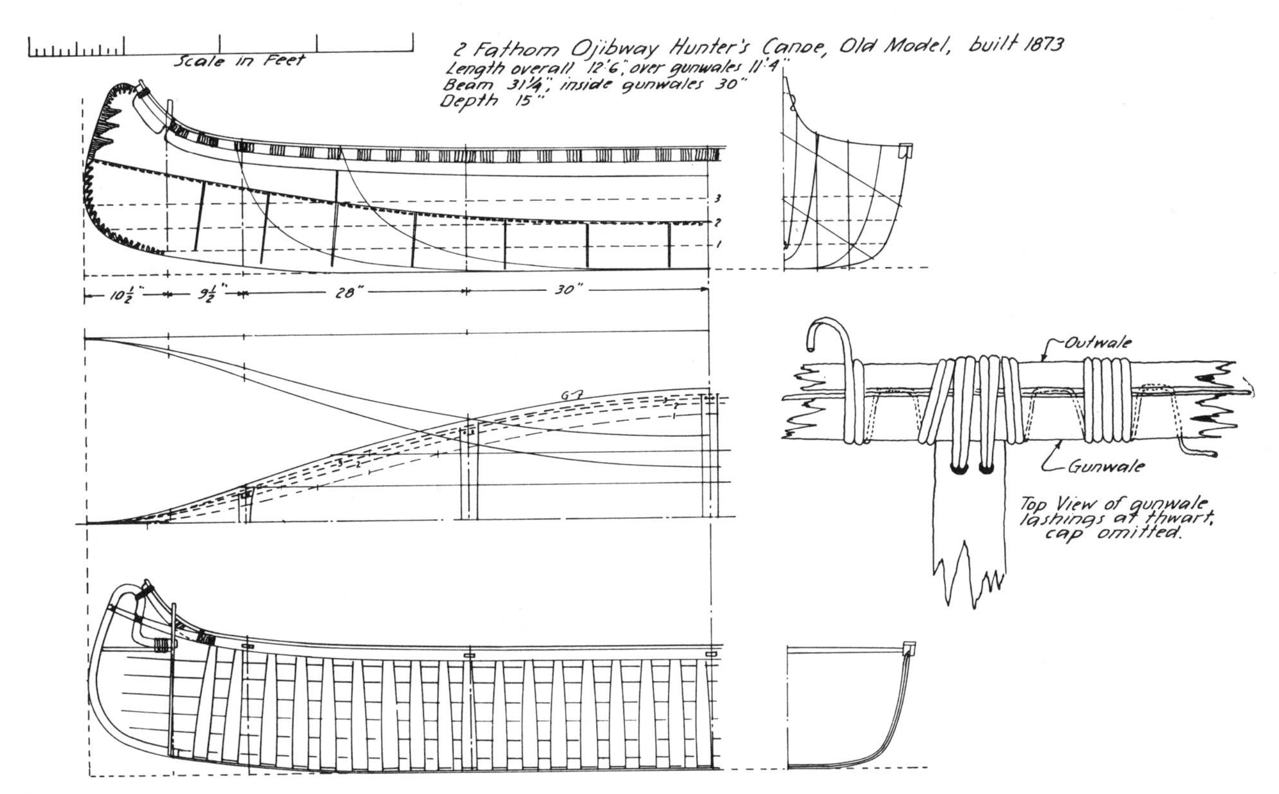


Figure 114

Ojibway 2-Fathom Hunter's Canoe, used by the eastern tribal groups. Probably the ancient model.

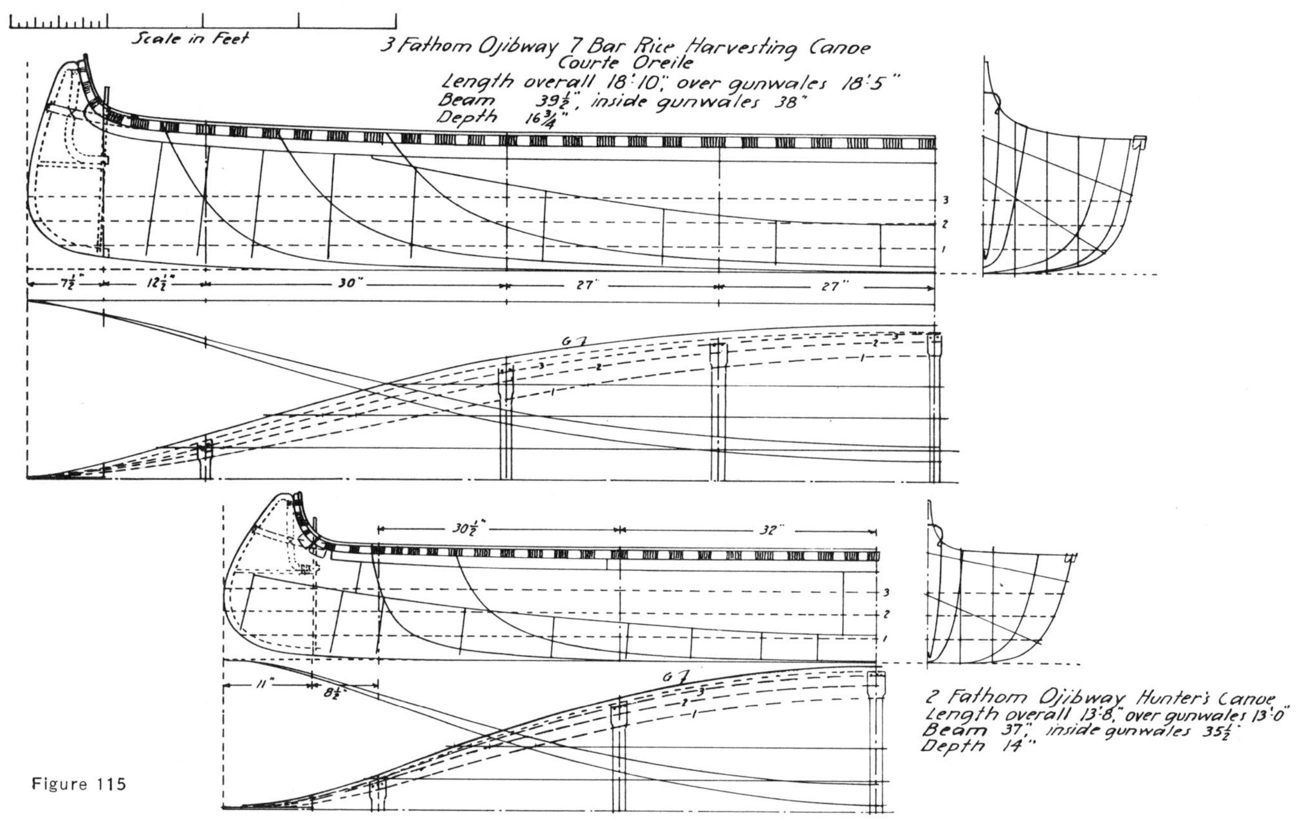


Figure 115

Examples of the Old Model Ojibway 3-Fathom rice-harvesting canoe (above), and 2-fathom hunter's canoe, showing the easy paddling form used.

Ojibway 3-Fathom Freight Canoe From Lake Timagami, apparently a hybrid based on canvas canoes.

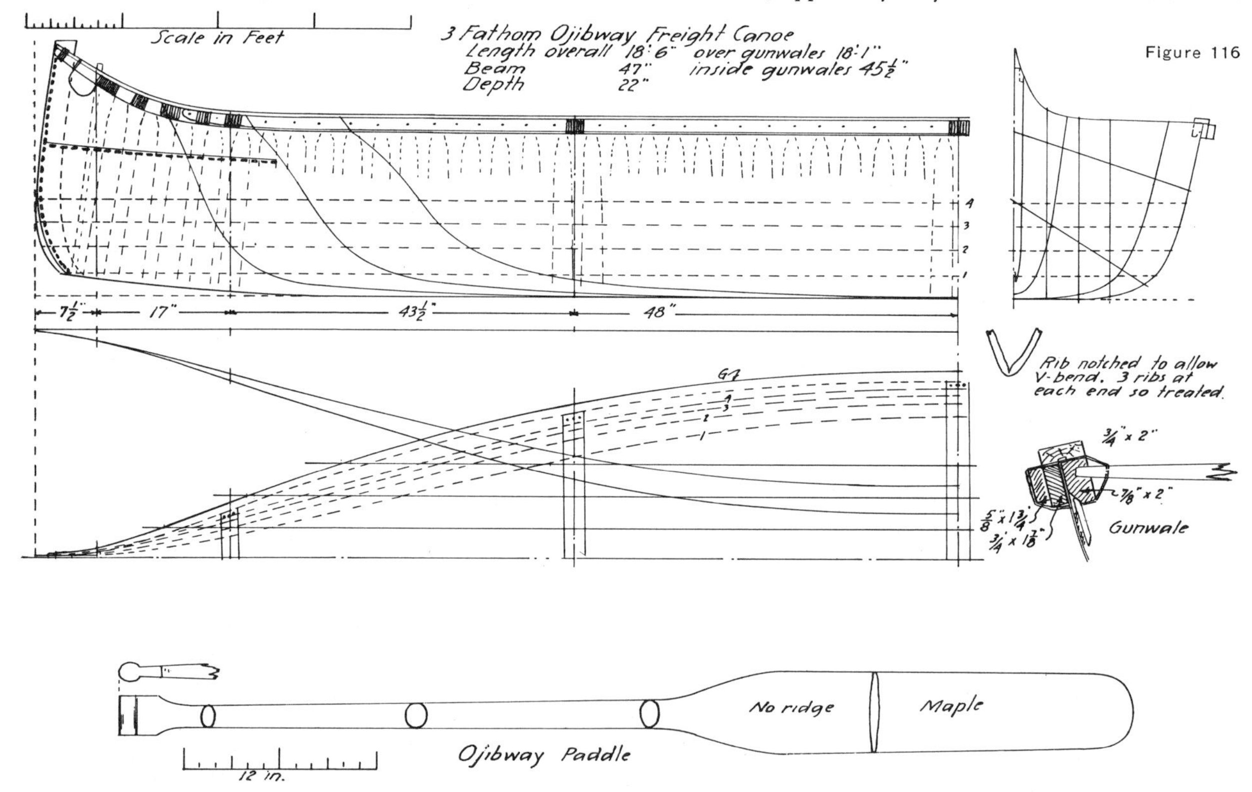


Figure 116

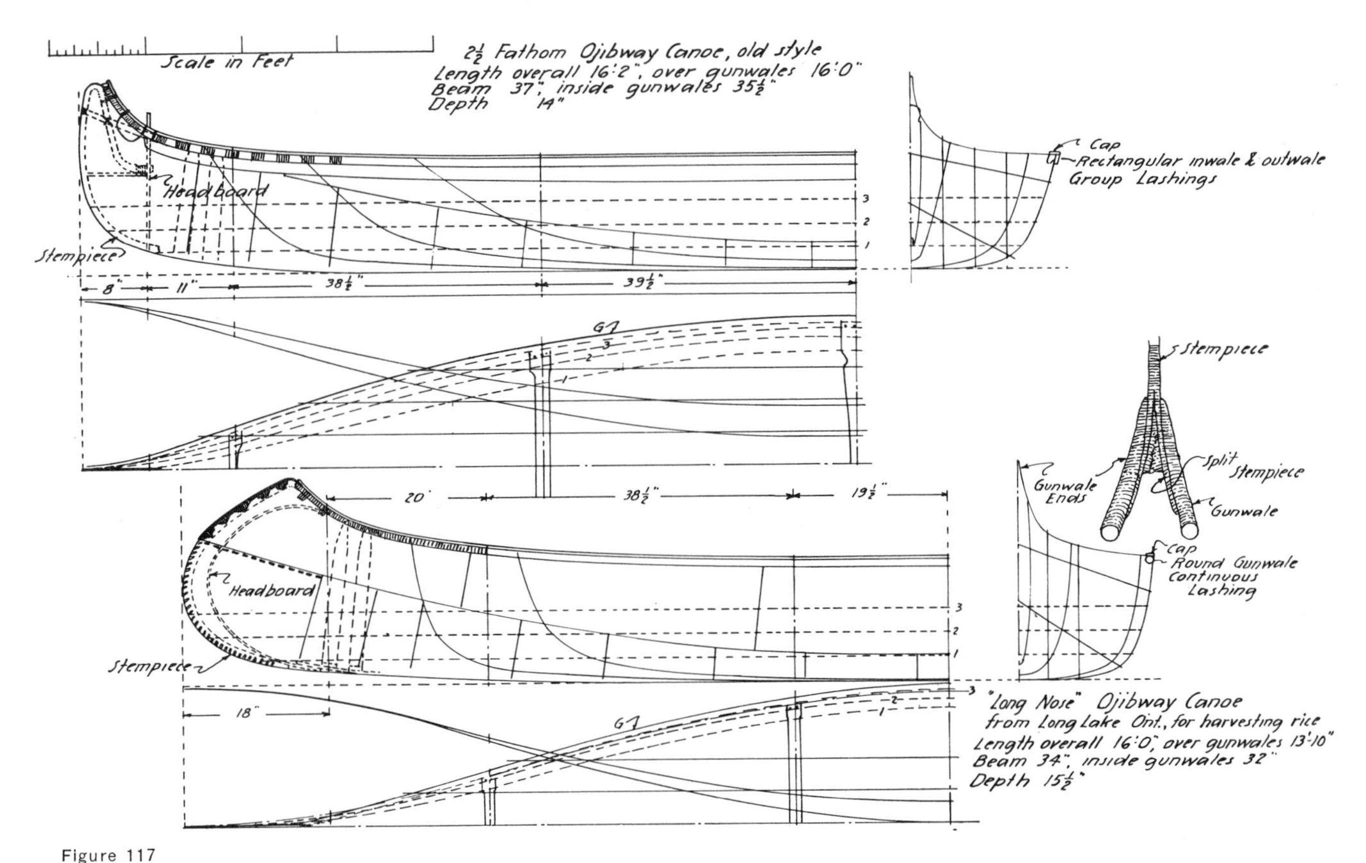


Figure 117

THE OLD FORM OF OJIBWAY 2½-FATHOM CANOE of the eastern groups (above), and the long-nose Cree-Ojibway canoe of the western groups.

At Lake Timagami, north of Georgian Bay in Ontario, the Ojibway used a low-ended canoe with a remarkably straight tumble-home stem profile; the forefoot had a very short radius ending at the bottom line with a knuckle, and the stem-head stood slightly above the gunwale caps. The stem-piece was made from a thin plank cut to profile; thus no lamination was necessary. The headboard stood straight, falling inboard slightly at the head. The midsection was dish-shaped, with a flat bottom athwartships and strongly flaring sides, the turn of the bilge being rather abrupt. The ends were strongly V-shaped in cross-section; a number of the frames there being "broken" at the centerline of the bottom. A canoe of this design was seen by Adney at North Bay, Ontario, in 1925, indicating that the design may have been used in some degree outside the Lake area in later years.

The most common Ojibway model used to the northwest and west of Lake Superior was the so-called "long-nose" form, a rather straight-sheered canoe. The bottom, near the ends, had a slight rocker, and the sheer turned up very sharply there, becoming almost perpendicular at the extremities, yet the ends were not proportionally very high. The end-profile came up from the bottom very full and round, then fell sharply inboard in a slightly rounded sweep to join the upturned sheer well inboard. The midsection was somewhat dish-shaped, but with well-rounded bilges, so that the flare of the topsides was rounded and not very apparent to the casual observer. The end section developed into a tumble-home form, so that a section through the top of the headboard was rather oval. As a result, these canoes appeared rather clumsy and unfair in their lines, but this apparently did not harm their paddling qualities or seaworthiness.

These canoes had narrow headboards that were sharply bellied, somewhat like those in the crooked canoes, and the belly was sufficient to allow the heel of the end-board to pass under the bottom sheathing and inside the bark cover so that two end ribs served to hold the heel in place. The inside stem-piece was

Figure 118

Eastern Ojibway Canoe, Old Form. (*Canadian Pacific Railway photo.*

Figure 119

Ojibway Long-Nose Canoe, Rainy Lake District.

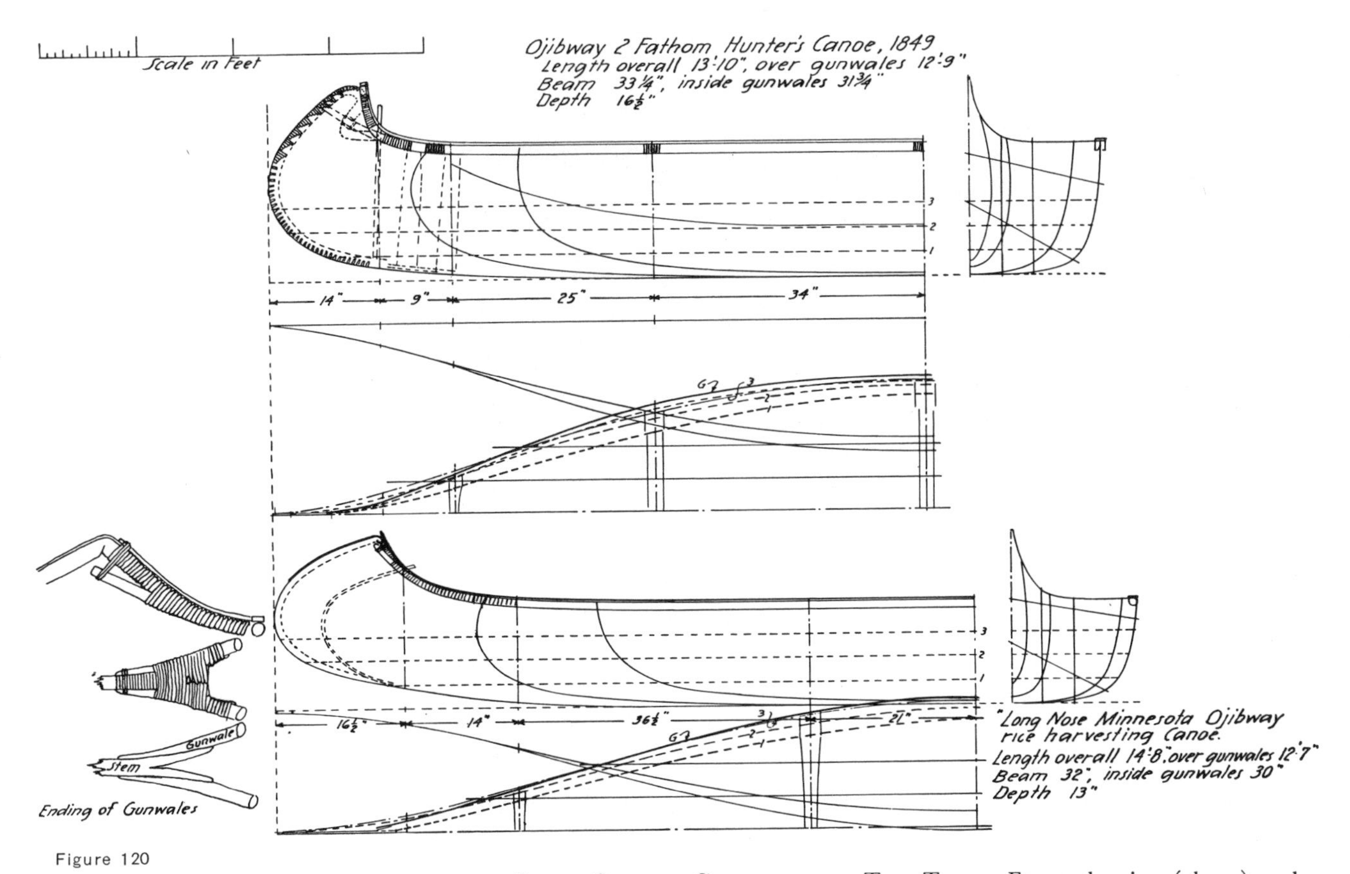


Figure 120

SMALL OJIBWAY CANOES OF THE TWO TRIBAL FORMS showing (above) early trend toward the long nose form, and the final Ojibway-Cree hybrid form combining flaring sides amidships with tumble-home sections at ends.

often no more than a light stick or rod bent to profile, with the head split and brought over the gunwale ends and down inside, between them. Each half of the split was then lashed to its neighboring gunwale member. A strip of bark was often placed over the end of the bark cover and carried down the face of the stem, under the sewing. The rail caps were then brought up over the tops of the gunwales and overlapped the top portion of the stem piece. The heel of the stem-piece was bevelled off on the inboard side so that it could be wedged under the headboard, inside the bark cover. These headboards, it should be noted, were no more than a thin, narrow batten, and in some canoes the head of this batten was lashed under the gunwale ends instead of coming up between them inboard, as usual. A variation in the fitting of the stem head was found in a canoe at Long Lake, Ontario; the stem head, instead of being split, was lashed between the gunwale ends and thus was brought inboard level with the top of the gunwales.

The cross section of the main gunwales was round or nearly so in nearly all long-nose canoes, and often a gunwale cap was fitted. The bark cover was secured to the gunwales by a continuous lashing, but in at least one example, from Minnesota, the gunwale wrappings were in groups over an outwale after the regular fashion to the eastward. The ends of the thwarts were wedge- or chisel-shaped and instead of being tenoned were forced into splits in the round gunwales. Many canoes had bark covers at the gunwale ends and vestiges of the *wulegessis* were to be seen.

All Ojibway canoes were built with a building frame, the bed being slightly higher at midlength than at the ends. The stakes were driven nearly perpendicular, instead of with heads slanted outward. It is apparent from observed examples that some canoes were built by the same procedure as the Algonkin, but that not all the long-nose canoes were built by spreading the gunwales; some were built using the methods of the St. Francis.

Figure 121

OJIBWAY CANOE BUILDING, LAC SEUL, 1918.

Preparing a building site or bed; building frame in place.

Bark set up; bark staked out on building bed.

Bark cover being sewn on building bed.

(See pp. 170–171 for more photos of Ojibway canoe building.)

Gunwales being lashed.

Securing gunwales.

Pitch being applied to seams.

Figure 122

LONG LAKE OJIBWAY LONG-NOSE CANOE. (*Canadian Geological Survey photo.*)

The lashing in the high-ended Ojibway canoes was about the same as that in the Algonkin canoes, but in the long-nose type the workmanship was often coarse. On many of the latter the stems were lashed by use of small groups in which two turns were taken through each of two closely spaced holes in the bark and the connection between the groups was made by a long spiral around the outside of the stem. This pattern was carried down from the stem-head to about the level of the midship sheer height; from there down around the forefoot the lashing consisted of a simple spiral. Another style was to use widely spaced groups made up of two or three turns through a pair of facing holes in the bark, one on each side and inboard of the stem. The turn went around the stem, and the last connected with the next pair of holes below. A few canoes of this style used closely spaced wrapping, as in the high-ended canoes.

The long-nose Ojibway canoe is surprisingly primitive by comparison with the graceful and well-finished high-ended model built after the Algonkin style. Adney believed that the long-nose type originated with the Sioux Dakotas, before the combined Ojibway and Cree movement forced them out of the forest lands to the west of Lake Superior. He considered it possible that both the Ojibway and Cree adapted the Dakota model, modifying it somewhat to their methods of construction. It is true that the western Cree built a long-nose canoe, but it had less chin than the Ojibway model. On the other hand, the Ojibway prebent ribs in pairs like the eastern Cree, and used spreaders in the end ribs while drying them, in exactly the same manner. A picture taken in 1916 shows the gunwales of a Cree long-nose canoe being set; it was laid on the ground and weighted along the midlength by stones laid on boards placed across the longitudinals. The ends had been sheered up and were supported at each end by a thong made fast to the gunwale end and then brought over a post, or strut, a few feet inboard and made fast to the middle thwart.

It is unnecessary to detail the construction of the Ojibway canoes, as they employed a building-frame, as the drawings on pages 123 to 127 show plainly enough the pertinent details of fitting and construction. It is important to observe that the wide variation in model and in construction details of the Ojibway canoes produced a variety of building procedures that in the main were like those of the Algonkin and Cree. Hence the older tribal method of construction cannot now be stated with any accuracy.

The paddle forms used by the Ojibway groups varied somewhat. Most were made with parallel-sided blades and oval tips. The hand grip at the top of the handle was rectangular and was large in comparison to the grip of the eastern Cree paddles. A few variations have been noticed; the blade of one was widest at the top, the tip was almost squared off, and the upper hand grip was much as in the factory paddle of today. This paddle, from an unknown locality, was used in 1849.

As in the case of the Algonkin, the eastern Ojibway built fur-trade canoes under supervision. Though these canoes differed somewhat from those built by the Algonkins, it is now impossible to say whether or not there was any real relationship between them and the small, high-ended "old-form" canoe. Likewise, the Ojibway built a version of the *wabinaki chiman* which seems to have influenced some types of their own, such as, for instance, the straight-stem Lake Temagami canoe.

Figure 123

NINETEEN-FOOT OJIBWAY CANOE with thirteen Indians aboard (1913).

## *Western Cree*

The western portion of the great Cree tribe appear to have occupied the western shore of James Bay and to have moved gradually northwestward in historical times. Their territory included the northern portion of Ontario and northern Manitoba north of Lake Winnipeg, and as early as 1800 they had entered northwestern Alberta. The line of division between the canoes of the eastern and western Cree cannot be strictly determined, but it is roughly the Missinaibi River, which, with the Abitibi River, empties into the head of James Bay at the old post of Moose Factory. The southern range of the Cree model was only a little way south of the head of James Bay, irregularly westward in line with Lake St. Joseph to Lake Winnipeg. To the west, the Cree type of canoe gradually spread until it met the canoe forms of the Athabascan in the Northwest Territories, in the vicinity of Lake Athabaska in north-western Saskatchewan.

The canoes of the western Cree, as has been noted, strongly resembled the long-nose Ojibway model except that they had less pronounced chin. But unlike those of the eastern Cree, their canoes employed an inside stem-piece that was sometimes a laminated piece and sometimes a piece of spruce root. The stem head was commonly bent sharply and secured between the gunwale ends at the point where the two longitudinals were fastened together, much as in some Ojibway long-nose canoes. The Cree canoe had basically the same dish-shaped midsection, but it had very full, round bilges and the flare was so curved in the topside that it was even less apparent than in the Ojibway model. The shorter chin of the Cree canoe also made tumble-home in the end sections unnecessary, and cross section near the headboards was given the form of a slightly rounded U.

The bottom had very little rocker at the ends, being straight for practically the whole length. The stem-piece if laminated (often in only two or three laminations) came up from the bottom in a fair round forefoot and then tumbled in by a gentle curve to the stem-head, where it was bent sharply to pass down between the gunwale ends as previously noted. But if the stem-piece was of spruce root, the profile was often somewhat irregular and the chin was more pronounced. In a common style the stem came fair out of the bottom in a quick hard curve, then curved outward slightly until the height of the least freeboard amidships was reached, at which height another hard turn began the tumble-home in a gentle sweep to the stem-head, where there was a very hard turn downward. The stem-head was often split, as in some Ojibway canoes, so that it came over the joined ends of the main gunwales and the two halves were then lashed to the inside faces of the gunwales.

Birch bark was often poor or scarce in the territory of the western Cree, as in that of their eastern brothers. As a substitute, they employed spruce bark and in general seem to have achieved better results, for their spruce-bark canoes had a neater appearance. If the canoe was built when or where root material was difficult to obtain, the western Cree used rawhide for sewing the bark cover. When the stems were lashed with rawhide, a stem-band of bark under the lashing was common.

The gunwales were round in cross section and were often spliced amidships. The bark cover was lashed to these with a continuous lashing, no caps or outwales being employed. As in the Ojibway long-nose canoe, the headboards were very narrow and much bellied. These canoes were built with four or five thwarts; the 4-thwart type was used for gathering wild rice, as was the Ojibway type, while the 5-thwart canoe was the portage model. The thwarts were sometimes mortised into the gunwales, but some builders made the thwart ends chisel-pointed and drove them into short splits in the gunwales before lashing them, one or two holes being drilled in the thwart ends to take the lashing thongs. When the thwarts were tenoned into the gunwales, the builders of course made the inside of the gunwales flat.

When spruce bark was employed, its greater stiffness made it possible to space the ribs as much as 10 inches on centers, but with birch the spacing was about 1 inch, edge to edge. The sheathing was in short splints and the inside of the canoe was "shingled" or covered irregularly without regard to lining off the strakes, a practice sometimes observed in Ojibway long-nose canoes. The much-bellied and narrow headboards were fitted as in the long-nose canoe, and the heel was secured under a piece of sheathing and held by it and the first two ribs.

Western Cree canoes were built with a building frame, and the bed was raised in the middle. The sewing varied. The ends were lashed with combinations of close-wrapped turns, crossed turns, grouped, and spiral turns; the lashing commonly went around the inside stem piece rather than through it. Side

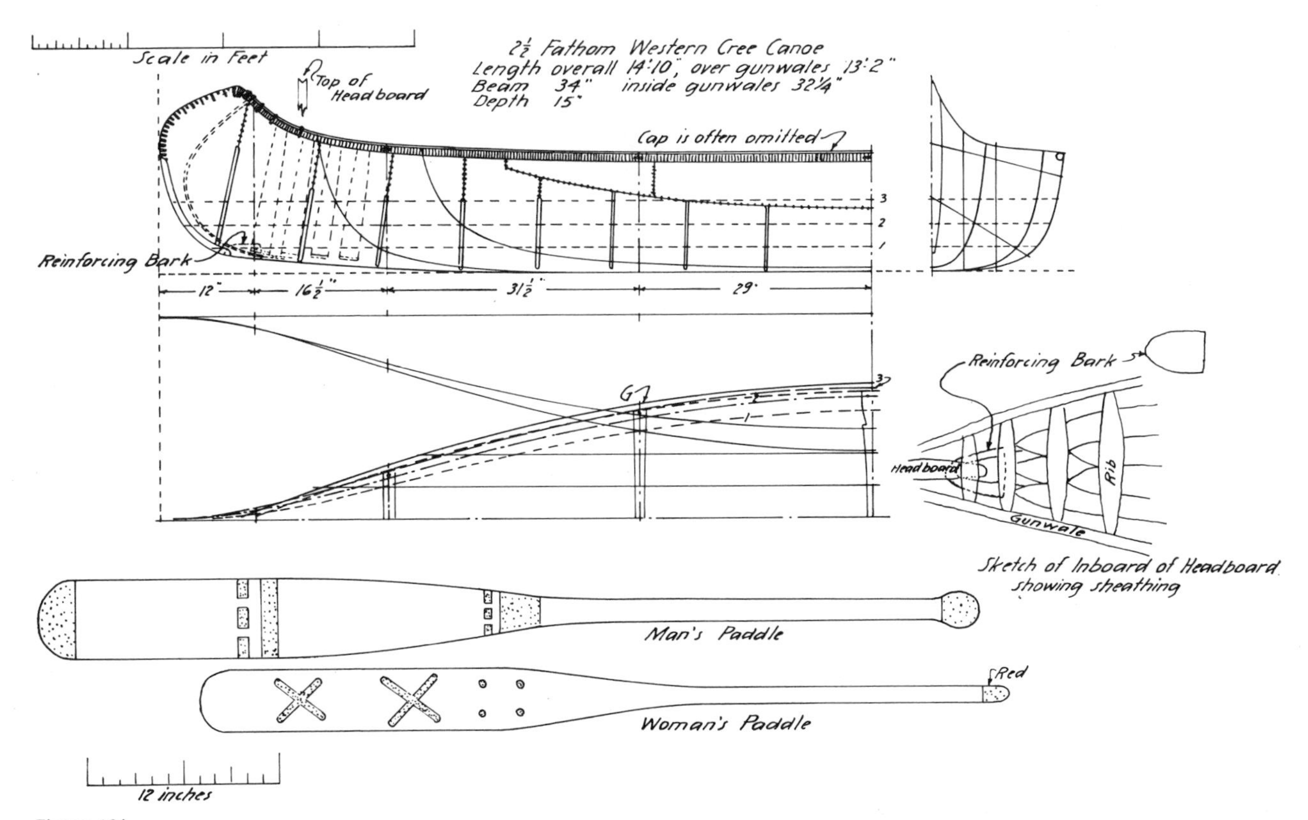


Figure 124

Western Cree 2½-Fathom Canoe, Winisk River District, northwest of James Bay. Built of either birch or spruce bark. Inside root stem piece, round gunwales, and much-bellied headboard are typical.

panels were sewn with in-and-out stitches or back stitches, and the gores with the usual spiral. Gumming as a rule was done with clear spruce gum tempered by repeated meltings.

The woodwork varied with the building site; some builders could use much cedar, but spruce was most common and the thwarts were usually of birch. When spruce bark was used it was never employed in a single large sheet, since it would have been impossible to mold it to the required shape. Hence the bark cover was pieced up, whether birch or spruce, as an aid in molding the form. Before the spruce bark was sewed and gummed, the edges of the pieces had to be thinned to make a neat joint. Furthermore, in the continuous lashing it was desirable to take two or three turns through one hole in the bark cover to avoid weakening the material with closely spaced holes.

The western Cree paddles had parallel-sided blades with rounded tips; the handle sometimes had a ball-shaped top grip and sometimes it was pole-ended. The blade did not have a ridge on its face near the handle. Old Cree paddles were often decorated with red pigment bands, markings in the shape of crosses, squares in series, and dots on the blades; the top grip might also be painted.

Many tribal groups in the western portion of the area have been mentioned—Teton, Sioux, Assiniboine, Illinois, Huron, and many others—but no record of their canoe forms has survived and the assigning of any model to them is pure speculation. The fur trade alone brought about a period of tribal movement among the Indians long enough to erase many tribal distinctions in canoes and to cause types to move great distances.

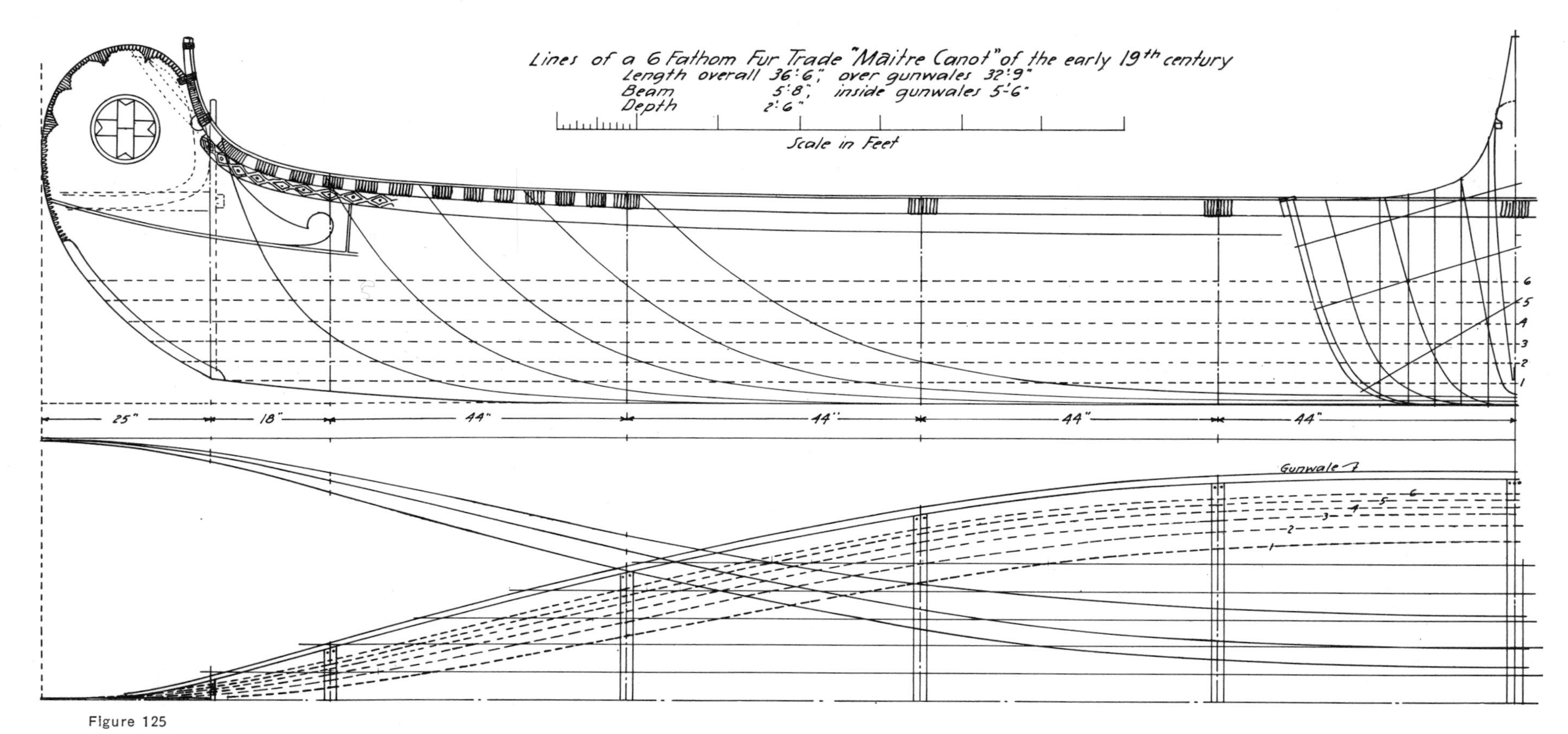


Figure 125

An Old 6-Fathom Fur-Trade Canoe, or "rabeska," used on the Montreal-Great Lakes run. Also called the Iroquois canoe, it approximates the canoes built for the French, at the Trois Rivières, Que., factory and is of the style used by the North West and Hudson's Bay Companies.

# *Fur-Trade Canoes*

Of all birch-bark canoe forms, the most famous were the *canots du maître*, or *maître canots* (also called north canoes, great canoes, or *rabeskas*), of the great fur companies of Canada. These large canoes were developed early, as we have seen in the French colonial records, and remained a vital part of the fur trade until well toward the very end of the 19th century—two hundred years of use and development at the very least. A comprehensive history of the Canadian and American fur trade is yet to be written; when one appears it will show that the fur trade could not have existed on a large scale without the great *maître canot* of birch bark. It will also have to show that the early exploration of the north country was largely made possible by this carrier. In fact, the great canoes of the Canadian fur trade must be looked upon as the national watercraft type, historically, of Canada and far more representative of the great years of national expansion than the wagon, truck, locomotive, or steamship.

Little has survived concerning the form and construction of the early French-colonial fur-trade canoes. Circumstantial evidence leads to the conclusion that the model was a development, an enlargement perhaps, of the Algonkin form of high-ended canoe as described on pages 113 to 116. The early French came into contact with these tribesmen before they met the Great Lakes Ojibway, the other builders of the high-ended model. It is known that the Indians first supplied large canoes to the French governmental and church authorities and that when this source of canoes proved insufficient, the canoe factory at Trois Rivières was set up and a standard size (probably a standard model as well) came into existence. As the fur trade expanded, large canoes may well have been built elsewhere by the early French; we know at least that building spread westward and northward after Canada became a British possession.

In the rise of the great canoe of the fur trade, the basic model was no doubt maintained through the method of training its builders. The first French engaged in bark-canoe building learned the techniques, let us say, from the original Indian builders, the Algonkin. As building moved westward, the first men sent to the new posts to build canoes apparently came from the French-operated canoe factory. It would be reasonable to expect that as building increased in the west, local modifications would be patterned on canoes from around the building post, but that the basic model would remain. This may account for the departures from the true Ojibway-Algonkin canoes seen in the *maître canots*.

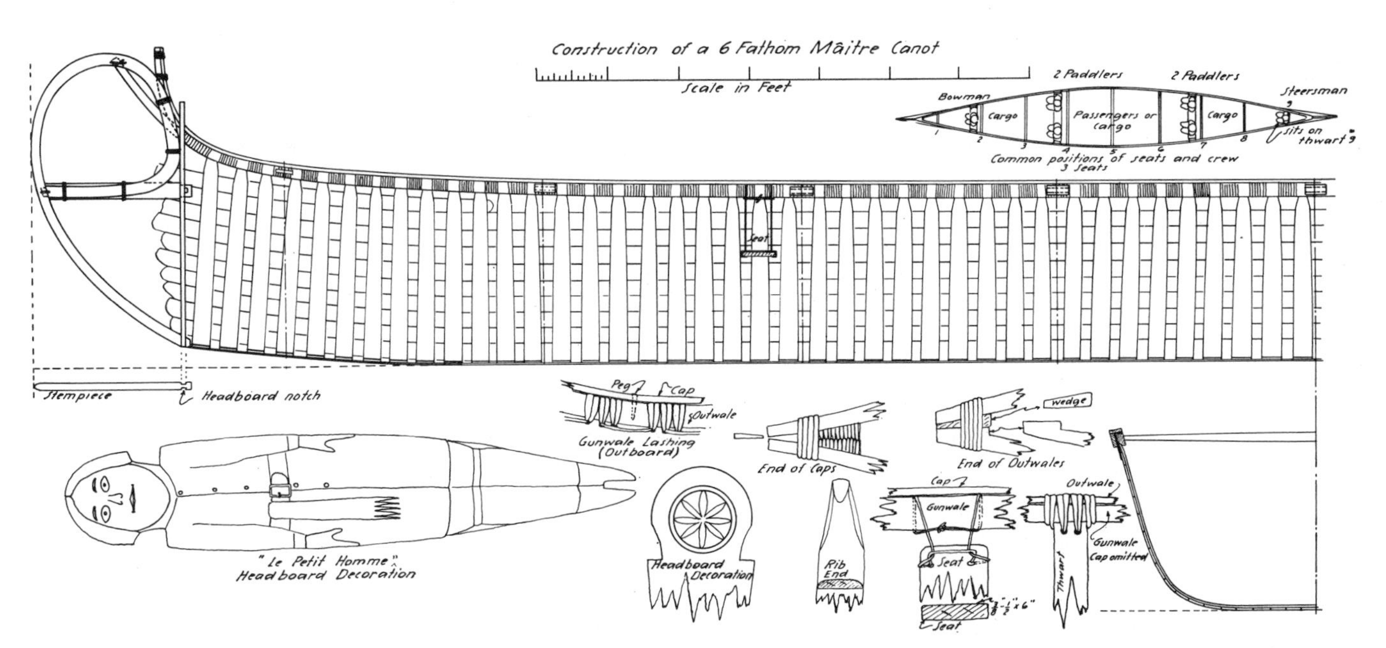


Figure 126

Inboard Profile of a 6-Fathom Fur-Trade Canoe, and details of construction, fitting, and decoration.

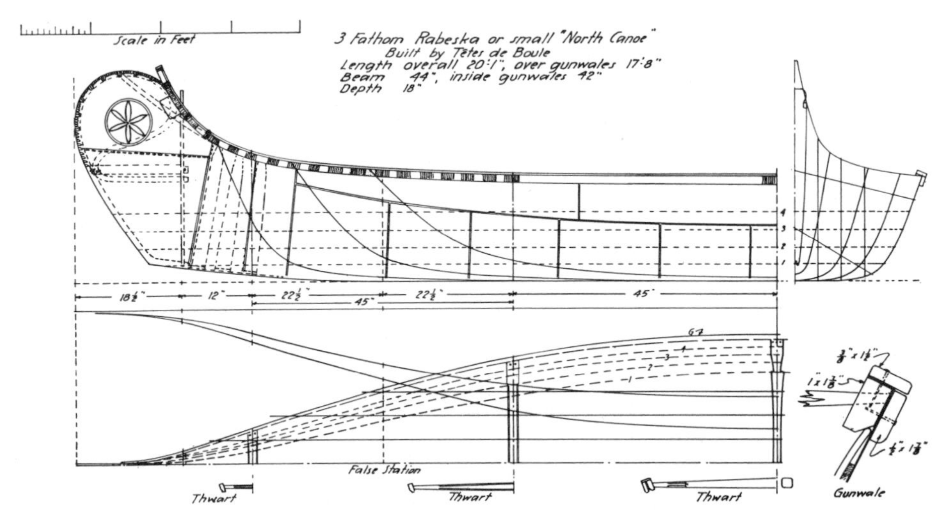


Figure 127

SMALL 3-FATHOM NORTH CANOE of the Têtes de Boule model. Built in the 19th century for fast travel, this Hudson's Bay Company canoe was also called nadowé chiman, or Iroquois canoe.

In model, all the fur-trade canoes had narrow bottoms, flaring topsides, and sharp ends. The flaring sides were rather straight in section and the bottom nearly flat athwartships. The bottom had a moderate rocker very close to the ends. In nearly all of these canoes, the main gunwales were sheered up only slightly at the ends and were secured to the sides of the inner stem-piece; the outwales and caps, however, were strongly sheered up to the top of the stem. The curvature and form of the ends, in later years at least, varied with the place of building.

After the English took control of Canada and the fur trade, a large number of Iroquois removed into Quebec and were employed by the English fur traders as canoemen and as canoe builders. Though the aboriginal Iroquois were not birch-bark canoe builders, they apparently became so after they reached Canada, for the fur-trade canoes built on the Ottawa River and tributaries by the Algonkins and their neighbors became known after 1820 as *nadowe chiman* or *adowe chiman*, names which mean Iroquois canoe. These "Iroquois canoes," however, were not a standard form. Those built by the Algonkin had relatively upright stem profiles, giving them a rather long bottom, and the outwales and caps stood almost vertical at the stem-heads; in contrast, the "Iroquois canoes" built by the Têtes de Boule had a proportionally shorter bottom than those of the Algonkin, because the end profiles were cut under more at the forefoot. Also, the outwales and caps of the Têtes de Boule canoes were not sheered quite as much as were those of the Algonkin.

It is supposed that the Têtes de Boule were taught to build this model by Iroquois, who had replaced the French builders subsequent to the closing of the canoe factory at Trois Rivières, sometime about 1820. After the English took possession of Canada in 1763, the old canoe factory had been maintained by the Montreal traders (the "North West Company"), and it was not until these traders were absorbed by the Hudson's Bay Company that canoe manufacture at Trois Rivières finally came to a halt, although it is probable that the production of canoes there had become limited by shortages of bark and other suitable materials. However, the North West Company had built the large trading canoes elsewhere, for many of its posts had found it necessary to construct canoes locally, and when the Hudson's Bay Company finally took over the fur trade it continued the policy of building the canoes at various posts where material and builders could be found. This policy appears to have produced in the fur-trade canoe model a third variant in which the high ends were much rounded at the stem head; this was the form built by the Ojibway and Cree (see p. 139). It must be noted, however, that the variation in the three forms of fur-trade canoe was expressed almost entirely in the form and framing of the ends; the lines were all about the same, though

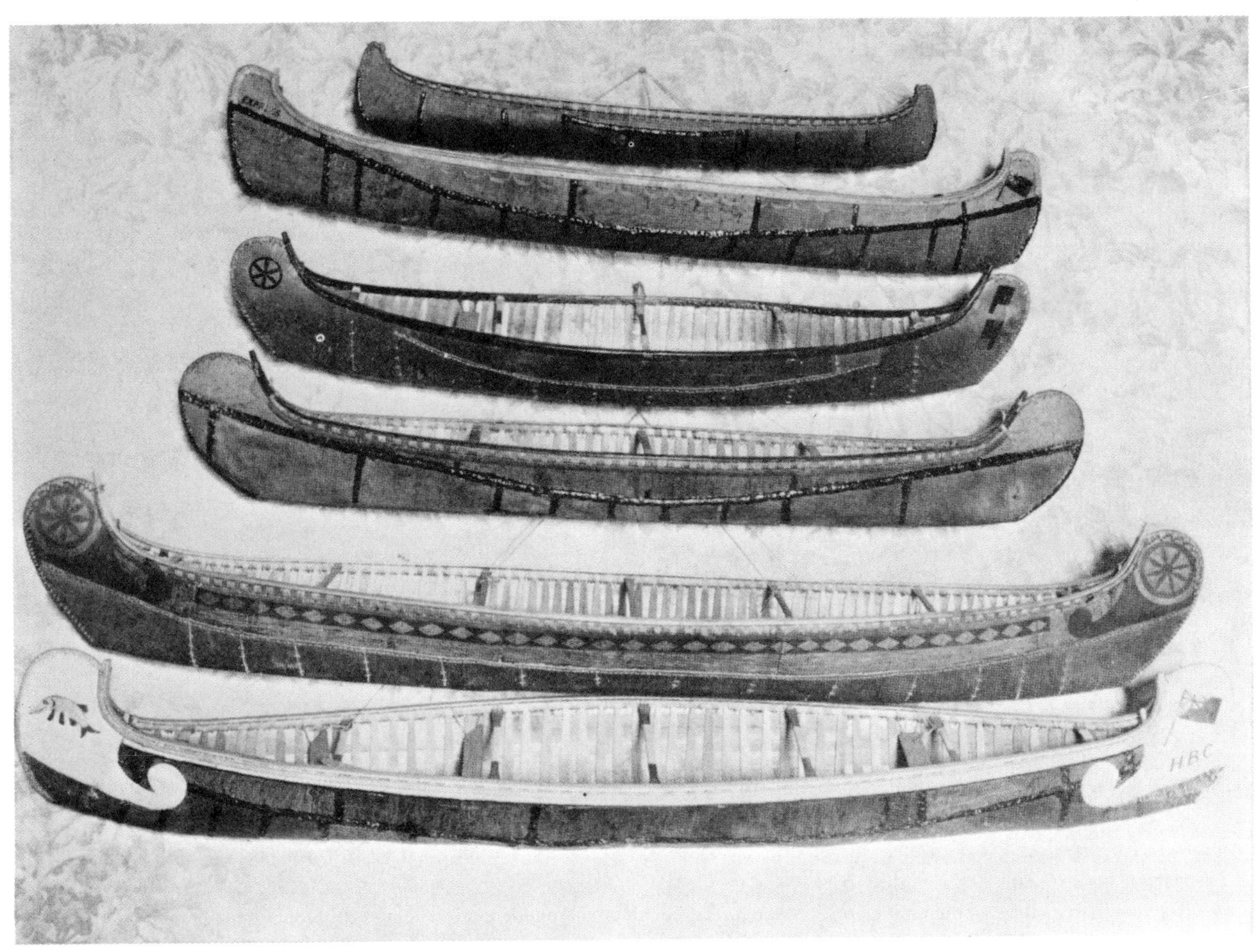

Figure 128

Models of Fur-Trade Canoes, top to bottom: 2½-fathom Ottawa River Algonkin canoe, Hudson's Bay Company express canoe, 3½-fathom Têtes de Boule "Iroquois" canoe, 3¾-fathom Lake Timagami canoe, 5-fathom fur-trade canoe of early type, and 5-fathom Hudson's Bay Company canoe built in northwestern Quebec Province.

small variations in sheer, rocker, and midsection must have existed.

Although no regulations appear to have been set up by the fur companies to govern the size, model, construction or finish of these canoes, custom and the requirements of usage appear to have been satisfactory guides, having been established by practical experience. As a result, the length of canoes varied and the classification by "fathoms" or feet must be accepted as no more than approximate.

The form of the canoe was determined by the use to which it was to be put, in trade or in travel. Fur-trade accounts often mention the "light canoe," or *canot léger*, often misspelled in various ways in early English accounts, and this class of canoe was always mentioned where speed was necessary. Commonly, the light canoe was merely a trade canoe lightly burdened. Due to the narrow bottom of these canoes, they became long and narrow on the waterline when not heavily loaded and so could be paddled very rapidly. It is true, however, that some "express canoes" were built for fast paddling. These were merely the common trade models with less beam than usual at gunwale and across the bottom. Some posts made a specialty of building such canoes, often handsomely painted, for the use of officials of the company, or of the church or government, during "inspection" trips. Not all of the highly finished canoes were of the narrow form, however, as some were built wide for capacity rather than for high speed.

Figure 129

"Fur-Trade Maître Canot With Passengers." From an oil painting by Hopkins (*Public Archives of Canada photo*).

The fur traders used not only the so-called fur-trade canoes, of course, but they employed various Indian types when small canoes were required. And in the construction of the high-ended fur-trade models, they did not limit themselves to canoes of relatively great length. Each "canoe road" forming the main lines of travel in the old fur-trade had requirements that affected the size of the canoes employed on it. The largest size of fur-trade canoe, the standard 5½-fathom (bottom length), was employed only on the Montreal-Great Lakes route, in the days before this run was taken over by bateaux, schooners, sloops, and later, by steamers. At the western end of this route, a smaller 4- or 4½-fathom canoe came into use. The latter was used on the long run into the northwest. Even smaller canoes were often employed by the northern posts; the 3- or 3½-fathom sizes were popular where the canoe routes were very difficult to operate. For use on some of the large northern lakes, the large canoes of the Montreal-Great Lakes run were introduced. Fur coming east from the Athabasca might thus be transported in canoes of varying size along the way.

In judging the size of the canoe mentioned in a fur-trader's journal, it is often very difficult to be certain whether the measurement he is employing is bottom or gunwale length. In the largest canoes, however, the 5½-fathom bottom-length was the 6-fathom gunwale length, and the use of either usually, but not always, indicates the method of measurement. This is not the case in the small canoe however, where the matter must too often be left to guesswork. To give the reader a more precise idea of the sizes of the canoes last employed in the fur trade, the following will serve. The *maître canot* of the Montreal-Great Lakes run was commonly about 36 feet overall, or about 32 feet 9 inches over the gunwales, and a little over 32 feet on the bottom. The beam at gunwale was roughly 66 inches (inside the gunwales) or about 68–70 inches extreme beam. The width of the building frame that formed the bottom would be somewhere around 42 inches. The depth amidships, from bottom to top of gunwale might be approximately 30–32 inches and the height of the stems roughly 54 inches. These dimensions might be best described as average, since canoes with gunwale length given as 6 fathoms were built a number of inches wider or narrower, and deeper or shallower. The earlier fur-trade canoes of the French and of the North West Company, for example, were apparently narrower than the above.

The 5-fathom size that replaced the larger canoe at the close of the bark-canoe period was about 31 feet long over the gunwales or 30 feet 8 inches in a

Figure 130

"Bivouac in Expedition in Hudson's Bay Canoe." From an oil painting by Hopkins (*Public Archives of Canada photo*).

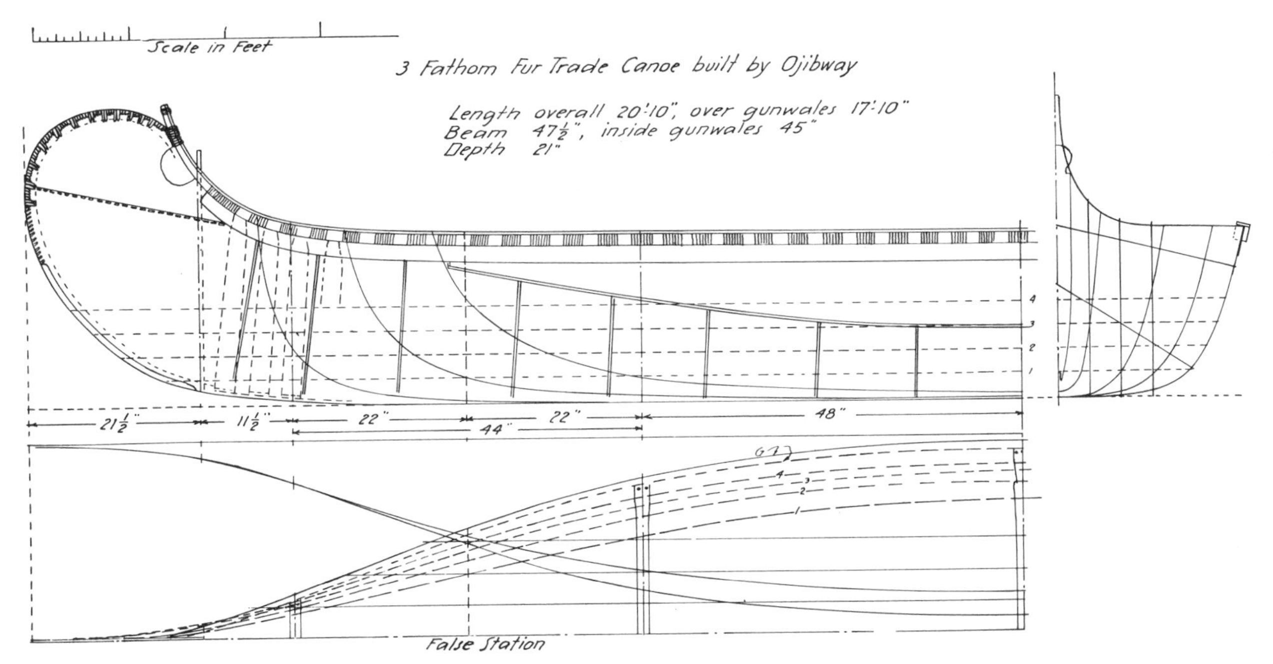


Figure 131

Ojibway 3-Fathom Fur-Trade Canoe, a cargo-carrying type, marked by cut-under end profiles, that was built as late as 1894.

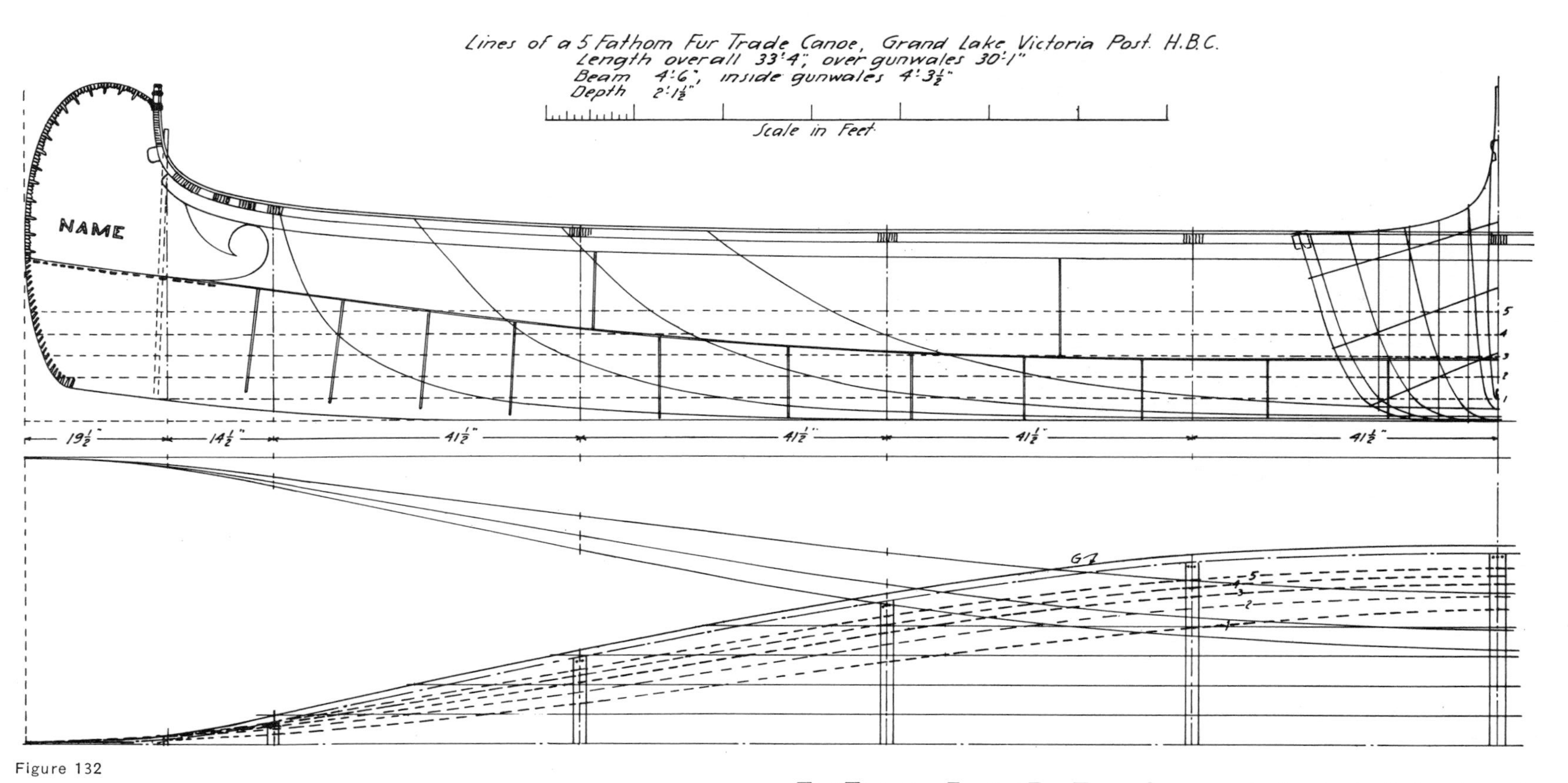


Figure 132

THIS TYPE OF 5-FATHOM FUR-TRADE CANOE was built at L. A. Christopherson's Hudson's Bay Company posts at Grand Lake Victoria, Lake Barrière, and Lake Abitibi. Called the Ottawa River canoe by fur-traders, it was used for fast travel and shows the upright stems of the northwest Quebec Algonkin.

Figure 133

"Hudson's Bay Canoe Running the Rapids." From an oil painting by Hopkins (*Public Archives of Canada photo*).

straight line from tip of upturned rail cap at one stem to the other. The beam inside the gunwales was 60 inches. The width of the building frame would be between 40 and 45 inches, and the frame when formed would be about 26 feet 8 inches long. The depth of the canoe amidships, from bottom to top of gunwale, was approximate 30 inches and the height of the stems about 50 inches. The overall length of such a canoe was about 34 feet 4 inches. An express canoe of this size would be about 56 inches beam inside the gunwales or even somewhat less, and the depth amidships about 28 inches or a little less.

A 4-fathom canoe measured 26 feet 8 inches over the tips of the upturned rail caps, and 29 feet 11 inches overall. The beam amidships was 57 inches inside the gunwales and the depth amidships to top of gunwales was 26 inches; the height of the stem was 53 inches.

A 3-fathom canoe was 19 feet 2 inches overall, 16 feet 8 inches over the ends of the gunwale caps, 42 inches beam amidships inside of gunwales, the depth of the canoe from bottom to top of gunwale amidships was 19 inches, and the height of the ends was 38 inches. The building frame for this canoe was 15 feet 8 inches long and 27 inches wide.

The canoes falling between the even-fathom measurements were often of about the same dimensions as the even-fathom size next below; a 3½-fathom canoe would have nearly the same breadth and depth as a 3-fathom; only the length was increased. The half-fathom rarely measured that—a canoe rated as 3½ fathom was actually only 20 feet 5 inches overall. One express canoe rated 3½ fathoms measured 20 feet 1 inch overall, 18 feet 3 inches over the gunwale caps, 44 inches beam inside gunwales amidships, and 21 inches deep, bottom to top of gunwale cap. The height of the ends was 39 inches. This example will serve to indicate how inexact the fathom classification really was. It should also be noted that the height of the ends varied a good deal in any given range of length, as this dimension was determined not by the length of the canoe but by the judgment and taste of the builder and his tribal form of end. Generally, however, small canoes had relatively higher ends than

Figure 134

"REPAIRING THE CANOE." From an oil painting by Hopkins (*Public Archives of Canada photo*).

large canoes, in proportion to length, because, as will be remembered, one function of the end was to hold the upended canoe far enough off the ground to permit the user to seek shelter under it.

Extremes of dimension appear to have been rare in fur-trade canoes; none whose length overall exceeded 37 feet have been found in the records, and the maximum beam reported in a *maître canot* was 80 inches. When canvas replaçed birch bark in the fur-trade canoes, the high-ended models disappeared; the canvas freight canoes were commonly of the white man's type having low-peaked ends, or a modified Peterborough type.

Before discussing the methods of construction, the loading and equipment of the fur-trade canoes should be described from contemporary fur trade accounts. The goods carried in these canoes were packed into easily handled bundles, or packages, of from 90 to 100 pounds weight. Wines and liquor were carried in 9-gallon kegs, the most awkward of all cargo to portage. In some cases the furs were packed into 80- or 90-pound bundles in the Northwest, and were repacked into 100-pound bundles before being placed on the large canoes of the Montreal-Great Lakes route, but bundles lighter than 90 pounds were made up for the shipment of small quantities of individual goods to isolated posts. The bundles, or packs, of furs were formed under screw presses so that 500 mink skins, for example, were made into a package 24 inches long, 21 inches wide and 15 inches deep, weighing very close to 90 pounds. Buffalo hides formed a larger pack, of course. In the canoe, packs were covered by a *parala*, a heavy, oiled red-canvas tarpaulin.

Boxes called *cassettes* were carried; these were 28 inches long and 16 inches in width and depth, made of ¾-inch seasoned pine dovetailed and iron-strapped, with the lid tightly fitted. The top, and sometimes the bottom too, was bevelled along the edges. The lids were fitted with hasps and padlocks and the boxes were as watertight as possible. Each box was painted and marked; in these were placed cash and other

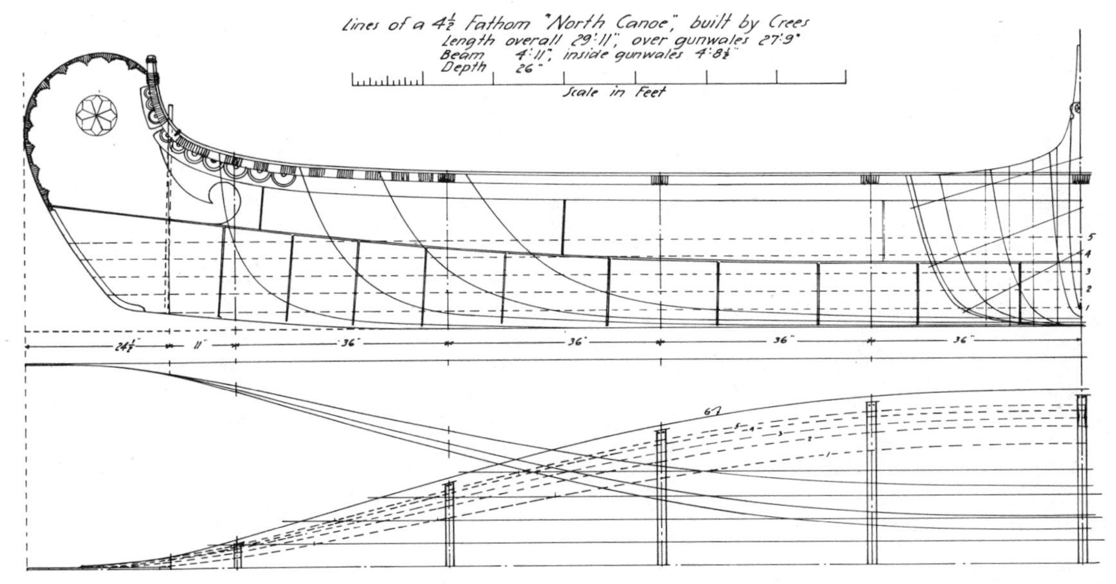


Figure 135

HUDSON'S BAY COMPANY 4½-FATHOM NORTH CANOE, of the type built by Crees at posts near James Bay in the middle of the 19th century, for cargo-carrying.

valuables. Also carried was a travelling case—a lined box for medicine, refreshments for the officers, and what would be needed quickly on the road.

Provisions such as meat, sugar, flour, etc. were carried in tins and were stowed in baskets which were usually of the form known to woodsmen as pack-baskets. Baskets also served to carry cooking utensils and other loose articles. Bedrolls consisted of blankets or robes, made up in a tarpaulin or oilskin ground-sheet and were used in the canoe as pads or seats. The voyageur's term for the canoe equipment—paddles, setting poles, sail, mast, and yard, and the rigging and hauling lines—was *agrès*, or *agrets*.

The term *pacton* was applied to packs made up ready for portage; they were ordinarily made up of two or more packages, so the weight carried was at the very least 180 pounds. No self-respecting voyageur would carry less, as it would be disgraceful to be so weak. The *pacton* was carried by means of a *collier*, or tump-line similar to that used to portage canoes (see p. 122). It was made of three pieces of stout leather. The middle piece was of stout tanned leather about 4 inches wide and 18 inches long, tapered toward each end, to which were sewn pliant straps 2 or 2½ inches wide and 10 feet long. These were usually slightly tapered toward the free ends. The middle portion of this piece of gear was of thick enough leather to be quite stiff, but the straps were very flexible. Sometimes the middle portion and 2 or 3 feet of the end straps were in one piece with extensions sewn to the latter. The *pacton* was lifted and placed so that it rested in the small of the carrier's back, with its weight borne by the hips. The ends of the *collier* were tied to the *pacton* so as to hold it in place, with the broad central band around the carrier's forehead. On top of the *pacton* was placed a loose package, *cassette*, or perhaps a keg. The total load amounted to 270 pounds on the average if the trail was good; the maximum on record is 630 pounds. With his body leaning forward to support the load, the carrier sprang forward in a quick trot, using short, quick paces, and moved at about 5 miles an hour over a good trail. A carrier was expected to make more than one trip over the portage, as a rule.

The traditional picture of the fur-trade voyageur as a happy, carefree adventurer was hardly a true one, at least in the 19th century. With poor food hastily prepared, back-breaking loads, and continual exposure, his lot was a very hard one at best. The monstrous packs usually brought physical injury and the working life of a packer was very short. In the early days, and during the time of the North West Company, the canoemen were allowed to do some private trading to add to their wages, but when the

Figure 136

FIVE-FATHOM FUR-TRADE CANOE FROM BRUNSWICK HOUSE, one of the Hudson's Bay Company posts.

Hudson's Bay Company took over this was not allowed and discipline became far more harsh. As a result, the French Canadians deserted the trade, to be replaced with Indians and halfbreeds. The paddling race against time, to reach the destination before the fall freeze, was labor comparable to that of a galley slave, but in a very harsh climate. Altogether, if the brutal truth is accepted, the life of the canoeman was far more hardship than romance.

The cargo of a fur-trade canoe was not placed directly on the bottom; light cedar or spruce poles were first laid in the bottom of the canoe and then the cargo loaded aboard. The poles prevented damage to the canoe by any undue concentration of weight. The weight of cargo carried varied with the size of the canoe and with the conditions of the canoe route. The canoes were usually loaded deeply, except in the case of the light express canoe, in which the cargo was reduced for sake of rapid travelling.

An account written in 1800 by Alexander Henry the younger gives the following list of cargo in a trade canoe on the run to Red River in the Northwest, where canoes under 4½ fathoms were generally used: General trade merchandise, 5 bales; tobacco, 1 bale and 2 rolls; kettles, 1 bale or basket; guns, 1 case; hardware, 1 case; lead shot, 2 bags; flour, 1 bag; sugar, 1 keg; gunpowder, 2 kegs; wine, 10 kegs. This totaled 28 pieces: in addition the crew had 4 bales (1 for each paddler) of private property, 4 bags of corn of 1½ bushels each, and ½ keg of "grease," plus bedrolls and the canoe gear. The trade goods carried to the posts included such items as canoe awls, axes, shot, gunpowder, gun tools, brass wire, flints (or, later, percussion caps), lead, beads, brooches, blankets, combs, coats, firesteels, finger rings, guns, spruce gum, garters, birch bark, powder-horns or cartridge boxes, hats, kettles and pans, knives, fish line, hooks, net twine, looking glasses, needles, ribbons,

Figure 137

FUR-TRADE CANOES ON THE MISSINAIBI RIVER, 1901. (*Canadian Geological Survey photo.*)

rum, brandy, wine, blue and red broadcloth, tomahawks or hatchets, tobacco, pipes, thread, vermillion and paint, and false hair.

The tarpaulins used to cover the cargo were 8 by 10 feet, hemmed and fitted with grommets around the edges for lashings. The cloth was treated with ochre, oil, and wax to give it a dull red color and to waterproof it. One of the tarpaulins usually served as the sail. The fur bales were each sacked, that is, wrapped in a canvas cover that was sewed on and stenciled with identification and ownership marks.

The cargo manifests were not always the same. Compare the previous list with this cargo, with which two light canoes were each loaded: 3 *cassettes*, 1 travelling case, 2 baskets, 1 bag of bread, 1 bag of biscuits, 2 kegs of spirits, 2 kegs of porter, 1 tin of beef, 1 bag of pemmican for officers and 2 for the crew, 2 tents for officers, cooking utensils, canoe equipment, and 1 *pacton* for each of the 9 men in each canoe.

The rate of travel varied a good deal, depending upon the condition of the waterway and of the men. Perhaps, as an average, 50 miles a day would be the common expectation during a 3-month run into the northwest. Traveling fast with good conditions, an express canoe might average as much as 75 or 80 miles a day, but this was exceptional.

The number of men required to man a fur-trade canoe varied with the use required of the canoe, with its load, and its size. There were rare occasions in which a *maître canot* had 17 paddlers and a steersman, but normally such a canoe was manned by between 7 and 15 men, depending upon how much space aboard was required by cargo or passengers and upon the difficulties of the route. An express canoe, traveling light and at high speed, was manned by 4 to 6 paddlers, one of whom acted as steersman or stern paddler, and one as the equally important bowman in river work.

The most valuable information on the construction methods of fur trade canoes was obtained in 1925 from the late L. A. Christopherson, a retired Hudson's Bay Company official. He had joined the Company in 1874 and retired in 1919, after 45 years service, 38 of which he had spent in western Quebec at the posts on

Figure 138

FUR-TRADE CANOE BRIGADE, CHRISTOPHERSON'S HUDSON'S BAY COMPANY POST, about 1885. Christopherson in white shirt and flat cap, sitting with hands clasped. Five-fathom canoes, Ottawa River type.

Lake Barrière and on Grand Victoria. These were canoe-building posts, and Christopherson had supervised the construction of both the 5- and 4½-fathom trade canoes. His posts had built the nearly vertical-ended *nadowé chiman*, the Iroquois, or Ottawa River, type of Algonkin canoe. The actual building was done by Indians, but the work was directed by the Company men.

In the building the eye and judgment of the builder were the only guides, aided by the occasional use of a measuring stick, and Christopherson made it abundantly clear that the Company had no rules or regulations that he knew of, regarding the size, model, and construction of the canoes, nor any standards for decoration. The model and appearance of the canoes were determined by the preferences of the builders and the size by the needs of the posts. For example, the 5-fathom canoe had been built at the Grand Victoria post until it was decided there that a 4½-fathom canoe would serve. The decoration, if any, was apparently according to "the custom of the post."

The method of construction described by Christopherson seems to be largely that of the Algonkin, modified slightly by Ojibway practices. The canoes were built on a plank building bed made of 2- or 2½-inch thick spruce; its middle was higher than the ends, as were the earthen beds used in the east, and holes were bored in it to take the stakes. A stake was placed near the end of each thwart and one between, along the sides of the canoe. The individual builders had their preferences as to the method of setting stakes; some set them vertically while others bored the bed so that the stakes stood with their heads pointed outward. A post might have two or more building beds, one for each size, or model.

Canoes were always built by means of a building frame. This was made with four or five crosspieces that determined the fullness or fineness of the bottom of the canoe toward the ends. By altering the lengths of the end crosspieces, the degree of fullness in the lines of the finished canoe could be predetermined. As a result the bed, which was usually about 18 inches wider than the building frame, might have the shape of its frame marked on it twice, with two sets of holes for stakes. Otherwise, the alteration in the building frame would require a special bed to be used. In addition to the alteration in the ends of the building frame, there could also be variations in its width amidships. Christopherson's posts commonly built canoes intended for fast travel, so most of them were narrower in beam at the gunwale and across the bottom than were the fur-trade canoes of the period, and the building frame was likewise narrower.

The length of the building frame used in these canoes was the same as the bottom length, or a little longer than the distance between the two headboards of the finished canoe. Thus, in a 5-fathom canoe the

Figure 139

Forest Rangers, Lake Timagami, Ontario. (*Canadian Pacific Railway Company photo.*)

bottom length would be 30 feet, and in a 4½-fathom canoe, 27 feet; the beds would be some 6 feet longer than these lengths.

As the canoes at Christopherson's were built for speed and rarely measured more than 48 inches beam between the gunwale members, the building frame was about 32 inches wide amidships, or approximately two-thirds the beam inside the gunwales in a 5-fathom canoe. The beam of his 4½-fathom canoes was less, say 42 inches inside the gunwales and 27 or 28 inches across the building frame, with a depth, bottom to top of rail cap, of between 19 and 21 inches. A 5-fathom canoe of this narrow model would carry nearly 2½ short tons with a crew of six, while the smaller model would carry nearly 2 tons. However, the capacity of a wide canoe was much greater. A 6-fathom canoe, the *Rob Roy*, built by another post about 1876 to bring in the bishop for the consecration of a church at the Lake Temiscaming post, was described by Christopherson as being about 6 feet beam on the gunwales. Considered a fine example of a freight canoe, the *Rob Roy* was afterwards loaded with 75 bags of flour, totaling 3½ tons deadweight, and carried as well a crew of seven and their provisions and gear.

The bark cover was commonly in two lengths on the bottom of the canoe, summer bark being used. The post maintained a supply of bark for canoe building and sheets 4 fathoms in length and 1 in breadth were not uncommon. Such sheets would have been ample for the cover of a small canoe but would not be expended so needlessly; hence, the canoes, large or small, had two lengths of bark in their bottoms. The lap was toward the stern. In what appears to have been a local characteristic of the canoes built at Christopherson's posts, the bows were indicated by making the thwarts toward that end slightly longer than those toward the stern, so that the forebody was fuller at sheer than the afterbody; the canoe master could thus instantly see which end was the bow without having to examine the bottom or the bark cover.

The two pieces of bark sewn together were placed on the building bed and the building frame placed on it and weighted down, in the usual manner. The stakes were then set in the holes in the bed and the bark secured to them with the usual inside stakes, as well as with the clothespin-like clamps used by the Algonkin and other Indian canoe builders. The end stakes were set in a peculiar manner: a short pair were set with their heads sloping inboard, for use later to support the sheering of the outwales, and a long pair were set raking sharply outboard to help support the bark required for the high ends. As the bark cover was made up, pieces were worked into the ends to allow the high ends to be made. The side panels often seen on the eastern Indian bark canoe were used, and the bark doubled at the gunwales. The doubling pieces were put on about 6 inches wide and trimmed off after the outwales were in place. The pieces were widest amidships, and when trimmed would extend about two inches or a little more below the outwales, narrowing somewhat toward the ends. Longitudinal battens to fair the bark along the sides were placed as usual in canoe building.

The main gunwales were originally made of white cedar, but when this became scarce at the posts, whipsawed spruce was used instead. The gunwales were rectangular in cross section, with the outer lower corner beveled off. The cross section of the inner gunwale member was smaller, in proportion, than the outwale, compared to a small eastern Indian canoe. The gunwales were bent "on the flat" in plan, and were sheered "edge bent." The tenons for the thwart ends were cut slanting, so that when the gunwales were made up they stood at a flare outward toward the top edge. The gunwales had much taper toward the ends as it was usual to work in some sheer in these members. The canoes built at Christopherson's posts, unlike some other trade canoes, had a good deal of sheer at the ends, as the main gunwales rose nearly to the top of the stem.

The manner of forming the gunwales varied somewhat. If the stakes around the building frame had been set to stand vertically, it was necessary to assemble the gunwales with temporary crosspieces, or false thwarts, each shorter by several inches than would be the finished thwart in their place, or twice the amount of flare desired. After the gunwale assembly had been set above the building frame on the usual posts to determine its height above the building bed, the bark cover would be lashed to each gunwale member. This done, each crosspiece would be removed in turn and replaced with its corresponding thwart. By this means the gunwales would be spread and, in the process, lowered in proportion to the change in beam. This would usually make too much sheer. Therefore, if the gunwales were to be spread as a result of the side stakes standing vertically, they had to be formed with some reverse sheer amidships. This was done as usual, by first treating each member with hot water and then weighting it on a long plank, or unused building bed, over a block placed under it at midlength. The height of the block would determine the amount the sheer was "humped" in the middle, usually only an inch or so. The gunwale ends were also treated with hot water and sometimes were split horizontally to get the required sheer there; they were then bent up and held, while drying and setting, by a long cord that was stretched between them and placed under tension by means of a strut, about 4 feet long, placed under the cord at midlength and stepped on the gunwale member being bent. However, if the side stakes were set sloping outward, it was unnecessary to hump the sheer amidships.

The reason why many builders preferred to set the stakes on the bed vertically was that it made easy the goring and the sewing of the bark cover side panels; if the bark available for the cover required little sewing, the sloping stakes might be preferred. It appears, however, that the usual procedure was to set the stakes vertically and to spread the gunwales, since good bark was usually available. A good deal of judgment was required to estimate the amount of hump or reverse to be worked into the gunwale members; too much would leave a hump in the sheer of the finished canoe and not enough would cause too much dip amidships. Before being bent to sheer, the gunwale members were worked smooth with a plane or with scrapers made of glass or steel. The building frame was taken apart and removed from the canoe after most of the thwarts were in place.

The ribs Christopherson called "timbers" and the sheathing, "lathing." The ribs, commonly of cedar, were usually ¼ to ⅜ inch thick, and were 2½ to 3¼ inches wide in most canoes, with a long taper so that near the ends the width was about half that at the middle, and at the ends they tapered almost to a point. Some large canoes had ribs 4 inches wide at the centerline, amidships, but these appear to have been unusual. The ribs were placed on the building frame at their proposed position and the width of the

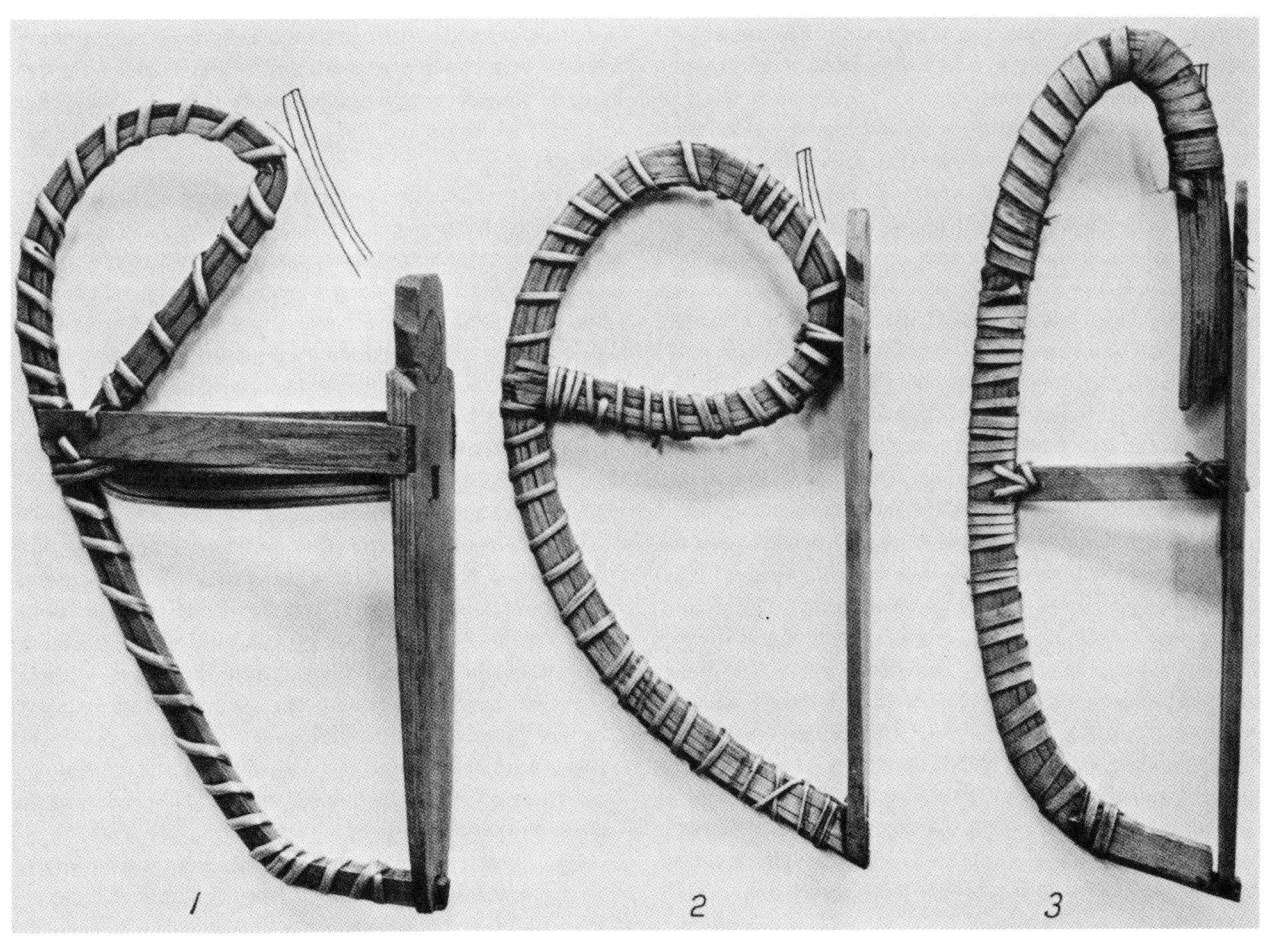


Figure 140

Fur-Trade Canoe Stem-Pieces, models made by Adney: 1, Algonkin type; 2, Iroquois type, Ottawa River, old French; 3, Christopherson's canoes.

frame at that point was marked on each. After being cut to about the required length and tapered, the ribs were then treated with hot water, and were then usually bent over the knee in pairs, the marks determining where the bending was to be done. In a freight canoe the ribs amidships would be nearly flat across the bottom but in a fast canoe they would be slightly rounded. The parts of the rib nearest the ends were not bent, and thus the rib would appear dish-shaped when in form. Each pair while drying was sometimes held by cords tied across the ends, or the ribs might be inserted in about their proper location in the unfinished canoe and held in place by battens and struts until they took their final set. The ribs at the extreme ends were often "sprung" or "broken" at the centerline to get the V-section required there, particularly in a sharp-ended express canoe.

The sheathing was about ¼-inch thick and was laid according to the tribal practice of the builder; Christopherson appears to have followed the Algonkin practices generally in this as in other building matters at his posts.

Whereas Malecite practice was to lash the bark cover to both inwale and outwale, in the western type of canoe the cover was lashed to the main gunwale first, owing to the spread gunwales, and the outwale was then pegged to the gunwale and also lashed, the ends being wrapped with figure-eight turns. All gunwale lashing in fur-trade canoes was in groups. Because of the sheer at the ends, the outwales were split horizontally into four or more laminae, and the splitting extended almost to the end-thwart positions. In a few canoes outwales were omitted or were short and did not extend beyond the end thwarts, but this practice was relatively uncommon. The outwales were usually rectangular in cross section and much tapered toward the ends.

The rail caps were also rectangular in cross section, but often they had the outboard upper edge rounded off or beveled. The caps were pegged at 1-foot intervals to the main gunwales, but at the ends they could only be lashed to the outwale, as both outwales and caps were so sharply upswept at the ends that they stood almost vertically. The ends were squared off and stood a little above the top of the stems, so that when the canoe was placed upside down as a shelter for the paddlers and packers it rested upon these members rather than on the sewing of the bark cover on the tops of the stems, as was usual with all the high-ended Algonkin and Ojibway canoes.

The stem-pieces and headboards were assembled into single units, as shown on pages 149 and 151, before being installed during construction. The stem-pieces were of white cedar, about four fingers deep fore-and-aft and laminated, and about ¾ to 1¼ inches wide, depending upon the size of the canoe and the judgment of the builder. In Christopherson's area the stem-piece was relatively short, the head coming up and around and ending at a point far enough under the rail-cap ends for it to be securely lashed to these members and to the outwale ends. It was bent by use of hot water and the laminae were secured by wrapping the stem piece with fine twine. The stem was stiffened by stepping the headboard on its heel in the usual manner, and the two were held in the required position by two horizontal struts, the outboard ends of which were lashed to the sides of the stem piece well up above the heel; the inboard ends were pegged at the sides of the headboard, in notches, or were passed through the headboards in slots and the strut ends secured with wedges athwartships on the inboard face of the headboard. The result was a rigid and strong end-frame. More complicated bending was employed at some posts, where the building of fur-trade canoes followed Algonkin or Ojibway practices. In these, as has been mentioned, the stem-pieces were brought down and around under the stem-head to the back or inboard edge of the stem-piece and lashed, then brought inboard horizontally to end in a hole in the headboard, between struts placed as in the Christopherson-built canoes. Another method was to bring the stem-piece around the stem head and down and around outboard to the inboard face of the stem, where the end was split and each half lashed to the sides of the stem-piece. In this case there was a lashing between stem-piece and the headboard, placed where the reverse was made, inboard and below the top of the stem, well up on the headboard. The heel of the headboard and stem-piece were pegged together.

Struts were not required with this construction, described earlier (on p. 123) as the Ojibway method. In bending the stem-piece, the reverse curve around the stemhead was formed over a short strut that was removed when the stem-piece was dried and set to shape. As a variety of forms were used in shaping these stem-pieces, it was the ingenuity of the builder that decided just how the end of the stem-piece was best secured and how the whole was to be braced. These details will be better understood by reference to the plans and illustrations on pages 134 to 151.

The headboards were not sprung or bellied, but stood nearly vertical in the canoes. The inboard face was often decorated; in the old French canoes and in those of the North West Company, the board was carved or painted to represent a human figure, *le petit homme*, which was often made in the likeness of a voyageur in his best clothes. In some canoes, only a human head was used, or the top of the headboard, or "button," was decorated with a rayed compass drawn in colors.

The thwarts were usually rather heavy amidships and were made in various forms to suit the taste of the builder. They were commonly of maple, but Christopherson's canoes had spruce or tamarack thwarts, the latter being his preference. These thwarts were not intended to be used as seats, though the sternman, or steersman, often sat on the aftermost one. The paddlers often used seats in the large canoes; these were planks slung from each end by cords made fast to the gunwales. These cords allowed the height of the seats to be adjusted; the paddlers usually knelt on the bottom of the canoe with hips supported by the seat. The seats were usually slung before the thwarts, except amidships, where the space was taken up by passengers or cargo.

The factors often took great pride in the appearance of the canoes from their posts and many, like Christopherson, had the craft gaily painted in a rather barbaric fashion. Christopherson's canoes did not use any of the circular decoration forms; his canoes usually had painted on them, he recalled, such names as *Duchess*, *Sir John A. MacDonald*, *Express*, *Arrow*, and *Ivanhoe*. The ends were often painted white, with the figures or letters on this background. The Company flag was often painted on the stern

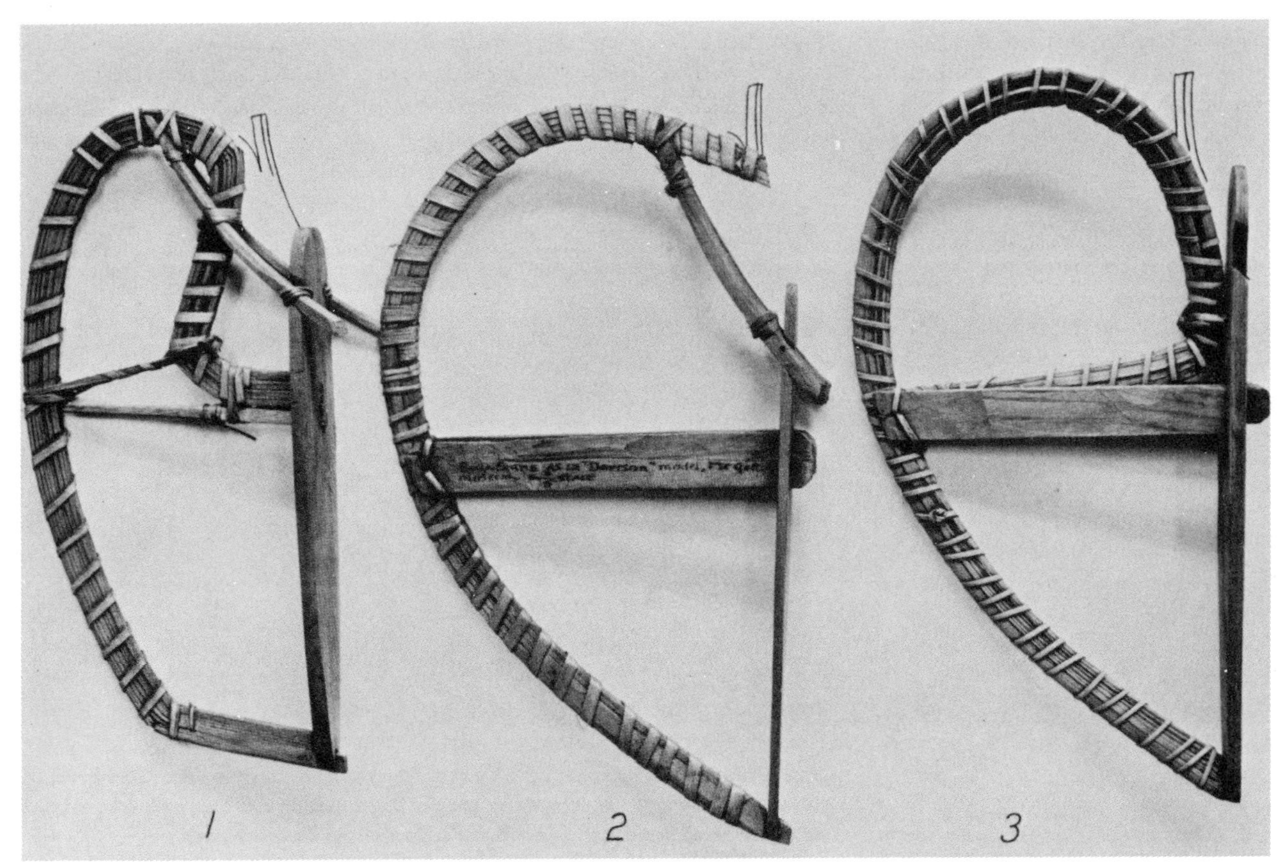


Figure 141

FUR-TRADE CANOE STEM-PIECES, models made by Adney: 1, Têtes de Boule type; 2, Ojibway form; 3, old Algonkin form.

with the initials of the Company, H.B.C., said to mean "Here Before Christ" by disrespectful clerks. Many posts used such figures as the jackfish, loon, deer, wolf, or bear, on the bow. The rayed circular devices appear to have been long popular and were said to have been introduced by the French. There is no record of any device being officially required in any district but the *cassettes* of certain districts were marked with distinctive devices at one time; Norway House used a deer's head with antlers, Saskatchewan two buffalo, Cumberland a bear, Red River a grasshopper, and Manitoba a crocus.

During Christopherson's long service he knew the canoes built in his vicinity at such nearby building posts as Lake Abitibi, Lake Waswanipi, and Kipewa, in western Quebec; and Lake Timagami (Bear Island), Matachewan on Montreal River, Matagama (west of Sudbury), and Missinaibi, in nearby Ontario. These were but a few of the building posts, of course, for canoes were built at numerous posts to the west and northward.

When portaged, the large canoes might be carried right side up or upside down, the former being more usual method. The *canot du nord* was often light enough to be carried by two paddlers, one under each end, with the canoe right side up and steadied by a cord tied to the offside gunwale and held in the carrier's hand. The *maître canot* required four men to carry it. Various methods were used. One was to lash carrying sticks across the gunwales near the ends and to carry the canoe right side up with a man on the end of each stick. Another way was for the men to distribute themselves along the bottom of the canoe, near the ends, and to use steadying cords. Or the canoe might be carried upside down with the men carrying it by placing one shoulder under the gunwales

Figure 142

PORTAGING A 4½-FATHOM FUR-TRADE CANOE, ABOUT 1902, near the head of the Ottawa River. Shows an unusually large number of carriers; four would be the normal number. (*Canadian Pacific Railway Company photo.*)

at convenient places. When a bad place in the portage was reached, the whole crew might have to turn to. The method of portaging had to meet the physical limitation of the portage path and the matter was not so much one of standard procedure as of improvisation of the moment.

The voyageur was particular about his paddle; no man in his right mind would use a blade wider than between 4½ and 5 inches, for anything wider would exhaust him in a short distance. The paddle reached to about the users' chin, when he stood with the tip of the paddle on the ground in front of him. Longer paddles, about 6 feet long, were used by the bow and stern men, the two most skillful voyageurs in the canoe and the highest paid. These men had, also, spare paddles whose total length was 8 feet or more; these were used in running rapids only. The paddles were of hardwood, white or yellow birch or maple, as hardwood paddles could be made thin in the blade and small in the handle without loss of strength, whereas softwood paddles could not. The blades were sometimes painted white, the tips in some color such as red, blue, green or black, but other color combinati ons were often used.

In Christopherson's service, sail was rarely used, as the canoemen were unskilled in handling it and loss had resulted. In early times, however, it appears to have been much used on the Great Lakes routes by the French and the North West Company. A single square-sail was the only rig employed; the canoes could not be worked to windward under fore-and-aft sails.

During the great seasonal movements the trade canoes moved in fleets called brigades, the usual brigade in early times being three or four canoes, but later, when the needs of the individual posts had grown, the brigade could be of any necessary number of canoes to carry in the required supplies and goods or to bring out the season's catch of furs. The leader of the brigade was the *conducteur* or *guide;* sometimes he was the post's factor. In French times the *maître canot* would be loaded with 60 pieces, or packs, to the total of about 3 short tons and half a ton of provisions, and eight men, each with an allowance of 40 pounds for gear, so that the whole weight in the canoe would be something over 4 short tons. An example of such a canoe measured, inside the gunwales, 5½ fathoms long and 4½ feet beam. The usual brigade of four of these canoes would thus carry roughly 12 short tons of goods.

The Company would send one brigade after another, at close intervals of time, until the whole seasonal movement was in progress. Those brigades going the greatest distance were started first. Although cargoes left the coast from early spring on to late summer, the great canoe movement took place towards the fall. Canoe travel north and northwestward from the Great Lakes had to be carefully timed, as goods had to be accumulated at the base posts on the Lakes and the brigades placed in movement at the last safe date which would permit them to reach their destination before the first hard freeze-up. The base posts were those where the run of the *maître canot* ended and that of the *canot du nord* began, the places where reloading for the individual trading posts in the Northland was necessary. The late start was usually desirable in order to await the arrival at the base posts of all the goods required, for movements of freight were uncertain before the days of railroads and steamers.

In the late 18th and early 19th centuries, before the whole canoe trade fell under the control of the Hudson's Bay Company, it was the custom to distribute 8 gallons of rum to each canoe for consumption during the run, and it was also the custom for all hands to see how much of this they could drink before starting out. This grandiose undertaking usually began as soon as the local priest, who gave his blessing to the canoemen, had left the scene. The magnificent drunk lasted one day and the next morning the crew had to be underway. The first day's run, old accounts repeatedly show, not only was short but was often beset by difficulties.

The era of the bark trading canoe did not close with a dramatic change. Its ending was a long, slow process. By the last decade of the 19th century the bark trading canoe had disappeared from most of the old routes, and even in the Northwest it had been almost wholly displaced by York boats, scows, bateaux, and canvas or wooden canoes of white-man construction. By the beginning of the first World War, the *maître canots* and *canots du nord* were finished, except as curiosities—hardly even as these, for not one was preserved in a museum.

Indeed, so complete was the disappearance of the fur-trade canoe that any attempt to record its design, construction, and fitting would have been almost hopeless, had it not been for the notes, sketches, and statements of such men as L. A. Christopherson, aided by a few models and pictures, and for the memories of a few Indian builders who had worked on the canoes.

Figure 143

DECORATIONS: FUR-TRADE CANOES. (*Watercolor sketch by Adney.*)

# *Chapter Six*

# NORTHWESTERN CANADA

INDIANS OF THE NORTHWEST TERRITORIES and the Province of British Columbia in Canada, and the States of Alaska and Washington, built bark canoes that may be divided into three basic models.

The first may be called the "kayak" model, a flat-bottom, narrow canoe having nearly straight flaring sides and either a chine or a very quick turn of the bilge. These bark canoes were low-sided and were usually partly decked. A number of tribal groups built canoes of this model, the variation being relatively minor. The rake and form of the ends varied somewhat as did the amount of decking; there were also some slight variations in structure and method of construction. While these bark canoes had some superficial resemblance in general proportions to the Eskimo kayaks, it is necessary to point out that they did not, particularly in Alaska, have the same hull form as the seagoing kayaks in that area. In fact, the single-chine form of the Alaskan version of this canoe appears only in the kayaks of northern Greenland and Baffin Island. The Alaskan seagoing skin kayaks are all multi-chine forms that approximate a "round-bottom" hull. It has been thought that the flat-bottom seagoing kayak form may have existed in the Canadian Northwest, at the mouth of the Mackenzie; a kayak so identified is in the collections of the U.S. National Museum (see p. 202), but there is now doubt among authorities as to the correctness of this identification. As will be shown later, it seems probable that it has been improperly assigned to the Mackenzie delta and is, in fact, an eastern Eskimo model.

The second model used in the Northwest area was a narrow-bottom flaring-sided bark canoe with elevated ends, having, perhaps, a faint resemblance to the Algonkin-Cree canoes of the old type. Here too there was some variation among the canoes of tribal groups, mostly in the shape and construction of the ends and in the fitting of the gunwales. Most of the canoes of this type had stem-pieces formed of a plank-on-edge, but in a few examples the stem-pieces were bent. This model was built by the same tribal groups in Canada that built the kayak form, the explanation being that the kayak form was the hunting while the second model was commonly the family or cargo canoe. In Alaska, however, only the kayak-form was used and the family, or cargo, canoe was merely an enlargment of it.

The third model may be called the "sturgeon-nose" type; in this the ends were formed with a long, pointed "ram" carried well outboard below the waterline as an extension of the bottom line of the canoe. Primitive in both model and construction, it was built in a rather limited area in British Columbia and in the State of Washington. The last canoes built on this form were canvas-covered; in earlier times spruce or pine bark was usually employed.

The birch in most of the Northwest is a small tree and the bark is of poor quality for canoe building; hence, in many areas spruce bark was commonly employed in its place; a single tribal group might build its canoes of either, depending upon what was available near the building site. However, near the Alaska coast, where kayak-form bark canoes were used and good birch was usually not available, some tribes used seal or other skins as a substitute. In the framework spruce and fir were most commonly employed, but occasionally cedar was available and was used.

The canoe-building Indians in northwestern Canada were mostly of the Athabascan family and included the Chipewyan or "Chipewans," the Slave or "Slavey" (=Etchareottine), the Beaver (=Tsattine), the Dogrib (=Thlingchadinne), the Tanana (=Tenankutchin), the Loucheux, the Hare (=Kawchodinne), and others. Some of these tribal groups built not only bark canoes but also dugouts. There were also some Eskimo people who built bark canoes for river service, as well as skin canoes, on the same model as the bark kayak-form.

In the vicinities of Lake Athabasca and Great Slave Lake, the Chipewyan employed not only their own models of canoes but also that of the western Cree. The latter had invaded Chipewyan territory before the arrival of the first white men in the Northwest and undoubtedly had influenced canoe-building technique during the long period of the fur trade that followed. It is therefore not possible to say where the influence of Chipewyan building techniques ends and that of the Cree and the eastern Indians, as introduced through the fur-trade canoes, begins. This raises the question whether the high-ended Athabascan canoe is itself the result of influence. One may infer from Samuel Hearne's description of his travels in this area, in his *Journey . . . to the Northern Ocean,** that only the Kayak-form then existed, for this type is the only one he describes, and he describes it in great detail. However, Alexander Mackenzie, in an entry in his journal for June 23, 1789, refers to the "large canoe" in a manner indicating that it was a local type. It may well be that then, as later, the kayak-form and cargo canoe existed side by side, or it may be that Mackenzie was referring to a large kayak-form canoe like the family canoe of the Alaska Yukon Indians. Perhaps the reason that Hearne did not mention the "large canoe" is that the people he met on his way to the Coppermine River, and on his way back by way of Lake Athabasca to Hudson Bay, did not then use canoes of the second model.

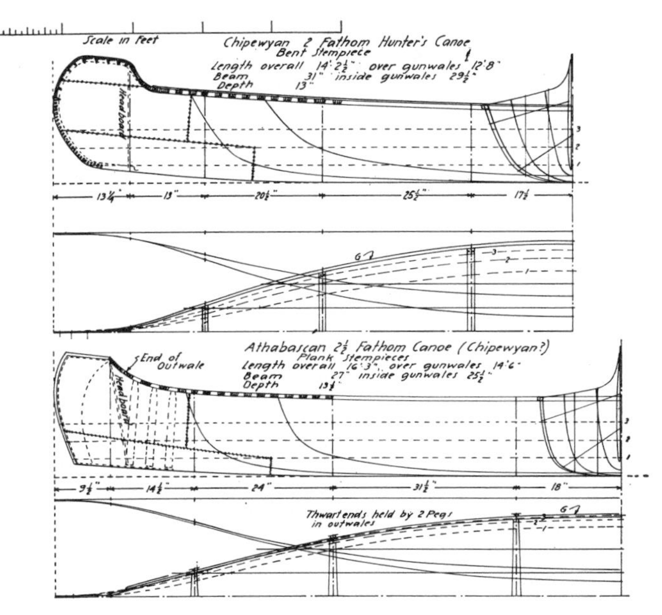


Figure 144

CHIPEWYAN 2-FATHOM hunter's canoe (top), with bent stem piece, and Athabascan 2½-fathom canoe with plank stem piece. Plank and bent stem pieces were both employed in Athabascan canoes. Spruce or birch bark were used without alteration of the design or basic construction methods.

## Narrow-Bottom Canoe

Because the variations in the second model, the Algonkin-Ojibway type, are relatively slight, it will be easiest to describe this first. The canoe is known to have been built extensively by the Chipewyan, Dogrib, and Slave. The sizes most common were 16 to 22 feet over the gunwales, with a beam of between 36 and 48 inches. The sheer was usually rather straight, the sharp upward turn to the end taking place very close to the gunwale ends. Most of the bottom was straight; the rocker, if existing, occurred close to the ends of the canoe and was moderate. The midsection was dish-shaped and nearly flat across the bottom, with a rather slack, well-rounded bilge and almost straight flaring sides, the amount of flare being usually great. The bottom apparently was never dead flat athwartships, for in all known examples it was somewhat rounded. Near the ends the sections were in the shape of a V with apex rounded; the form of the ends was sharp and without hollow either at the gunwale or at the level lines. The ends of the canoes were never lofty and many had end profiles that were very long fore-and-aft and showed a marked angularity. Inwales and outwales formed the gunwale structure; some canoes also had gunwale caps which stopped well short of the end profiles. The ends of the inwales were carried to the stem-pieces; they were sharply tapered and curved to sheer, and were elaborately cross-wrapped to secure them there. The end profiles were formed of a thin plank-on-edge in most canoes, but some had stem-pieces split into laminae in the usual fashion and bent. In all cases headboards were employed; the heads were forced under the inwale ends and against the inside face of the stem-piece. The gunwale lashings were in groups, although some canoes exist in which

*See bibliography.

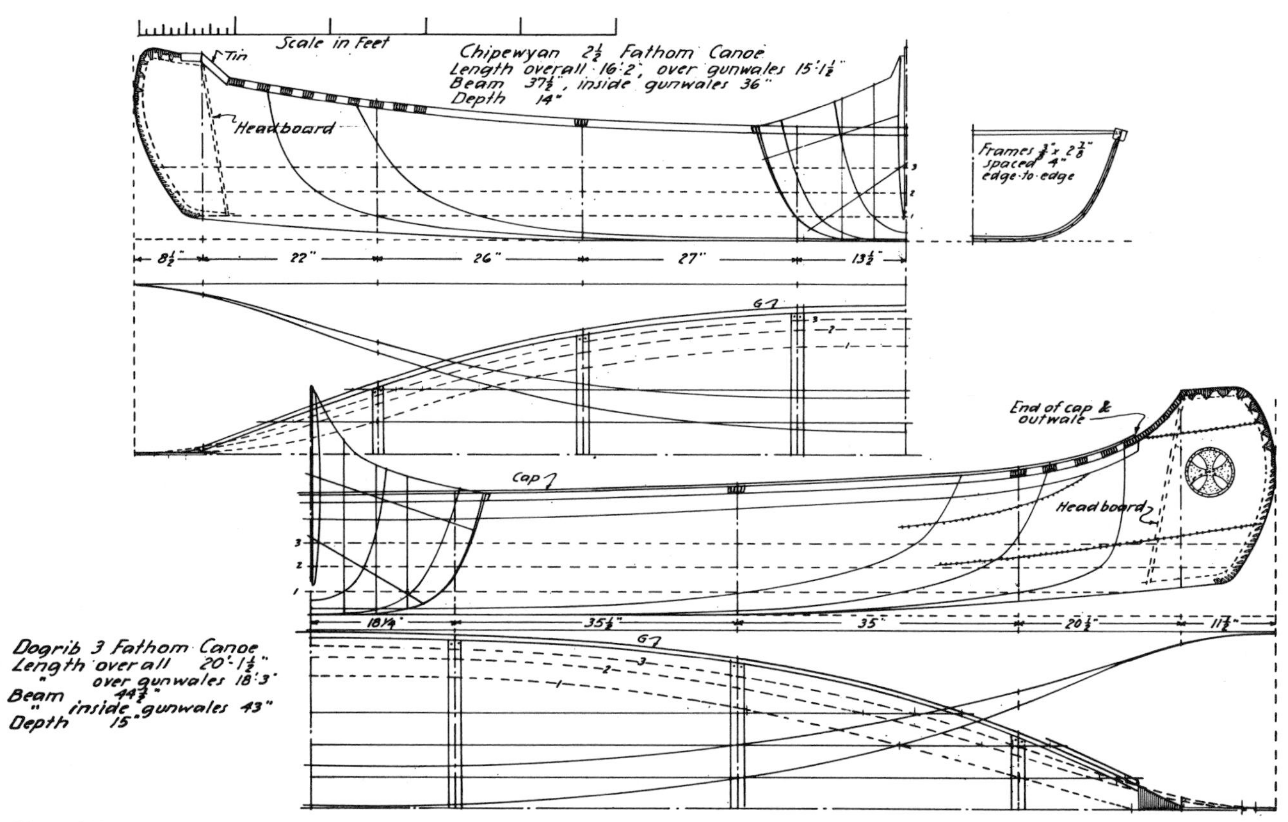


Figure 145

ATHABASCAN CARGO OR FAMILY CANOES WITH BENT STEM PIECES, Chipewyan 2½-fathom (top) and Dogrib 3-fathom. These canoes were covered with spruce or birch bark.

the outwale was omitted and the lashing was continuous; these canoes usually had laminated bent stem-pieces and their stem lashing was identical with that of the Algonkin-Ojibway fur-trade canoes. This departure, it is reasonable to assume, was the result of outside influence on the Athabascan technique. When the stem-piece was of thin plank, the bark was usually fastened to it by multiple turns of two thongs passed, one from each side, through the bark and through holes bored in the stem.

The end profile varied with the tribe of the builder. Chipewyan canoes had a very long end profile fore-and-aft; the heel was angular, and the outline of the stem then swept forward in an easy curve to a height about two-thirds the depth of the canoe amidships, then began to tumble in a little, the curve becoming gradually sharper until the head was reached. The stem-head in its fore-and-aft length was almost one third the height of the ends and was roughly parallel to the bottom of the canoe directly beneath it. Because of the rocker of the bottom, the after end of the head was thus lower than the fore end. The sheer was faired up to the after end of the head in a short, quick curve. Usually the outwales were cut off short of this point, but in some canoes they were brought up along with the inwales to the stem-head. Wedges were used in making up the gunwale-end lashings in both the Chipewyan and Dogrib canoes; these served to tighten the lashings and formed a sort of breasthook. In a few examples of the Athabascan type, the stem-pieces were of cedar root without lamination; this use of the roots enabled the angular form of the plank-on-edge stems to be retained. It cannot be determined whether the root stem-pieces were part of the old Athabascan technique or were an importation from the western Cree. The lashing in these canoes followed the forms used in the fur-trade canoes—long-and-short turns in

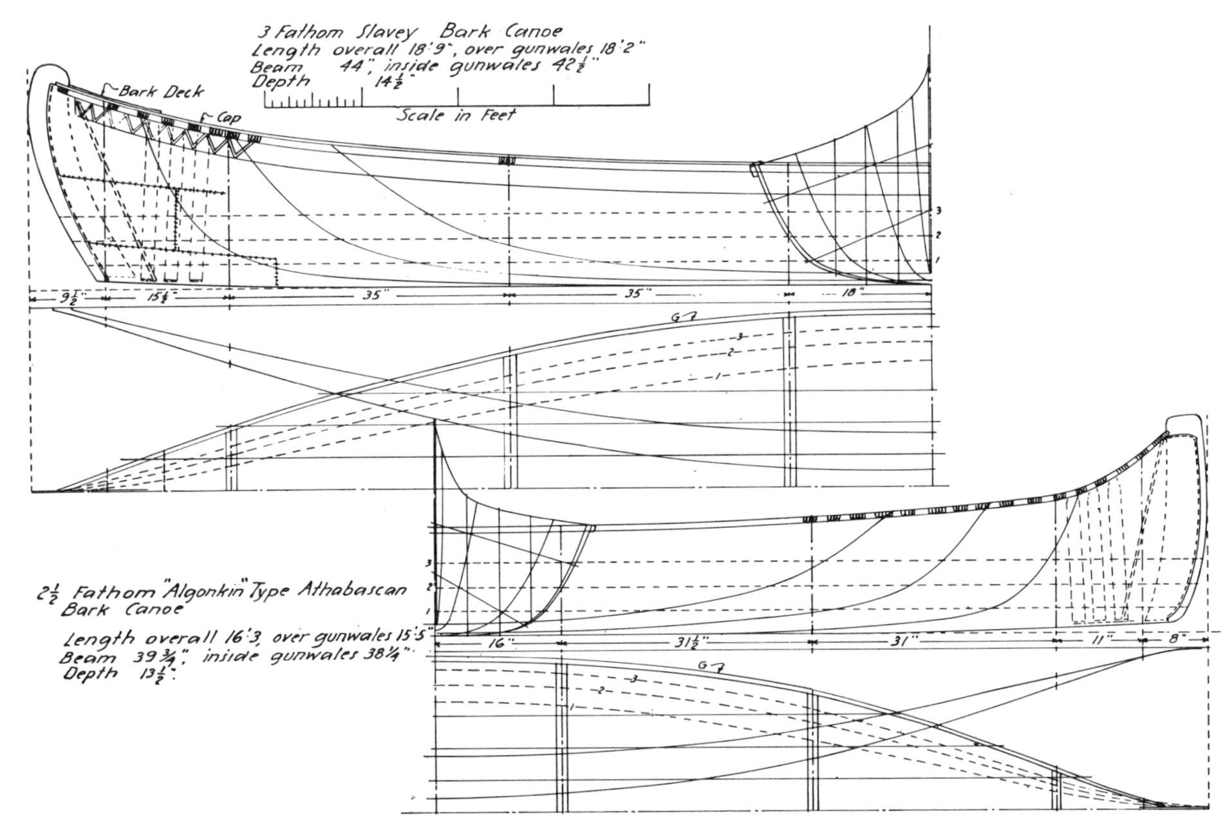


Figure 146

Plank-Stem Canoes of Hybrid Forms, 3-Fathom Slavey (top) and 2½-fathom Algonkin-type Athabascan, probably the results of the influence of fur-trade canoe-building.

groups generally triangular in shape, with a spiral turn between groups.

The canoes of the Dogrib were practically identical with those of the Chipewyan except that the end profiles were usually slightly deeper fore-and-aft; also the Dogrib canoes were perhaps more often of birch bark, judging from the remaining canoes and models. The form of the ends in the Dogrib canoes was such that they often appeared higher than they really were, as the stem-heads stood some distance above the ends of the sheer, an effect which was heightened by the small fore-and-aft depth of the stem-heads.

The large canoes of the Slave had the same hull characteristics as the others but differed in end profiles and did not have rail caps. In the Slave canoe, the ends were formed of thin plank and in profile were almost upright and slightly curved. The stem line came out from the bottom in a sharp, almost angular curve and ascended with a slight sweep to a point about level with the gunwale amidships (in some, to within a few inches of the stem-head); from there a tumble-home carried it to the stem-head, which was short fore-and-aft and slightly crowned, the inboard end dropping vertically downward inside the gunwales. The headboards were under the gunwale ends. Inwales and outwales were both carried to the stems but the end lashings were quite short. There were no rail caps. The bark cover was lashed to the stem with an in-and-out stitch from side to side through holes in the plank. The sheer was brought up nearly to the top of the stem in a rather long, easy sweep beginning inboard at the endmost thwart.

The gunwale members in all these Athabascan canoes were quite light compared with their Eastern counterparts. A reinforcing strip of bark was placed under the outwales so as to hang down below them some four inches or so amidships and less toward the ends; this was sometimes decorated with a painted zig-zag stripe or with widely spaced circles. The end lashings of the gunwales were protected by short bark deck pieces inserted under the caps. The edges of these deck pieces were trimmed flush with the outboard edges of the caps, so that no *wulegessis* resulted.

In spruce-bark canoes, because the bark was stiff the ribs were spaced 6 to 8 inches, whereas in birch-bark canoes the ribs were spaced about as usual, 1 to 2 inches edge to edge. In the Dogrib and Slave canoes the ribs were without taper; in the Chipewyan there was usually a slight taper from the bottom to the gunwale end. The ends of the ribs were forced under the gunwales in the usual manner employed in the east, the gunwales being rectangular in cross-section, with the lower outboard corner beveled.

The thwarts were all parallel-sided, but tapered toward the ends, in elevation. The thwart ends were tenoned into the inner gunwale and usually had two holes in each end for the lashings.

In the bark cover the horizontal sewing was often over root battens. In many canoes rawhide was used in much of the lashing and sewing, and in the last-built bark canoes the end lashings of the gunwales were often protected by a decking formed of a small triangular sheet of metal, obtained from a large can and crimped along its edges so as to clamp the bark and main gunwales. When this metal deck-piece was used, the cap and outwale ended against the inboard edge of it.

For use in open water these canoes were often fitted with a blanket square-sail. The sapling serving as a temporary mast stood in a hole in the second thwart, and was stepped on a block, or board, pegged or lashed to the ribs.

The sheathing of all canoes of this class was of the same form—wide, short strakes amidships, narrower short strakes afore and abaft. The midship strakes were often quite short and their ends were over the longer end strakes. The end strakes were, of course, tapered toward the stems. The placing of the strakes was often irregular, with the result that the butts were somewhat staggered. Some canoes had four strakes to the length, but three appears to have been most common.

The large canoe was employed on the large lakes of the Mackenzie region; smaller canoes of the same general form, 14 to 16 feet in length and 30 to 40 inches in beam, were used on the large rivers and streams. In the smaller canoes of this class, the flare of the topsides was often less than in the larger craft. The Cree in this area, particularly to the south of Great Slave Lake, also employed the Athabascan form. This class of canoe, in general, appears to have been strongly affected by outside influence; consequently this description must be understood to cover existing canoes and models, not pure Athabascan canoe building.

The usual construction methods were employed in building this class of canoe; the stakes around the building frame were set vertically, and when the bark cover was lashed to the gunwale members (inwale and outwale together) the gunwales were spread and the thwarts inserted in their tenons. Skill was required in preshaping the gunwale members, which, as in the fur-trade canoes, had to be arched in sheer amidships to allow for the change in sheer caused by spreading the gunwales in construction. The building bed was also arched at midlength to allow for the lifting of the ends that occurred in spreading the gunwales with the bark cover attached.

A typical large Chipewyan canoe of this class was 21 feet 4 inches in overall length, 43 inches beam and 14 inches in depth amidships. A smaller Dogrib canoe of the same class was 14 feet 7 inches in overall length, 31¼ inches beam, and 11½ inches in depth. However, these smaller canoes appear to have been relatively uncommon, and the average large canoe was about 20 feet long.

## *Kayak-Form Canoe*

The kayak-form canoe was widely employed in the Northwest and was highly developed in both model and construction. It was essentially a portage and hunting craft, ranging in length from 12 to 18 feet and in beam from about 24 to 27 inches, with a depth between 9 and 12 inches. In areas where the kayak form was used as a family and cargo canoe, the length would be as great as 20 or 25 feet and the beam might reach 30 inches. Except in the family or cargo canoe, which had none, there was usually some decking at the ends, most of it forward. Some tribal groups built the kayak form with its greatest beam at midlength, but the most common form had its greatest beam abaft midlength and its greatest depth there likewise. Many of the kayak forms had unlike end profiles, so that there was a distinct bow in appearance as well as in fact.

There was much variety in end profile, and the canoes of each tribal group were usually identifiable by this means. The kayak-form bark canoes of the

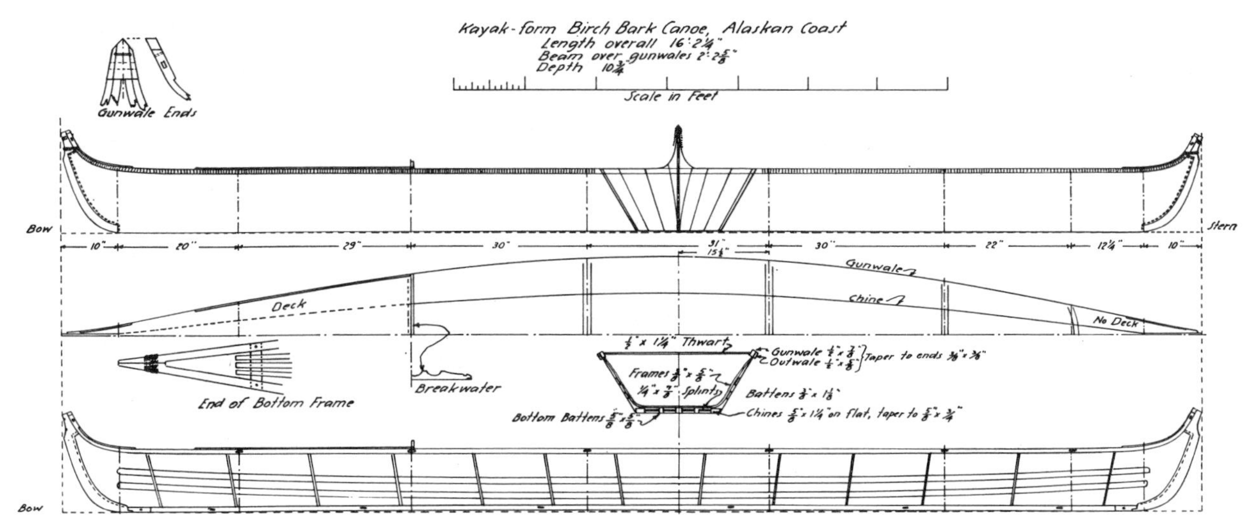


Figure 147

Eskimo Kayak-Form Birch-Bark Canoe From Alaskan Coast, with long foredeck batten-sewn to the gunwales, no afterdeck, and rigid bottom frame.

lower Yukon and neighboring streams had a short overhang, formed in a curved rake and alike or very nearly so, at bow and stern. On the upper Yukon and adjoining streams the canoes had much rake at both ends, the rake being straight from the bottom outward for some distance, then curving rather markedly. The bow rake was usually greatest, but the stern might be higher by one or two inches. The bottom was without rocker, being straight or even slightly hogged in most of these canoes. The sheer was straight to the point where the rake began, then rose in a easy sweep to the ends. The end decks on the upper Yukon canoes were short, those on lower Yukon canoes were much longer; on the latter the bow deck was nearly a third the length of the canoe, on the former about a fifth. In the Mackenzie Basin, the kayak-form canoes had a moderate rake, curved in profile, at bow and stern and a rather low stem-head; the depth at the stern was noticeably greater than at the bow, and the deck forward was commonly a little less than a fourth the length of the canoe. In these canoes the greatest beam in most cases was abaft midlength, and this was also true of the lower Yukon canoes. On the upper Yukon and in some of these canoes on the lower Mackenzie, the greatest beam was amidships and the depth at bow and stern were equal.

The variation in depth at bow and stern in some of the kayak-form canoes seems to have been related to the position of the greatest beam; when the beam was abaft the midlength, the greatest depth was aft, whereas when the greatest beam was amidships, the depth at the ends was equal. With the beam abaft midlength, the weight of the paddler trimmed the canoe by the stern somewhat, hence greater depth aft than forward was necessary to make the canoe run easily and turn readily in smooth water. In the sea kayaks of the eastern Eskimo, on the other hand, the depth and the draft were greatest forward, to bring them head to the sea when paddling ceased. The Alaskan sea kayaks were commonly of equal draft at bow and stern or might have a slightly greater draft aft than forward.

A third variation of the kayak form existed in British Columbia in early times, and apparently was employed by the Beaver, Nahane, and Sekani. It was an undecked bateau-shaped canoe having a fair sheer in a long sweep from end to end, the stem profiles were nearly straight, the ends were raked rather strongly, and the bow was somewhat higher than the stern. The beam was greatest slightly abaft midlength. It is estimated that canoes of this type, which has long been extinct and now can only be reconstructed from a model, were about 14 feet 8 inches long and 30 to 36 inches in beam, and probably were built of both spruce and birch.

The gunwales of the kayak-form canoes were formed by inwales and outwales; no caps were employed. In the Alaskan types and in the extinct British Columbia bateau variation, the gunwale lashings were contin-

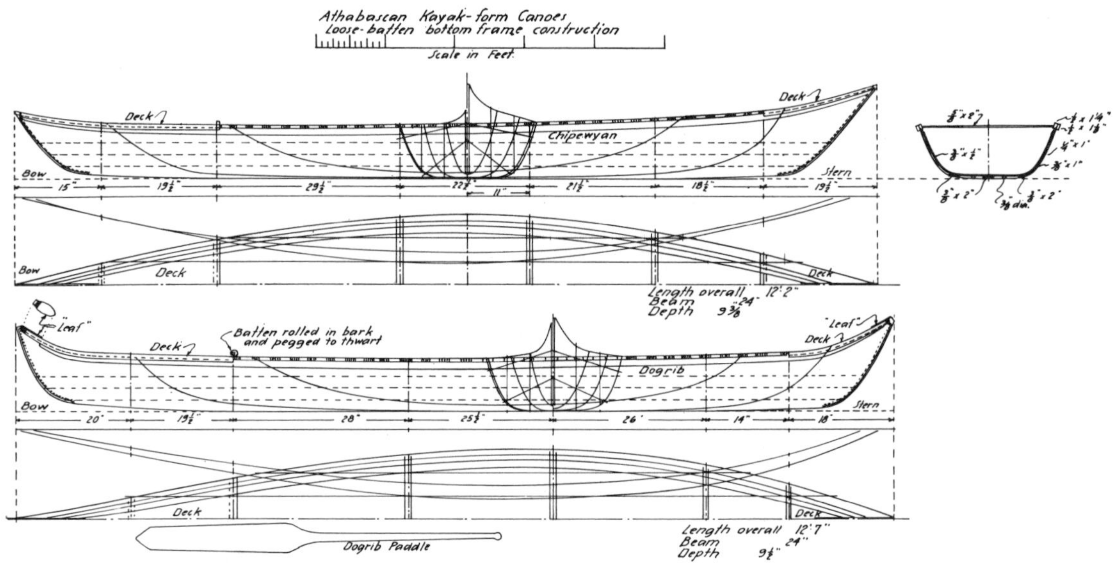


Figure 148

Athabascan Hunting Canoes of the Kayak Form, showing characteristic hull shape. These canoes were light, handy, and fast.

uous, but in the Mackenzie models the lashings were in groups. Inwales and outwales in all the kayak forms ran to the stem-pieces, which were plank-on-edge of a thickness that varied according to tribal practice. No headboards were employed. The gunwale members were rectangular in cross-section and were bent square with the flare of the sides. The ends sometimes were swelled and rounded, and in the bateau variation the gunwales, in cross section, appear to have been rounded. Six thwarts appear in most of the kayak forms but the Loucheux model had five and the bateau variation seems to have had but three.

Reinforcing bark was placed under the outwales in all Mackenzie Basin canoes, but not in the Alaskan or in the bateau variation. The ribs in all these canoes were small, usually about ½ inch square, and widely spaced, about 9 to 14 inches on centers. No ribs were placed in the rake of the ends. The ends of the ribs were chisel-pointed and were forced between the inwale and outwale, against the inside of the bark cover. In some canoes, however, the ribs near the ends of the canoe were forced into short splits on the underside of the inwale. The thwart ends might also be forced into short splits on the inside face of the inwales or might be tenoned there; in any case a single lashing was used at the thwart ends. Thwarts were parallel-sided in plan and slightly tapered toward the ends in elevation; no shoulders were used. In the bateau variation, a heavy thwart was placed directly under the middle thwart with its ends against the side battens, apparently to act as a spreader. Each end was notched over the side battens and was held by two lashings to the bottom crosspiece below it. This structure was probably made necessary by the fragile construction of this form of canoe. In all kayak forms there was no complete sheathing—the one, two, or three narrow battens to a side above the chine were held in place only by the sprung ribs (without lashings); in the bateau form, however, the side batten was lashed to each frame after the manner of of an Eskimo sea kayak.

The characteristic detail in the structure of the bark kayak-canoe, including the bateau variation, was the bottom framing. It was variously formed, according to tribal designation. The bottom framing was made up of five or six longitudinal battens (four in one extinct form of canoe). In the Yukon canoes six rectangular battens, all of about the same cross section, were used with the narrow edge outboard. These battens were held rigidly to form by thin crosspieces, or splints, about ¼ by 1 inch forced athwartships through short splits in the battens and pegged at the ends on the chine battens. The ends of the four inner longitudinals were cut off on the snye to bear on the inside face of the chine battens (in some instances they were cut short of this). The chine ends were beveled together or lashed to the sides of the stem-pieces.

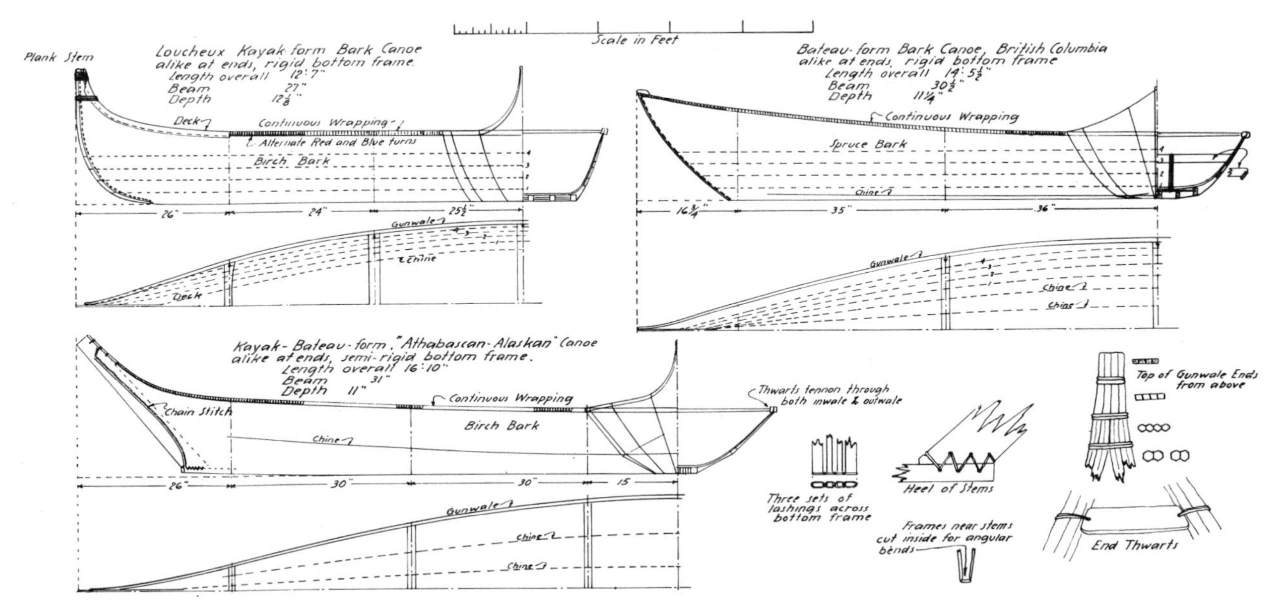


Figure 149

Extinct Forms of Canoes Reconstructed From Old Models, showing variations in the bottom frame construction and the effects of hull form. Dimensions are estimated from the sizes of canoes in the area of each example.

But in the Mackenzie form of canoe, the longitudinals had no cross-members and, like the side battens, were held in place by the pressure of the sprung ribs against the bark cover. There was a difference in the form of midsection: in the Yukon canoes the bottom athwartships was flat, but in the Mackenzie canoes there had to be some rounding there. At least one exception existed in the Mackenzie Basin, where the Loucheux canoe was formed on the Yukon bottom. Another is to be seen in an old model of an extinct Athabascan kayak form, which has only four longitudinals and chine members that are very wide and rounded only on the outboard face. Between the chine battens are two light rectangular battens. These are all held together by a few splints and by lashings which pass around each individual batten, thus serving both as lashing and spreader. This canoe has what is apparently a very narrow bottom compared to known types. In some of the Eskimo-built birch kayak forms, the separators between the bottom battens were rectangular blocks held in place by a thong threaded through two holes in each batten and block, to make a round turn, and tied at one chine.

In some bateau variations of the kayak-form canoe, the longitudinals were secured by crosspieces, the ends of which were tenoned into the inside faces of the chine battens. The three inner battens were below the cross pieces. As a result, their bottoms were slightly below the bottom of the chine members, so that in this canoe two chine lines show through the bark cover on each side of the canoe.

From tribe to tribe the method of building the kayak-form canoe varied somewhat, but generally the following procedure was employed. On a smooth, level piece of ground the form of the canoe was staked out in the usual manner, using a building frame, with the stakes sloped outward at the top to match the desired flare of the sides.

Stem and stern posts were shaped of cedar by charring and scraping. The gunwales were made in the same manner and were then lashed at the desired heights on the stakes. Next, the bark cover was formed, usually of two or more sheets sewn together. This was placed inside the stakes and the building frame was forced down on it and weighted with stones. The ends were then trimmed and the sides were gored, sewn, and trimmed to fit the gunwales, to which the bark was laced. The stem and the stern post were then placed and lashed to the gunwales and secured to the bark by lashing, in some instances through holes in the posts. The bark at this stage was usually quite dry and stiff and the gunwales could be freed from the side stakes.

The bottom frame, assembled before other construction had started, was hogged; the middle was placed on a log or block and the ends weighted. Hot water was often applied to set the bottom frame.

Next, the bark cover was thoroughly wetted with boiling water to make it pliable and elastic. The building frame and stones were now removed, the bottom frame was substituted, and its ends fastened or engaged to the heels of the stem and stern posts. The bottom frame was then forced flat and held there by stones. This stretched the bottom bark longitudinally, and increased the sheer slightly toward bow and stern. The hogged bottom frame was known as a "sliding bottom" by some Indians.

The transverse frames, or ribs, had been prebent in the usual manner before assembly began; a few of these were now put in place, the ends being forced under the gunwales between their outer faces and the bark, or into a grove on the underside of the gunwale. This stretched the bark transversely and vertically. Once the bark had been forced into form by this method, the remaining ribs were added, and these now held the hogged bottom down so that the weights or stones could be removed. The canoe was then turned over, the seams gummed, and any tears or rents repaired.

This method of building usually produced a slight hogging in both bottom and in the sheer amidships, but when the canoe was afloat and loaded the light, flexible construction caused the hogging to disappear. The kayak-form canoes of the Dènè tribe appear to be the most highly developed of all in this type.

The decks of many of the kayak-form canoes were made of a triangular sheet of bark cut with the grain of the bark running athwartships, so that it could be held in place by the curl of its edges, which clamped under the outwales, as well as by three lashings. The edges were curled by passing a glowing brand along them. One lashing was around the stem-head and two were at the inboard end of the deck, around inwale and outwale. If the inboard end of the deck was not on a thwart it was stiffened by a batten lashed on top of the deck athwartship, at the deck end, to serve as an exterior deck beam and breakwater in one. If the deck end was on a thwart, a batten might be pegged athwartship on top of the deck; sometimes this batten was rolled in a sheet of bark first. Another method was to use a small sheet of bark tightly rolled, with its free edge tucked under the deck end and secured at the ends of the roll by the deck-gunwale lashings there. Some canoes had their decks lashed over battens for a short distance along the gunwales. In some Mackensie Basin kayak forms, the end of the deck at the stem-head was protected by a small paddle- or leaf-shaped piece of bark placed under the lashing there and shaped to reach a little over onto the stem piece so as to seal the seam.

The fitting of the bark cover of the kayak-form canoes was not the same in all types. In the Mackenzie canoes the bottom, which might be in three, four, or five pieces sewn together, was alike on both sides; to it the side pieces were sewn at, or just above, the chines. The sides were made up of deep panels, five to nine to a side. There were no horizontal seams other than the one near the chines.

In some Yukon canoes, however, the bottom sheet was often made of three pieces and covered not only the bottom but also a portion, such as the after two-thirds, of one side. The forward portion of that side would then be covered by a single large panel or perhaps two, so that the horizontal seam on that side would run from the stem aft to the inboard end of the foredeck and would be just above the chine. On the opposite side a sheet would cover the bottom there and the bow topside from the stem aft for a short way. Deep panels would then cover the rest of that side to the stern, so that the horizontal seam there began forward at the sheer, some feet abaft the bow, and swept downward in a gentle curve to near the chine and then ran aft to the stern in a long sheered line just above the turn of the bilge, rising slightly as it neared the stern. Hence the foremost of the panels on that side was nearly triangular and the others were nearly rectangular. Inside, at the chine, was placed a reinforcing strip of bark wide enough to reach 3 inches beyond both sides of each chine longitudinal and running the length of the bottom; or if a seam near the chine permitted, the side and bottom pieces were overlapped. As has been noted, in the Yukon canoes a reinforcing piece at the outwale was not used, but was in the Mackenzie canoes; it extended down the side about 3 inches below the underside of the outwale amidships and ran to the ends of the canoe, or nearly so, tapering with the outwales to a width of about 1½ inches at bow and stern. In these canoes much of the lashing at stem and stern was double-thong; the longitudinal sewing was often over a batten in the usual spiral stitch, and a simple spiral stitch was also used to join the panels, although in-and-out stitching might also be seen in some canoes.

In many of the kayak-form canoes two ribs often stood noticeably close together amidships, and the rest stood parallel to the rake of the end on their side,

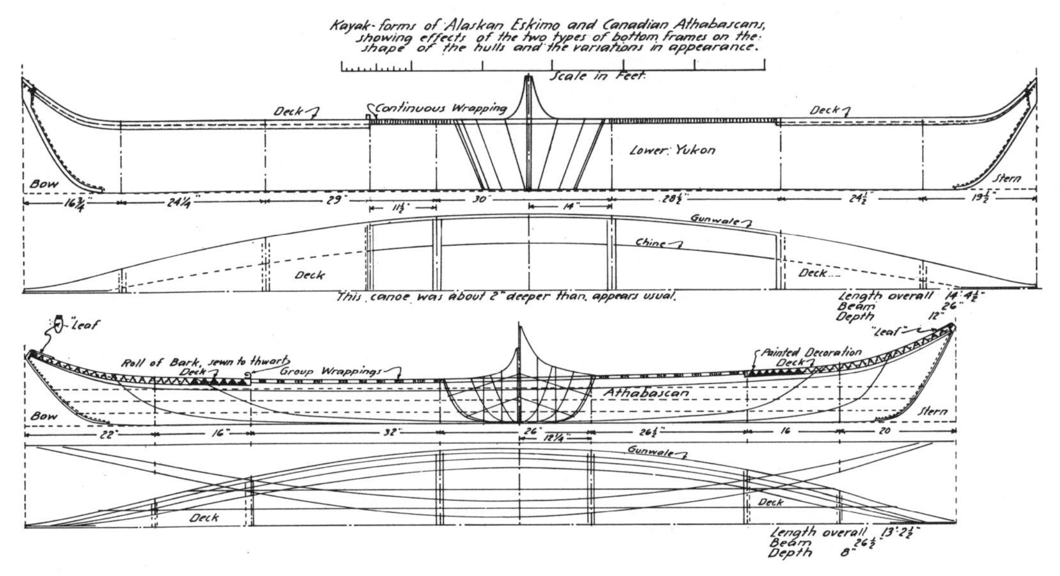


Figure 150

Kayak-Form Canoes of the Alaskan Eskimos and Canadian Athabascan Indians: chine form of Eskimo birch-bark canoe (above) and the dish-sectioned form of the Canadian Athabascans.

respectively, of the middle ribs. However, not all these canoes had such double ribs; some were framed out in the usual manner, with the ribs widely spaced and canted toward their respective ends of the hull, away from the midship of the canoe.

In most of these canoes the paddler sat on a sheet of bark secured on the bottom; this was held in place by one or two false ribs having their ends under the inner gunwales and their middle forced down against the bark on the bottom framework. In place of bark, some Eskimo builders of the type used thin splints of wood laced together by two or three lines of double-thong stitching athwartships, which was passed through two holes in each splint. This might be loose or held in place by a false frame.

The paddle was single-bladed and the same as that used with the second class of Mackenzie Basin canoe (fig. 151). The blade was parallel-sided with the point formed in a short straight-sided V-form; The blade of Yukon paddles was often taper-sided toward the point, which was a rounded V. Other variations in blade form existed, however, and the narrow leaf-shaped blade was used in some areas in Alaska. In the Mackenzie paddles the handle ended in a knob, but in Alaskan versions it ended in a cross-grip like those of paddles used with some Alaskan sea kayaks. The Eskimo double-blade paddle was used with the kayak-form canoe by some paddlers; Hearne mentions its use.

Some of the kayak-form canoes were decorated; in Alaska this decoration often took the form of a line of colored beads sewn along each side of the afterdeck at the gunwale, or it consisted of a few oval panels of red, blue, or black paint along the sides or centerline of the afterdeck. In some Mackenzie kayak forms the decks were painted in various designs; a rather common one seems to have been two or more bands of paint around the deck edges, along the gunwales, ending at bow and stern with a full round sweep. Painted disk designs appeared on some of the large Algonkin-Ojibway canoes of the second type.

A number of kayak forms became extinct before any accurate, detailed records of their shape and construction had been made; models of some of these canoes exist but are not to scale and are untrustworthy as to detail, since they are often simplified. One model of the extinct British Columbia bateau form, for example, showed but three longitudinals in the bottom, though the probable size of the canoe undoubtedly would have required a greater number. On the other hand, the model may have represented a spruce-bark canoe constructed for temporary use, in which case a simplified construction might have been employed. One can only speculate which it was. Models of some kayak-form Yukon canoes show the decks lashed to the gunwales with a very coarse spiral stitch not recorded for any of the observed full-size canoes; thus it may be a model-maker's

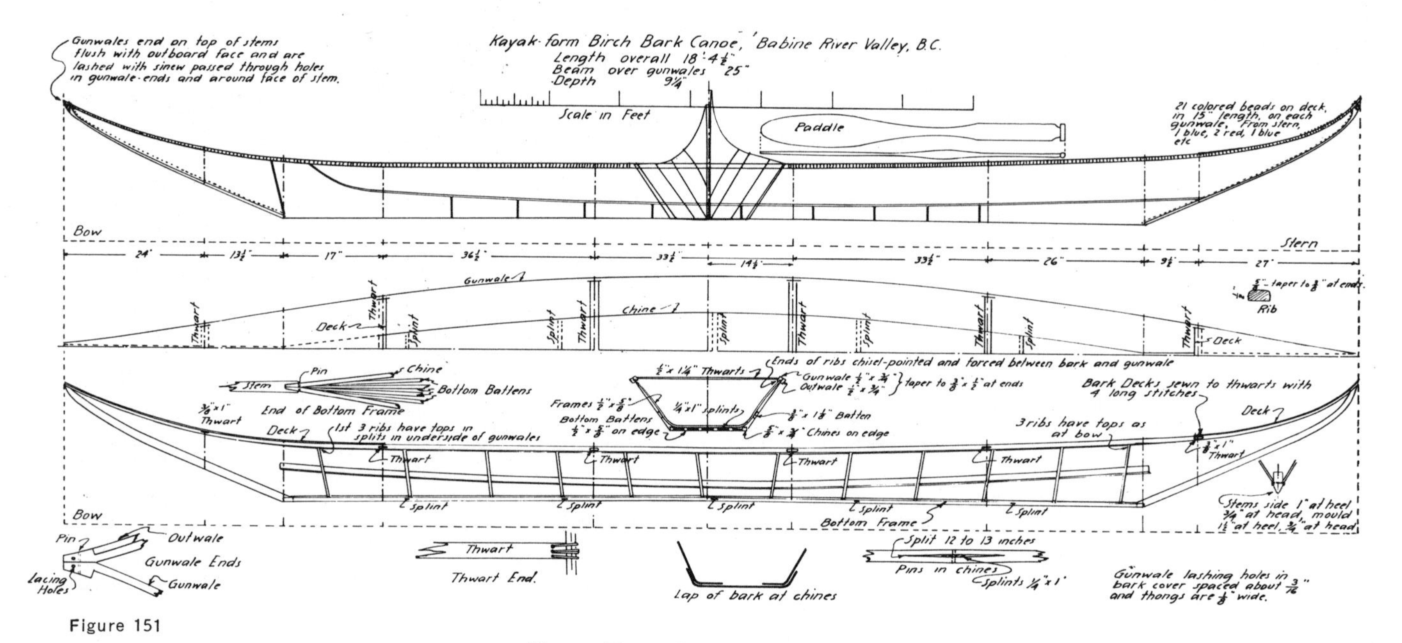


Figure 151

KAYAK-FORM CANOE OF BRITISH COLUMBIA and upper Yukon valley. Shows hogged bottom, usual in the type with a rigid bottom frame, which becomes straight or cambered when canoe is afloat and manned. Original in the Museum of the American Indian, New York.

method of securing the decking firmly rather than an actual practice used on full-size canoes.

It now remains only to give short descriptions of the various kayak-form canoes that have been observed.

The ends of the Eskimo-built canoes of the lower Yukon had a short rake, the heel of the end profile breaking out of the bottom line at a slight angle and sweeping upward and outward in a gentle curve, often becoming almost straight near the stem head. The bow and stern were nearly the same height, the bow being a little higher, about half the midship depth above the sheer amidships. The sheer at each end was almost dead straight until within a few inches of the end; thence it swept up sharply with the inner gunwale ends, broadened, resting on the inboard side of the stem piece. The extreme ends of the inner gunwales were thus at the extreme stem-head. The stem-pieces were of plank, the cutwater portion outside the bark cover being sharpened the full height of the stems. These lower Yukon canoes had three side battens above the chine piece, but not all ran the full length in one piece; some were in two, in which case the ends merely ran past one another for a few rib-spaces and were neither butted nor lapped. The forward deck extended nearly one-third the canoe's length and had a batten across the inboard deck-end; the after deck reached to the after thwart. Adney's model of such a canoe shows the after deck lashed to the gunwales with spiral turns over a batten along the deck edges and finished toward the stern with chain stitching, but no such arrangement was seen in any full-sized canoe.

The form of these Eskimo-built canoes was nearly that of a double-ended flat-bottom skiff; the bottom being flat athwartships and without rocker fore-and-aft. The sides flared and were nearly straight. The turn of the bilge was quite sharp, the chine having a very short radius. In plan, the canoe showed no hollow in the ends, which were convex both at gunwale and on the bottom frame. In some of the full-sized canoes inspected there appeared to be a slight hog ranging from ¼ to ⅜ inch in the bottom, but there was no evidence to suggest that this was a result of the drying and shrinkage of the canoe structure with age. Hearne's drawing of a kayak-form canoe shows an impossible amount of hog in the bottom, and he indicates that some hog was intentional in building. This would disappear when the canoe was loaded afloat owing to the light and flexible structure, and it is evident that the builders usually sought to have the bottom slightly hogged.

Figure 152

CONSTRUCTION OF KAYAK-FORM CANOE of the lower Yukon, showing rigid bottom frame. (*Smithsonian Institution photo.*)

The kayak-form canoes of the lower Yukon and neighboring streams all appear to have been small canoes "tailored" to their owner's weight and height: 14 to 15 feet in overall length, 2 to 2¼ feet wide, and 10 to 12 inches deep. The bottom frame was from 12 to 14 inches wide amidships.

The kayak-form canoes of the upper Yukon Valley and those used in northern British Columbia and in Yukon Territory had ends with a long rake that came up in a straight line from an angular break at the bottom line to the height of the sheer amidships or thereabouts; there a gradual upward curve continued to the stem-head. The stern was 2 inches or so higher than the bow, and the rake of the latter was usually about an equal distance longer than that of the stern. The sheer was nearly straight, with only about 2 inches of sag from the heel of the stem to that of the stern. Beyond the heels, the sheer lifted in a fair sweep, becoming sharper toward the ends, where the broadened inwales were secured on top of the stem and stern pieces. There was no rocker in the bottom, and some examples showed as much as ⅜ inch of hog amidships. The bottom was flat athwartships and the almost straight sides flared a good deal. The turn of the bilge was on a very small radius and in some canoes appeared angular. The bow deck was usually just under one-fifth the length of the canoe. Most of the canoes did not have a stern deck, at least on the Yukon headwaters, but on those that did, it was about one-ninth the length of the canoe. The greatest beam was abaft amidships and the canoe was usually about 1½ inches deeper at the heel of the sternpost than at the heel of the stem. In plan, the ends (at gunwale and bottom frame) were convex; the gunwale ends alone might appear slightly hollow close to the posts in some examples. The canoes in Alaska and British Columbia and at the headwaters of the Yukon had a rigid bottom structure, with the splint spreaders usually numbering five.

The 1-man hunting canoes were commonly 18 to 19 feet long, 24 to 27 inches beam, and usually 10 to 11 inches deep amidships. The single example of a family or cargo kayak-form that has been measured from this area was 20 feet 1 inch overall and 30¼ inches beam over the gunwales. It was 18 inches wide on the bottom frame, 13 inches deep amidships, 14 inches deep at heel of stem, and 16 inches at heel of stempost. Height of the stem was 29 inches, of the stern 30½ inches, the after rake was 38 inches, and the fore rake 40½ inches. The canoe had no decks and was rather sharp-ended.

The kayak-form canoe of the Athabascan Loucheux had a rigid bottom-frame; the bottom was flat athwartships and it had no fore-and-aft rocker. The sides were flaring and slightly curved. Both ends were alike, and the canoe was unusual in having only five thwarts, with one amidships. The stem was short in rake and curved; the stem profile came out of the bottom line in a fair, quick curve which became vertical at a height of little more than two-thirds the depth amidships of the canoe. The height of the stem was almost twice the midship depth. Between the end thwarts the sheer was straight, thence it swept upward in a gradually sharpening curve to the inboard stems; the inwale ends stood vertical on the face of the stem, with their ends brought to the top of the stem-head. The stem-pieces were of unusually thick plank, with the head broadened and the cutwater part outside the bark cover sharpened until near the head, where it gradually became as wide as inboard. The gunwales were lashed with continuous turns, as in the Alaskan canoes. In plan, the gunwales and bottom frame were full-ended and convex. These canoes were decked equally at both ends. The deck extended inboard far enough to just cover the end thwart, to which, in the example seen, it was lashed with four simple in-and-out passes of rawhide thong. The chine-pieces of the bottom were lashed to the sides of the stem-pieces. The covering was birch bark. Two battens on each side were employed with the usual six longitudinals in the bottom frame. These canoes were well-built and their ends resemble those of the seagoing kayaks used at the mouth of the Mackenzie, but these for at least the last 70 years of their use were round-bottomed. The Loucheux canoes were small, usually about 15 feet long, 30 inches wide, and about 12 inches deep amidships.

The Chipewyan kayak-form canoe was of loose-batten bottom frame construction, with its beam well aft of amidships. Its bottom was slightly rounded athwartships, with a slight rocker fore-and-aft; the sides flared outward and were nearly straight; and the turn of the bilge was almost angular. The bow and stern were of the same general shape; the end profile came out of the bottom line with a quick hard curve and then fell outboard in a long sweep that gradually straightened near the head. The rakes were short, however, and the stem was noticeably lower than the stern, the difference being as much as 6 inches in some canoes. The sheer was nearly straight to the end thwarts and thence it curved up in an easy sweep to the ends of the canoe. The canoes were markedly deeper at the stern than at the bow; the difference being as much as 1½ inches in some examples.

This kayak-form was very sharp-ended; the gunwales in plan often showed a slight hollow and the chine members came to the posts in an almost straight V. As a result, the end ribs were often intentionally "broken" to form a narrow-based, angular U. In some Eskimo-built kayak forms, a similar result in hull section was obtained in the endmost frames by stepping short struts in splits, or tenons, on top of the chine members and on the underside of the main gunwales. This construction was occasionally found in some of the lower Yukon kayak forms. The Chipewyan kayak forms were decked at both ends. The fore deck was slightly more than one-fourth the length of the canoe and extended inboard to the second thwart; the after deck was about one-tenth, and came inboard to the end thwart. No breakwater batten or bark was employed. There were two battens on the sides, above the bilges.

The gunwale wrappings were in groups. The bark cover was not folded over the top of the inner gunwale but, as usual in the Northwest canoes, was trimmed evenly with the top of the inwale and outwale. Reinforcing bark along the gunwales extended downward about 1½ inches below the bottom of the outwales amidships and about 1 inch at the ends. Of the bottom longitudinals, the keel and chine-pieces were roughly rectangular in cross-section, laid on the flat, and the intermediate two battens were round; the ends of the keel piece were merely butted against the stems, no lashing being used. The stem piece was thick plank and was sharpened outside the bark cover to form a cutwater. The stem lashing was of the usual two-thong form, and a batten was used in the longitudinal seams of the bark cover. The thwarts, six in number, were tenoned through both inwale and outwale and pegged between them. No thwart lashings were used. The decks often were not lashed into place, being held only by the curling of the edges of the bark sheets.

This canoe was a very good one; it was light and was fitted to the owner's build. In size it would be between 12 and 14 feet long and 20 and 24 inches wide over the gunwales, and the width of bottom

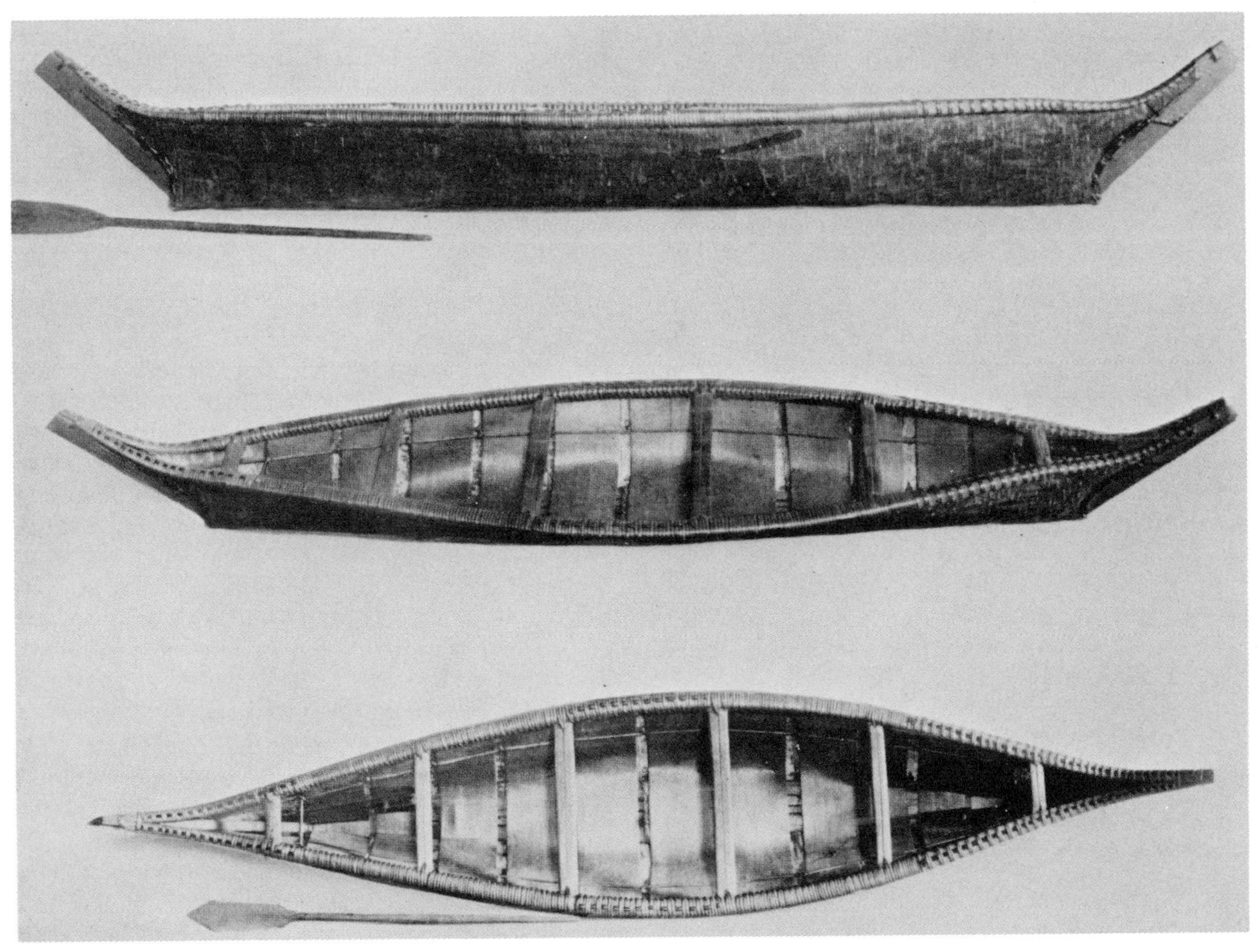

Figure 153

MODEL OF AN EXTINCT FORM OF BIRCH-BARK CANOE, Athabascan type, of British Columbia. In Peabody Museum, Harvard University, Cambridge, Mass.; entered in the museum catalog as of 1849.

over the chine members amidships would be 11 to 12 inches. The greatest beam would occur 7 to 8¼ feet abaft the stem. The depth at heel of stem would be 8½ to 9½ inches and at heel of stern, 10 to 11 inches. The amount of bottom rocker would be between ¾ and 1 inch, with its low point about amidships. The cover was usually birch bark, but sometimes spruce bark was used.

Another kayak-form canoe of unknown tribal designation from the Mackenzie Basin was 13 feet 3 inches long, 27 inches beam over the gunwales, 8½ inches deep amidships, 8¾ inches deep at heel of stem, 10 inches deep at the aftermost thwart, and with about ⅜ inch of rocker in the forebody, none in the afterbody. The greatest beam occurred 7 feet 2 inches from the stem. The width amidships of the bottom framework of loose longitudinals was 13 inches. The length of the rake foreward was 12 inches and aft, 12 inches. The fore deck extended inboard to the second thwart, where a roll of bark formed a breakwater. The after deck extended inboard to the aftermost thwart. Between the end thwarts the sheer was practically straight; at the ends it rose gently, becoming almost a straight line as it came to the stem and stern, and without the usual upward hook in the ends of the gunwales.

This was a very light and well-built canoe with a birch-bark cover, a slightly rounded bottom athwartships, slack bilge, and flaring sides showing some curve in cross-section. The ends were rather sharp, the gunwales coming in to them almost straight, in plan, as did the chine members. The stem and stern pieces were of wide plank sharpened along their outboard edge outside the bark cover, for their whole height, to form cutwaters. The stem and stern profiles were about the same as those of the Chipewyan canoes.

An old model in the Peabody Museum of an undecked kayak-form canoe of Athabascan construction represents a high-ended canoe having ends with a slight rake and a straight cutwater. This form of canoe has long been extinct, and no description of an actual canoe of the form exists. Judging by the model it had a very narrow flat-bottom and rounded flaring sides.

The extinct bateau variant has already been described (pp. 159–161); it might be considered a primitive form of the kayak-form bark canoes, were it not that no intermediate type, between the bateau and the later and highly developed bark kayak-form, has been found; as a result, any such statement can be no more than speculation.

## *Sturgeon-Nose Canoe*

In southern British Columbia and in northern Washington, the ram-ended or sturgeon-nose canoes were built. These were the canoes of the Kutenai, also spelled "Kootenay," and of the Salish tribal groups. Used on rivers and lakes, they were constructed of the bark of birch, spruce, fir, white pine, or balsam, whichever was available at the building site. Wherever possible a panel of birch bark was worked in along the whole length of the gunwales. The hull form of these canoes varied somewhat, perhaps by decision of the builder, or perhaps by local tribal custom. The ends were formed with a marked "ram," the stem profiles running down and out to the "nose" in a straight or nearly straight line. In some examples the stem profiles were in a hollow curve, starting down from the gunwales rather steeply and then curving outward more gently to the nose. Most examples had a bottom that was straight or slightly hogged, while those with the hollow curve in the ram often had a slight rocker. It is believed that the intention was always to have the bottom straight but that in construction the center of the canoe lifted somewhat, thus showing a slight hog in the bottom line. The effects of loading and use on the light and flexible structure of these canoes would cause the bottom to rocker and the outboard ends to lift, thus causing the hollow in the ram profiles. These effects of loading are confirmed by tests with models of this form of canoe.

The midsection was usually quite round, almost U-shaped, on the bottom, but some canoes showed the bottom slightly flattened and the sides flared out somewhat. Toward the ends, the U-shape became marked, and near the gunwale ends the sides of the U fell inboard slightly as they came to the gunwales, the bottom of the U having a hard turn. In plan, the gunwales approached the stems without hollow, being nearly straight or even slightly convex. The ram was long and sharp in its lower level lines and this, with the form of midsection, made this model a fast-paddling canoe, though rather unstable. Most of these canoes had but one thwart, placed at mid-length, but some have been found with three thwarts and a thong tie across the gunwales, close to the stems, as well.

No stem-pieces were used; the bark ends were closed by two outside battens, one on each side, whose heads were carried some 3 inches above the gunwales. A cutwater batten was placed over the edges of the bark between the battens, and the three were lashed together, with the bark, by a coarse spiral wrapping or by group ties. The bark cover was not sheathed inside; instead, six battens, ⅜ by 1½ inches, were placed on each side of the keel piece, which measured about ½ by 3 inches and tapered toward the ends. The battens, widely spaced, ran well into the ram ends, and were held in place, like sheathing, by the pressure of the ribs. The ribs, spaced 8 to 12 inches on centers, were often split saplings; sometimes they were shaped to approximately ¼ by ¾ inch. The batten nearest the gunwale on each side was lashed to every rib. In some canoes the heads of the ribs were brought up between the inwale and outwale, inside the bark cover, with their ends against the cap. The stitching of the longitudinal seam of the topside panel was passed around these frames and so helped to secure them. In one example, the ribs were passed through the bark cover just below the horizontal seam of the topside panel; there a turn of the stitching was passed around each rib; then the rib was brought inboard again in the seam by being passed between the edges of the bark cover and the panel. In many canoes there were no ribs in the ram ends, but this was not universal practice;

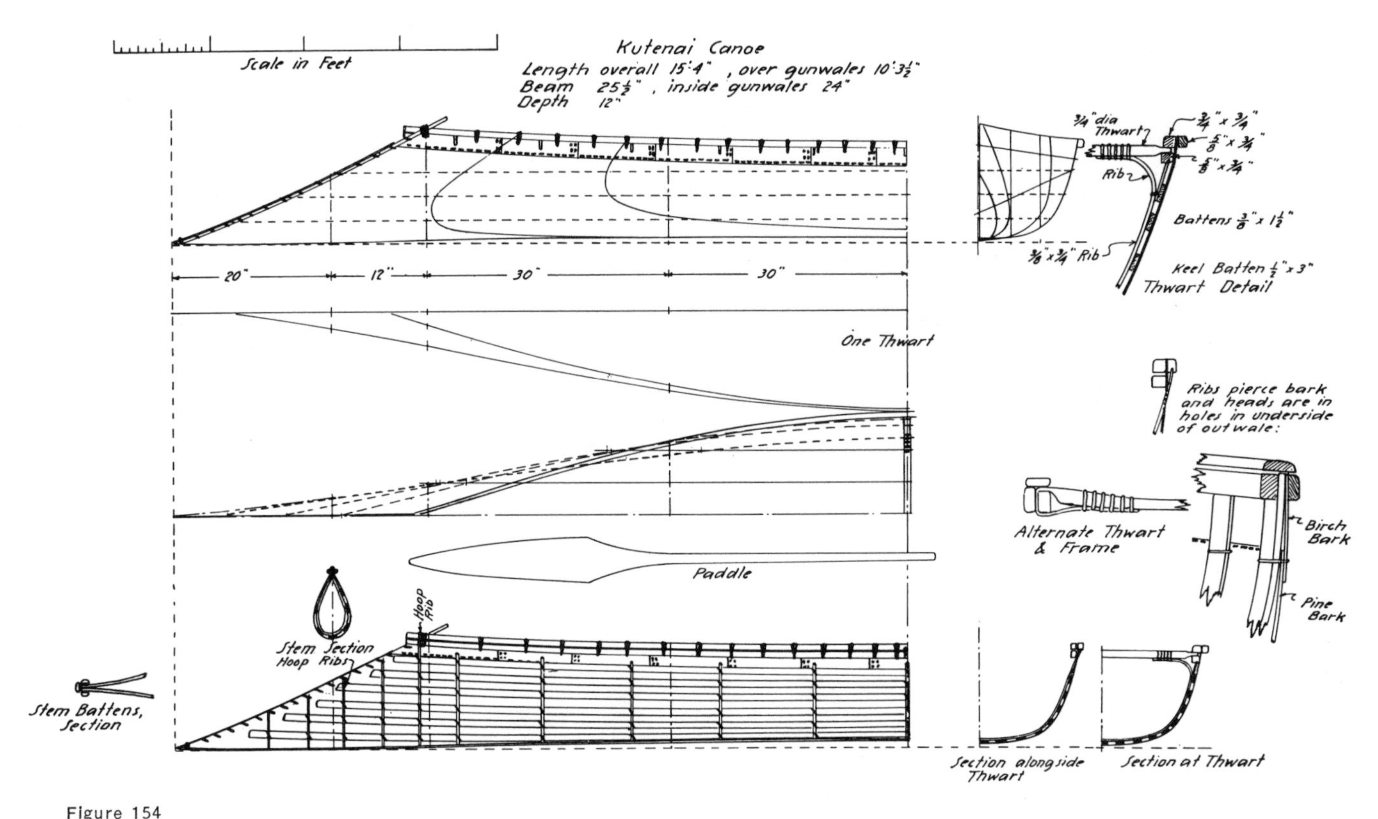


Figure 154

BARK CANOE OF THE KUTENAI AND SHUSWAP, about average in size and proportion. Original in the Museum of the American Indian, New York.

small light ribs were sometimes placed there, with their heads caught in the closure lashing of the end.

The canoes had 3-part gunwales consisting of inwale, outwale, and cap, but in many the arrangement of these was such that this nomenclature is misleading. In the latter construction, a lower inwale was used, as in the above drawing; rather small in cross section, it was almost square, with rounded edges. The rib ends, after passing through slits in the bark cover below the lower inwale, continued upward past it, outside the bark cover. Above the lower inwale and inside the bark cover was a larger upper inwale; this was flat on the outboard and bottom sides, the top and inboard sides being rounded into one another. The outwale, roughly rectangular in cross section, clamped the bark cover and heads of the ribs between it and the upper inwale. The ribs and bark were trimmed off flush with the tops of the outwale and upper inwale. The thwart amidships was caught, at the ends, between the lower and upper inwales. The gunwale members and bark cover were secured by group lashings of small extent and rather widely spaced.

The methods of fitting the thwarts differed in this class of canoe, and it cannot be determined with certainty whether this variation was tribal or the choice of the individual builder. In canoes having the lower inwale arrangement there was but one thwart amidships. As has been said, its ends were caught between the upper and lower inwales. Directly beneath it was a rib whose head was not brought up outside the bark cover but, after being secured to the uppermost sheathing batten, was brought around inboard in a quick hard turn and secured along the underside of the thwart with a close spiral lashing. Under this rib at the topmost batten was secured a short false rib head by forcing the beveled foot of the false rib between the batten and the true rib, after lashing; the head of the false rib was then brought up through and outside the bark cover in the customary manner, or it might be forced under the lower inwale, inside the bark cover. In this construction, the endmost ribs were at the gunwale ends, and the heads of these were lashed to the stem battens outside the gunwale ends, on the outside of the bark cover.

Figure 155

OJIBWAY CANOE CONSTRUCTION.
(See pp. 122–131.)

Peeling bark.

Staking out bark.

Assembling bark over on building site.

*(Canadian Geological Survey photos.)*

Making root thongs.

Setting ribs inside bark cover with a mallet.

Fitting gunwale caps on new canoe.

In canoes having the usual gunwales of inwale, outwale, and cap, the inwale and outwale were roughly rectangular, with their top sides horizontal, and the cap, very small and light, was flat on the bottom and rounded on top. In this construction, the rib heads usually were clamped between the inwale and outwale, inside the bark cover.

The ribs of the ends were lighter than those of the main body and more closely spaced, say 2 or 3 inches apart. These began about 8 or 9 inches inboard of the gunwale ends; the heads did not reach the gunwales, but instead were caught in the horizontal seam of the side panel and then cut off. Usually three ribs were so fitted. The rest of the end ribs, usually eight in number, either had their heads caught in the stem lashings or were made up as hoops with the heads overlapped and lashed together, the ribs being placed so that the overlap came to one side or the other of the canoe. Each hoop was usually caught by a turn in the end-closure lashing.

To strengthen the ram, the lower ends of the three stem battens were lashed to the extremities of the inside keel-piece, which was brought through the bark cover at this point. The opening resulting from this was sealed with gum or pitch. Minor variations in construction have been noted in the canoes exhibited in museums; in one, for example, only every fourth rib was caught in the topside panel stitching.

In canoes having the usual arrangement of gunwale members, with the cap over the ends of the ribs, the ends of the thwart were sometimes carried some 6 to 8 inches beyond the gunwales, at each end, and much reduced in thickness by cutting away about half the depth of the thwart. This part was then wrapped tightly around the inwale, brought inboard along the underside of the thwart, and there lashed. Examples show that the amount of end brought inboard under the thwart varied with the builder. It should be added that the thwarts were usually no more than barked saplings and were obviously installed in the canoe when green and treated with hot water so they would not break when wrapped around the inwales. In canoes having three thwarts, all were fitted in this manner, but often the thwarts on each side of the middle were also wrapped in a long spiral with a thong whose ends were tied to each gunwale. In 3-thwart canoes, there was commonly a cross tie, located roughly 12 inches from the gunwale ends and consisting of three or more turns of cord, or thong, around the gunwale members on each side and athwartships, secured by turns of the ends around the cross tie. In one canoe there was a thwart amidships and one at one end, about halfway between the middle thwart and the gunwale ends; at the other end were two cross ties, one replacing the thwart and another a foot inboard of the ends of the gunwales. In this canoe the ribs at the gunwale ends were hoops and there were only three hoop ribs in the ram ends.

One canoe, from Stevens County, Washington, had a peculiar double framing. The sheathing battens, instead of being on the inside of the bark cover, rested on light ribs, spaced about 6 inches apart, that ran only far enough up the sides to have their ends caught in the stitching at the bottom of the topside birch-bark panel along the gunwales. The longitudinal battens were placed inside these, with the batten nearest the gunwale lashed to the light ribs. Inside these battens and spaced about a foot apart was another set of ribs whose heads were secured between the inwale and outwale inside the bark cover; each of these inside ribs was also lashed to the uppermost batten. Only the keel batten was under the small ribs. The thwart ends were wrapped around the main gunwale members, and the stem battens were secured to the birch topside panels by but one group lashing, near the gunwales. The bottom cover was stiff pine bark.

The topside panel of birch bark was placed in these canoes so that its grain was horizontal instead of the usual vertical. Presumably this was done as a maintenance solution: the panel was much easier to repair or replace than the bottom bark; and by having the panel placed in this weak mode, it would split before the bottom bark if too much pressure were brought on the framework in loading.

These canoes paddled well in strong winds and in smooth water, and worked quietly in the marshes where they were much used. Canvas canoes of the same model replaced the bark canoes, indicating that the model was suitable for its locality and use. These sturgeon-nose canoes were so different from other North American bark canoes that they have been the subject of much speculation, particularly since ram-ended canoes, though of different construction, existed in Asia.

The size of the Kutenai-Salish sturgeon-nose canoes varied; the most common size appears to have been between 14 and 20 feet over the ends of the rams, 24 to 28 inches beam, and with a depth ranging from 12 to 13 inches amidships and from 14½ to 17½ inches at the ends of the gunwales. However, records exist that show rather large canoes were built on this model, 24 feet over the rams, 48 inches beam and 24 inches depth.

The building methods of this type of canoe have never been reported. Probably some kind of a rough building frame was used. Perhaps this was comprised of a couple of the battens and the keel piece, weighted with stones. The building bed was probably level. The main gunwale members were apparently made up temporarily and the bark cover shaped and staked out. From that point the work may have followed the usual canoe-building practices except that the ends could not be closed until the framing there was complete, otherwise it would have been impossible to fasten the small ribs in the rams. The structure of these canoes appears to have been almost entirely cedar, except for the bark and lacings which, in some instances, were partly some bark fiber as well as roots. In general, the construction of this class of canoe did not match in quality that of the other bark canoes of the Northwest.

Figure 156

INDIANS WITH CANOE at Alert Bay, on Cormorant Island, B.C. Dugout shown.

# *Chapter Seven*

# ARCTIC SKIN BOATS

*Howard I. Chapelle*

Among the three primitive watercraft of North America (the others being the dugout and the bark canoe of the American Indians), the Arctic skin boats of the Eskimos are remarkable for effective design and construction obtained under conditions in which building materials are both scarce and limited in selection. The Arctic skin boat is almost entirely to be found in the North American Arctic from Bering Sea to the East Coast of Greenland. In Russian Siberia, only in a small area of the eastern Arctic lands adjacent to the North American continent are any employed.

These craft, an important and necessary factor in the hunting lives of most Eskimo tribal groups, have long attracted the attention of explorers and ethnologists, and many specimens have been deposited in American and European museums. Like bark canoes, they have unfortunately proved difficult to preserve under conditions of museum exhibit. As a result, examples of once numerous types have become so damaged that they no longer give an accurate impression of their original form and appearance, and some have so deteriorated that they have had to be destroyed. Among the latter may have been examples of types long since out of use. One such type was represented by a single kayak, now destroyed; as a result this form has become extinct, and only a poor scale model remains to give a highly unsatisfactory representation of it.

In 1946 the late Vilhjalmur Stefansson, who was then projecting his *Encyclopedia Arctica*, asked me to prepare for it a technical article on the Arctic skin boat. The decision of the sponsors to discontinue the publication, after the first volume had appeared, prevented appearance of the article, but in 1958, through the kindness of Dr. Stefansson, it was returned to the author for publication by the U.S. National Museum. I have since revised and added to it, after receiving criticisms and suggestions from Henry B. Collins, of the Smithsonian's Bureau of American Ethnology, from John Heath, and from other authorities.*

The object of the study, as will be seen, was to measure the skin boats and to make scale drawings that would permit the construction of a replica exact in details of appearance, form, construction, and also in working behavior. Special regard was given to the diversity of types with respect to hull form and construction methods; but questions of ethnic trends, tribal migrations, and such matters, being outside the scope of the study, were not considered. Wherever possible, full-size craft were used as the source, but where only fragments existed, these had to be supplemented by reference to and interpretation of models of the same type.

In spite of the difficulty of locating skin boats of some Arctic areas, examples of most of those mentioned by explorers since 1875 have been found and recorded, so that, as far as possible, every distinctive tribal type of Arctic skin boat which in 1946 was represented by museum exhibits in the eastern United States is represented in plans here.

With the material available it was not possible, of course, to explore all the individual types and forms in full; hence, the geographical range of a type can be stated only approximately, owing to the overlapping of tribal groups and the almost constant

*For their aid to him the author takes this occasion to extend particular thanks. He also thanks his Smithsonian Institution colleagues in the Division of Ethnology, U.S. National Museum; members of the staffs of The American Museum of Natural History and The Museum of the American Indian in New York, of the Peabody Museum at Harvard, and of the Stefansson Library at Dartmouth; and the Washington State Historical Society and Museum, and others in the Northwest who gave both aid and encouragement.

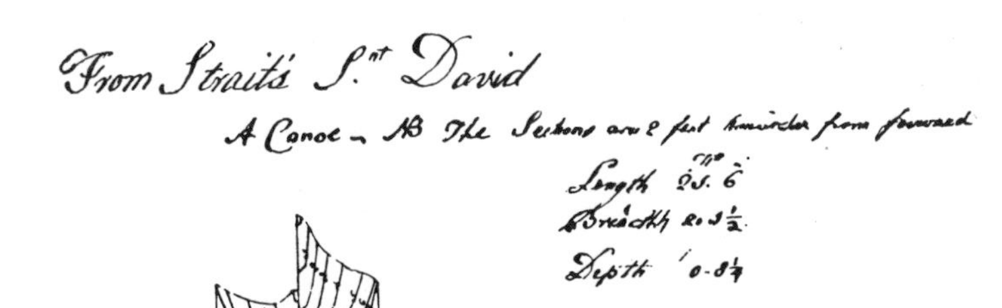


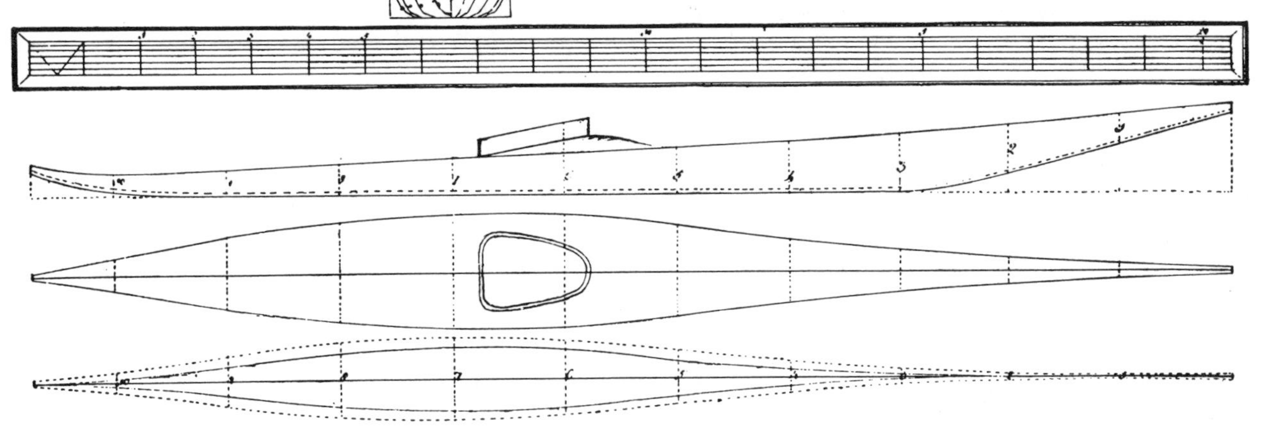

Figure 157

EIGHTEENTH-CENTURY LINES DRAWING of a kayak, from Labrador or southern Baffin Island (according to Dr. Kaj Birket-Smith of the Danish National Museum). Note the long stem that is characteristic of present day kayaks from Labrador. The lettering apparently reads:

From Strait's S.nt David
A Canoe—N.B. The sections are 2 feet asunder from forward
Length 21′–6″
Breadth 2′–1½″
Depth 0′–8¼″

(*Courtesy National Maritime Museum, Greenwich, England.*)

migratory movement of the Eskimo. Originally the 2- and 3-cockpit kayaks of Russian colonial Alaska had been omitted as being probably the results of Russian influence. John Heath, however, believing attention should be given to this type, has very kindly prepared for me a fine draught of such a kayak, or "baidarka" (other spellings of this name are common); this is shown on page 197.

Although the scale drawings accurately represent the form and details of construction, they necessarily idealize somewhat the primitive boat design. Also, in showing the hull-form, the usual method of projecting the "lines" of the hull was discarded as unsuitable. Instead structural features have been emphasized, with the result that "round"-bottom kayaks appear as multichine hulls, as they properly are. In view of the fluid state of design in Eskimo craft it is obvious that the examples shown represent the stage of development at the given date, though the alteration in most designs has been so gradual that the representation could serve to illustrate with reasonable accuracy a tribal or area type for a decade or more.

The Eskimos have produced two types of skin boats that have proved remarkably efficient craft for small-boat navigation in Arctic waters: an open boat ranging from about 15 to approximately 60 feet in length for carrying cargo and passengers for long distances, and a small decked canoe developed exclusively for hunting. With few exceptions these Arctic skin boats are wholly seagoing craft.

The open boat, called the umiak, is propelled by paddles or oars or sail or, in recent years, by an outboard gasoline engine, or it may be towed. While fundamentally a cargo carrier the umiak has been employed by some Eskimo in whaling and in walrus hunting. For these purposes a faster and more developed design is used than that used only to carry families, household goods, and cargo in the constant Eskimo search for new hunting grounds.

To a far greater degree than any other boat of similar size, this Eskimo boat is characterized by great strength combined with lightness.

The decked hunting canoe, the kayak, is propelled by paddle alone when used for hunting and fishing, but is occasionally towed by the umiak when the owner travels. The kayak is perhaps the most efficient example of a primitive hunting boat; it can be propelled at high speed by its paddler and maneuvered with ease. These hunting kayaks are commonly built to hold but one person, though one group of Eskimo built the kayaks to carry two or three. The kayak, remarkable for its seaworthiness, lightness and strength, has been perhaps one of the most important tools in the Eskimo fight for existence. Few tribes have been unacquainted with its use. Because of its employment, the kayak often has to be designed to meet very particular requirements and so there is greater variation in its form and dimensions than in the umiak.

Seagoing skin boats have not been common outside the Arctic in historical times. In fact only the European Celts are known with certainty to have used such craft. The Irish, in particular, employed large seagoing skin boats as late as the reign of Queen Elizabeth of England; a drawing of one preserved in the Pepysian Library was reproduced in the *Mariner's Mirror* (vol. 8, 1922, facing p. 200). Although there can be little doubt that large seagoing skin craft had been more widely used in prehistoric times, the perishable nature of the skin covering and the light framework probably account for the lack of any archeological remains that would indicate its range. The availability of the materials required in its construction, however, suggest that its use could have been very widespread. The long voyages made by the Irish, in the dawning of recorded history, could well have made its design and construction known to others.

There are still many skin boats in use by primitive people and even a few survivals in Europe, but with the exception of the Irish "curragh," these craft are designed for inland waters and are either rather dish-shaped, or oval in plan, like half a walnut shell. In design they are related to the coracle of ancient Britain rather than to a seagoing skin boat of the Irish or Eskimo type. Both the Irish curragh and the British coracle, now, of course, are covered with canvas rather than hide.

Traditions of long voyages by the ancient Irish in the skin-covered curragh make it apparent that such voyages were relatively common, and the design and construction of existing models of the curragh and umiak indicate that these voyages could have been made with reasonable safety. Compared to the dugout canoe, the skin boat was far lighter and roomier in proportion to length and so could carry a far greater load and still retain enough freeboard to be safe. The size of the early skin boats cannot be established with certainty; the modern Irish curragh is probably debased in this respect, but early explorers of Greenland reported umiaks nearly 60 feet in length and there is no structural reason why the curragh could not have been as large or even larger.

Compared with the curragh, the umiak is lighter, stronger, and more resistant to shock. The curragh was built with closely spaced bent frames and longitudinal stringers to support the skin cover, whereas the umiak has very widely spaced frames and few longitudinals, giving the skin cover little support. The difference in construction is undoubtedly a result of the type of covering used, for the curragh was covered with cattle hides, which were less strong than the seal or walrus skins used by the Eskimo. The strong and elastic skin cover of the umiak and the lack of a rigid structural support gives this boat an advantage in withstanding the shocks of beaching or of working in floating ice; and because of its relatively light framework and the method of securing the structural members, its frame is far more flexible than that of the curragh, adding to this ability.

The skin cover of the curragh was made watertight by rubbing the hides with animal fat, and the sewn seams were payed with tallow. The Eskimo soak the skin cover of the umiak with animal oil and pay the seams with blubber or animal fat. Both treatments produced a cover initially watertight but requiring drying and reoiling to remain so. Under most climatic conditions in the North Atlantic or Pacific the oiled skins remain watertight from four days to a week. This period can be lengthened by various methods; skin boats travelling in company can be dried out in turn by unloading one and placing it aboard a companion craft. There is evidence of other methods of treating the skin covering; waterproofing it with melted tallow, for example, or with a vegetable gum or a resin such as pitch, would enable it to remain watertight for a much longer time, though such treatments would make the covering less elastic. Pitch was also used at one time in curragh building, and it would be unwise to assume that the

Western Alaskan Umiak with eight women paddling, Cape Prince of Wales, Alaska, 1936. (*Photo by Henry B. Collins.*)

Figure 158

Western Alaskan Umiak being beached, Cape Prince of Wales, Alaska, 1936. (*Photo by Henry B. Collins.*)

Figure 159

Figure 160

Repairing Umiak Frame at St. Lawrence Island, Alaska, 1930. (*Photo by Henry B. Collins.*)

Figure 161

Eskimo Woman Splitting Walrus Hide to make umiak cover, St. Lawrence Island, Alaska, 1930. (*Photo by Henry B. Collins.*)

oil treatment used by the Eskimo was their only method of producing watertight skin covers in the period before they were first observed by Europeans.

The fundamental difference between the construction of the curragh and that of the umiak lies in the type of longitudinal strength members and the transverse framing used. The curragh, like the birch-bark canoe, depended entirely upon its gunwales for longitudinal strength, whereas the umiak has a strong keel, or, properly, a keelson since the keel was inside the skin cover. The curragh used longitudinal battens to support the skin cover. The umiak, on the other hand, has in its chine timbers rather strong longitudinal members that give additional strength to the bottom. Its transverse frames, unlike those of the curragh which were continuous from gunwale to gunwale, are in three sections, two side pieces and a floor, or bottom, member and the frame members are joined to gunwale, chines and keelson by lashings of sinew, whalebone, or hide, a method that, together with three-part frames, gives great flexibility to the framework. The frame of the early curragh may have been lashed, but because of the other fundamental differences in design and construction it was less flexible than that of the umiak.

The basic features of the umiak frame are not found in the kayak, the structure of which in most types approaches that of the curragh. The gunwale is the strength member in the kayak, and some types have a rather extensive longitudinal batten system as well. In only a few types of kayak is the keelson an important strength member, and even here the gunwales are of primary importance. The hypothesis has been offered that this indicates a different parentage for the kayak than for the umiak, and that the umiak represents the earlier type, it being argued that this type of boat was the one more required in migratory periods, and so would be first developed. Such

Fitting Split Walrus-Hide Cover to umiak at St. Lawrence Island, Alaska, 1930. (*Photo by Henry B. Collins.*)

Figure 162

Outboard Motor Installed on Umiak, Cape Prince of Wales, Alaska, 1936. (*Photo by Henry B. Collins.*)

Figure 163

Launching Umiak in Light Surf, with crew of 12 men. (Note outboard motor attached), Cape Prince of Wales, Alaska, 1936. (*Photo by Henry B. Collins.*)

Figure 164

theories should be accepted with caution, however, as the fundamentally different use requirements for the two types of craft might readily explain the variation in their principles of construction. Hunting would also have been necessary during migrations, as existence depended upon food; the earlier appearance of the umiak cannot be assumed on such limited grounds.

Eskimo skin boats possess remarkable advantages for their employment and conditions of use. Their hulls are light in weight, simple to build, and relatively easy to repair, yet they are highly shock resistant. They can carry large loads, yet are fast, they are capable of being propelled by more than one means, and they are exceptionally seaworthy.

Floating ice is considered a major hazard to craft of all sizes, but the umiak, for example, can resist the shocks of ramming the ice to a degree beyond the tensile strength of the skin covering, by reason of the method of attaching the skin cover to the framework of the hull, and to some extent the form of the boat itself. The skin cover of the umiak is not rigidly attached to the frame in a number of places, but rather is a complete unit secured only at the gunwales and to the heads of stem and stern. This permits the skin cover to be greatly distorted by a blow, so that the elasticity of the material at point of impact is assisted by the movement of the whole skin cover on the frame. Also, the frame itself is flexible and allows distortion and recovery not only within the limits of the elasticity of the wooden frame but also by the movement of the lashed joints in the transverse frames. Some kayaks have similar characteristics, though their small size and the light weight of both boat and loading make its resistance to shock of far less importance than that of the umiak.

Light weight is a highly desirable characteristic for small craft in the Arctic, since it permits the boat without the aid of skids or other mechanical contrivances to be removed from the water and carried over obstructions, and to be transported either by sledge or by manual portage over long distances. Lightness is obtained in the Eskimo skin boats by the small number and small size of the wooden structural members used in their construction. The resulting light weight hull permits heavy loading in proportion to the size of the boat, and it allows building with a minimum of material, in a country where such materials as wood are scarce and hard to obtain.

For all small craft in Arctic waters, where distances between sources of supply may be great and the time that the water is open to navigation is relatively short, speed is an important and desirable attribute that permits movement with a minimum of effort. The exigencies of Arctic travel make it further desirable that small craft be capable of propulsion under paddle, oars, sail, or low-powered gasoline motors. The umiak, because of its form and weight, can be modified to meet this requirement without loss of other desirable attributes, and to a slightly lesser degree, the same may be said of the kayak.

Simplicity in construction and repair are also basic requirements in the Arctic, where an emergency may make it necessary to repair or rebuild a damaged boat out of materials available nearby with the minimum of tools and under adverse weather conditions. The Eskimo has produced a boat construction that, as will be seen from the descriptions that follow, to a high degree meets this requirement.

Exceptional seaworthiness is required, as most Arctic waters are subject to violent storms; the Arctic skin boats have been developed with forms and proportions to meet this condition. In this matter, the light and flexible hull structure gives a special advantage. The kayak, in its highest state of evolution and in skillful hands is perhaps the most seaworthy of all primitive small craft. The umiak is a close second, but of the two, the kayak is safer under all conditions of Arctic travel.

The load-carrying capacity of skin boats has been mentioned. The Eskimo umiak is notable in this respect, exceeding the curragh and even craft produced by modern civilization. The umiak possesses this advantage because of its very light hull weight in combination with a nearly flat bottom and flaring sides. The resulting hull-form allows heavy loading with relatively little increase in draft, as the flaring sides cause the displacement to increase rapidly with the slightest increase in draft. Though a similar form exists in the lumberman's drive boat, the greater hull weight of this type makes it inferior to the umiak. Light draft when loaded has very definite advantages in the Arctic, for it allows loading and unloading on the beach or afloat, and allows the boat to be beached at points where this would not be possible with a deeper hull. The light draft also makes the umiak easy to propel manually.

The imperative need for very efficient watercraft has made the Eskimo seek improvements, and as his needs altered, so have his skin boats. Con-

Figure 165

UMIAKS ON RACKS, in front of village on Little Diomede Island, July 30, 1936. (*Photo by Henry B. Collins.*)

sequently the designs of these craft have gone through numerous changes since the first of the types were placed in American museums. It is noticeable that, among other changes, the amount of freeboard of umiaks has been altered as their owners met new conditions imposed by longer voyages, heavier cargo, and the outboard motor. The high-sided umiak, while suited for heavy loads and very seaworthy, was almost impossible to paddle or even row against a strong gale. When this condition had to be met, the freeboard and flare were reduced to minimize the windage. In recent years umiaks have appeared with round bottoms to give greater speed under paddle, the resulting boat being an enlarged kayak in construction. These changes to meet differing use requirements are not necessarily basic improvements, for they result in the sacrifice of some of the other qualities of the type. Nevertheless, they indicate the fluid state of primitive boat design in the Arctic, a condition that has been accentuated in most areas by the increasing influence of white men, their boats and their motors.

## *The Umiak*

The umiak was undoubtedly more widely employed by the Eskimo before the coming of the white man than existing records indicate. It was a type of boat most necessary for family migration by sea, and with it the early Eskimos could establish themselves on islands far from the mainland and could cross large bodies of water. From some areas where early explorers mention having seen the type, the umiak has disappeared; this suggests the possibility that tribes now unacquainted with the umiak had at some time in the past reached a location where such a boat was no longer necessary.

The umiak was common in open waters and was found from Kodiak Island through the Aleutians and north and eastward along the west and north coast of Alaska to the mouth of the Mackenzie River. On the Siberian coast, opposite Alaska and for a short distance westward, the umiak was also employed. From the Mackenzie eastward to Hudson Bay the umiak has not been employed in recent times, though it is highly probable that it was used in the migrations that populated this part of the Arctic coast with Eskimo. Early explorers found umiaks in use along the northwestern coast of Hudson Bay and Foxe Basin; the umiak disappeared from these areas during the last century, but its use continued in Hudson Strait and in Greenland, where it became highly developed.

Among the various tribes of Eskimo known to have employed the umiak in the last century, the form of the hull varied a good deal, as did its dimensions. In general its form was something like that of the lumberman's "drive boat," except that most umiaks had a slight V-bottom and were quite different from it in the shape of the bow and stern. The size of the umiak does not seem to have been established by a set of measurements as distinct as that used in the building of kayaks, but rather as the result of utilizing material available locally, with due regard to the intended use of the craft for relatively heavy transport. Such matters as the flare of the sides, rake and shape of bow and stern, and width varied from tribe to tribe. The Asiatic and Alaskan umiaks were usually rather sharp-ended, with little spread to the gunwales at bow and stern; one of the Asiatic types has the gunwales brought round in a full curve at the ends of the boat. In the East the umiaks have rather upright bows and sterns and the gunwales are often rather wide apart to form square ends to the hull. Some of the western umiaks were navigated with paddles only; with others, before the appearance of the Russians in the area, both oar and sail may have been used. In the East the umiaks were being paddled, rowed, and sailed when white men reached the Arctic in the 17th century.

The Greenland umiak frame is much heavier and more rigid than the Alaskan. In comparing eastern and western umiaks the frame of the eastern umiak seems to be somewhat better finished, but the models of the western umiak are undoubtedly the better. The eastern umiak is not intended for use in hunting but is primarily a cargo carrier; its use has been confined to women and its chief employment is moving the family and household effects from one hunting ground to another. While it is highly probable that this condition is the result of the disappearance of whaling in this region, the use of the umiak as a hunting boat ceased so long ago that the eastern umiak model may have degenerated to a great degree. It has been otherwise in the western Arctic where the use of the umiak in hunting has continued and the boats have been managed, to a very great extent, by the men. As a result, the boats are held in greater respect by their builders and the better models have survived. The tribal distinctions between the western umiaks are therefore more marked than in the east; including Siberia, at least three basic models and a very large variety of tribal variations, are to be found, as can be proved by existing models. In the east only two basic and distinct umiak models are known to have existed, the Baffin Island type used on both the north side and on the Labrador side of Hudson Strait, and the Greenland type. In the latter, there were slight tribal distinctions it is true, but these were minor.

The Asiatic umiaks may be classed into two types, the Koryak type of Eastern Siberia and the Chukchi model of the Siberian side of Bering Strait. The Koryak umiaks illustrated by Jochelson show a highly developed boat, rather lightly framed compared to boats on the American side. In profile the bow has a long raking curve and the stern much less; as a result the bottom is rather short compared to the length over the gunwales. Viewed in plan, the gunwales are rounded in at bow and stern to form almost a semicircle. At the bow the gunwales are bent around a horizontal headboard tenoned over the stem head but at the stern there is no headboard. The sheer is moderate and very graceful. The flare of the sides is great and there appears to be a little V in the bottom transversely. There is also a slight fore-and-aft rocker in the bottom. The construction is similar to that of the Alaskan umiaks except that the Koryak umiaks have double-chine stringers and also a double riser, or longitudinal stringer, halfway up the sides. The riser is not backed with a continuous stringer, as is the chine; instead three short rods are lashed inside the side frame members. The side stringers do not reach bow and stern. The four thwarts are located well aft, and between the first and second thwarts is a larger space than between the others, for cargo. The boats are rowed, two oarsmen to a thwart. The cover was formerly walrus hides split and scraped thin but more recently the skin of the bearded seal has come into use. A rectangular sail of deer skin is sometimes lashed to a yard and set on a tripod mast about amidships. Two legs of the mast are secured to the gunwale on one side, the remaining leg is lashed to the opposite gunwale. Judging by the drawing made by Jochelson* this umiak is perhaps the most graceful of all those known today.

*Reproduced in JAMES HORNELL, *Water Transport* (Cambridge: University Press, 1946), p. 160.

Figure 166

Umiak Covered With Split Walrus Hide, Cape Prince of Wales, Alaska. The framework can be seen through the translucent hide cover. (*Photo by Henry B. Collins.*)

The Asiatic Chukchi umiak is somewhat similar to that used on the American coast but with less beam in proportion to its length and less flare to the sides. The skin cover is of bearded seal. Bogoras measured an example and found her 35 feet 9 inches long, 4 feet 6 inches wide amidships, 2 feet 6 inches wide on the bottom over the chines. (An Alaskan umiak measured 34 feet 9 inches long, 8 feet 2 inches wide at gunwales and 2 feet 8 inches over the chines.) The Chukchi also use a very small hunting umiak, 15 to 18 feet long and having two or three thwarts, much like the small hunting umiaks once used in the Aleutians. The larger Chukchi umiaks have rectangular sails set on a pole mast; some boats carry a square topsail. The sails are lashed to their yards and the lower sail, or "course," is controlled by sheets and braces. The topsail, when used, has braces only. The sails were formerly of reindeer skins, but now drill is used. These umiaks were formerly paddled, as indicated by their narrow beam, but since the advent of the white man oars have come into use, and it is quite certain that the topsail also is the result of white man's influence, if not the whole rig.

In stormy weather some of these umiaks and also some of those in Alaska employ weather cloths, 18 or 20 inches high above the gunwales, raised on short stanchions lashed to the hull frames. The ends of the stanchions are inserted in slits in the top of the weather cloth, and in fair weather the cloths are folded down inside the gunwale out of the way. Also in some of these Asiatic and Alaskan umiaks, inflated floats, of seal skin, are lashed to the gunwales to prevent capsizing in a heavy sea.

The Alaskan umiaks varied much in size but are rather similar in form. The small hunting umiaks used by the Aleuts are about 18 feet long, while the large cargo carrying umiaks range up to about 40 feet long, so far as available records show. They are marked by heavily flared sides and often have a rather strong sheer; a few, however, are rather straight on the gunwales. Nearly all existing models and boats were built since 1880; and no information is now available on the forms and dimensions of earlier craft.

On page 184 is a drawing of a small umiak, used in walrus hunting, from the Alaskan coast in the neighborhood of the Aleutians. In the U.S. National Museum are the remains of a similar boat obtained in 1888 from Northern Alaska. This type of small umiak is also employed in fishing and is rather widely used as a passage boat for short voyages along shore. These craft, propelled by paddles, are primarily fast, handy hunting canoes rather than boats for migration or cargo-carrying. For this reason they are quite sharp-ended and shallow. The construction of this example will serve to illustrate the methods common to this type.

The umiak shown is 20 feet 8½ inches over the headboards, 4 feet 9½ inches extreme beam and 17⅜ inches depth—apparently an average-sized boat of her class. The width of the bottom over the chine

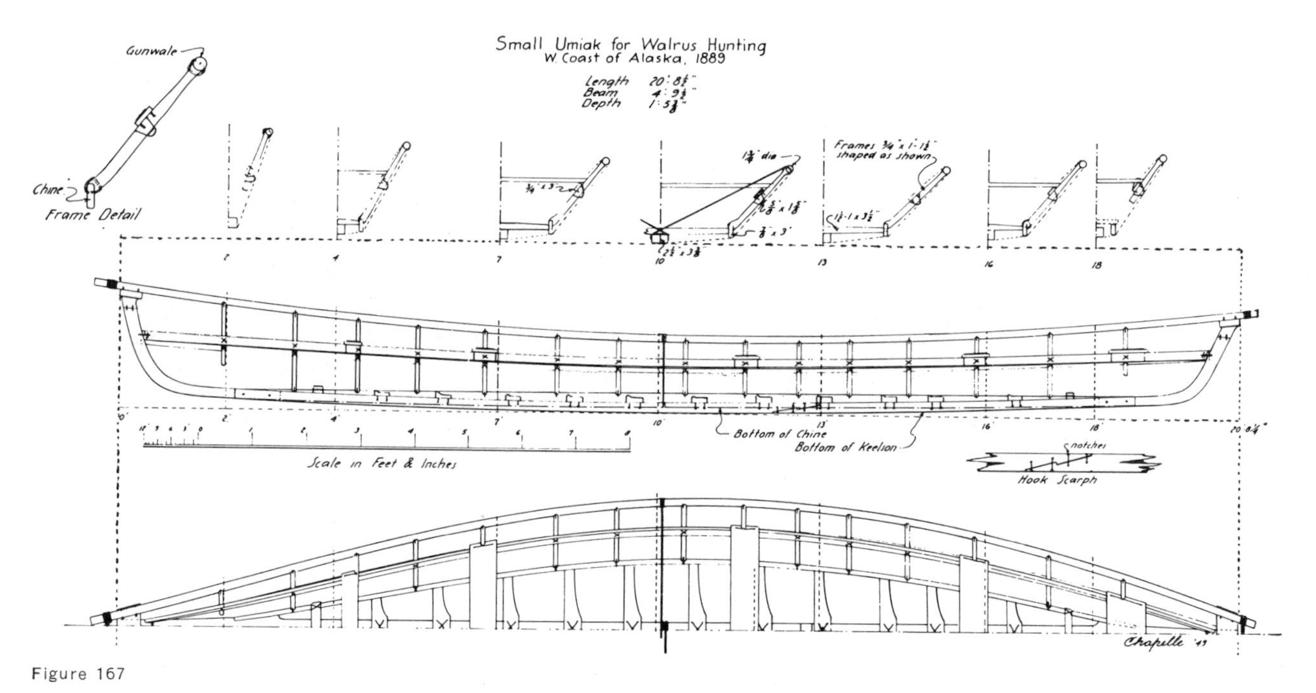


Figure 167

SMALL UMIAK FOR WALRUS HUNTING, west coast of Alaska, 1888–89. Reconstructed from damaged umiak formerly in U.S. National Museum, and from models.

members is 2 feet 7 inches. The keelson is rectangular in section and in two pieces, hooked-scarphed together; each piece is shaped out of the trunk of a small tree with the root knees employed to form the bow and stern posts. The floor timbers are quite heavy and support the chine members by having the floor ends tenoned into the chine pieces. At bow and stern the chines are joined to the keelson in a notched scarph; at these places the keelson is sided rather wide to give good bearing. It is evident that this portion of the boat's structure is the first built and forms a rigid bottom to the hull. The floor timbers are lashed to the keelson by lacings of sinew, whalebone, or hide, passed through holes bored in both, as indicated in the plan. The ends of the floors are pegged where they tenon into the chines and the ends of the chines are pegged to the keelson, but this was evidently not a universal practice, as there are models showing lashings at floor ends and at chine ends. The headboards are carved out of blocks in a T-shape and are stepped on top of the stem and stern posts and lashed. The fit is extremely accurate. The bow headboard is narrower athwartship than the stern headboard. The detail of the hook scarph in the drawing shows a method of lashing that is widely used.

Because of the manner in which the keelson is cambered and the floor fitted, the bottom of the covered hull shows in cross section a slight V, reducing toward the bow and stern, that is typical of the Alaskan umiak. The amount of deadrise seems to have been determined by the manner of fitting the floor timbers and it helps the boat to run straight under paddle and oars. In present day umiaks the amount of V in the bottom is slight; too much would make the boat difficult to sledge overland without employing chocks to steady the hull. Perhaps in the past, where sledging was not required, the deadrise was greater, as indicated by some old models.

After the chines and floor are fitted to the keelson, the frames at the thwarts are made and set up at the desired flare and height, being held in place by temporary spreaders lashed or braced. These are sometimes stiffened by thongs from frame head to keelson at each pair, to steady the frame while the gunwale is being bent. As the lengths of the thwarts are controlled by the fairing of the gunwales, the thwarts are not fitted until after the latter are in place. As shown in the figure above, the gunwales are round poles, slightly flattened on the lower side at the headboards, where they are secured by lash-

Figure 168

Umiaks Near Cape Prince of Wales, Alaska, showing walrus-hide cover and lacing. Frame lashings are walrus-hide thongs. (*Photo by Henry B. Collins.*)

ings. In building, the gunwales are shaped and secured by lashing them to those side frames selected to shape the hull. The lashings that secure the side frames to both gunwale and chine are passed through holes in each member and are hove taut by means of a short lever with a hole bored in it to take the end of the lashing, which is also wrapped around the lever to give temporary purchase. The side frames have saddle notches to bear on the chine and gunwale. All lashings in the frame, it will be noted, pass through holes bored in the members and in some cases the lashings are let in, so that the sinew is flush with the surfaces of the members, to prevent the lashing from being damaged by chafing.

With the gunwales faired, the remaining frames are then put in position and lashed to the gunwales and chines. An outside batten is run along each side and lashed by turns of sinew over the batten and around the side frames, with the lashings let into each member to prevent slipping and chafing. The batten is lashed at bow and stern in some umiaks, but in many it is stopped just short of coming home on the posts. Next, the short frames at bow and stern are put in place and the risers secured inside the side frames, then, with the thwarts fitted and lashed to the risers, and the ends of the gunwales are lashed together at bow and stern, the boat is ready to be covered. When ready to cover, the frame is stiffened by diagonal thong ties, each of which has one end secured by turns around the gunwale, with the other end passed through holes in the keelson and secured. These are commonly found in western umiaks; the small umiak has but one pair placed amidships. The timber used in such craft is fir, spruce, and willow, and is usually driftwood obtained at river-mouth.

When this umiak was examined, the skin cover was in such a condition that the number of hides used could not be determined, but it probably is comprised of three sea-lion skins sewn together. New skin covers are made by removing the hair and fat from the skins and then sewing them together by the method illustrated on page 186, to obtain proper

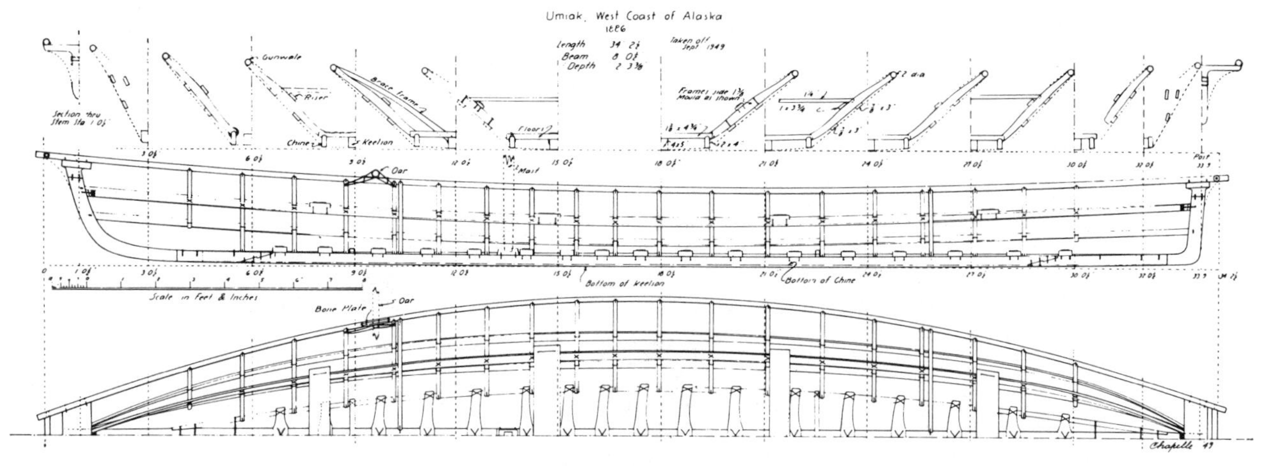


Figure 169

UMIAK, WEST COAST OF ALASKA, King Island, 1886. Taken off umiak at Mariner's Museum.

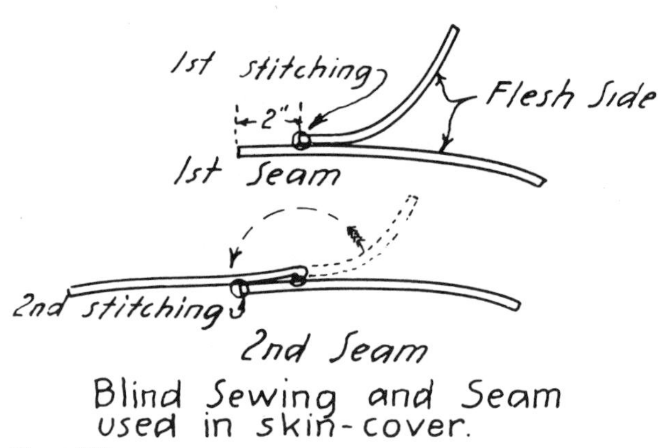


Figure 170

MAKING THE BLIND SEAM: two stages of method used by the Eskimo to join skins together. The edge of the skins are placed flesh side to flesh side with one overlapping the other about 2 inches. Then, by means of a thin needle and slender sinew, the skins are sewn together, with an over-and-over stitch, care being taken not to penetrate through the lower skin. When this is completed the skins are opened out and the second seam made on the grain side to complete a double seam without penetration of either skin. The width of the seam varies somewhat.

dimensions. Green skins are generally preferred, since they stretch into shape better than partly or wholly cured ones. Once stretched to shape and cured, the cover can be readily removed and replaced, without resewing. In fitting a new skin cover the skins are first thoroughly soaked in seawater. The cover is then stretched over the frame and worked taut by lacings. It is wide enough to reach from gunwale to gunwale and a little down inside the boat on each side, and is laced to the rising batten with turns of rope spaced 3 to 5 inches apart amidships and closer together in the ends of the hull. At the headboards the cover is laced around the gunwales and through holes in the headboards, two independent lacings of two turns each being used on each side. At the extreme bow and stern the cover is laced to the gunwale lashings. Where the cover will not stretch smooth in fitting, gores appear to have been cut out and the skin resewn. After being laced, the cover is allowed to shrink until it becomes smooth and tight, then it is heavily oiled and the seams rubbed with tallow or blubber. This treatment is repeated at regular intervals. While the boat is in service care is taken to dry out the skin cover once a day, if possible.

The sequence of construction described is not followed universally; sometimes spreaders are fixed between the gunwhales, which are then sheered by thongs to the keelson, after which the side frames are put in and the side and rising battens, and finally the thwarts, are added. Judging by the numerous models seen, the small hunting umiaks varied a good deal in the rake and sweep of the bow and stern, even in the same village. These hunting umiaks worked with kayaks in Aleutian walrus and sea-lion hunting; a practice that seems to have once been common along

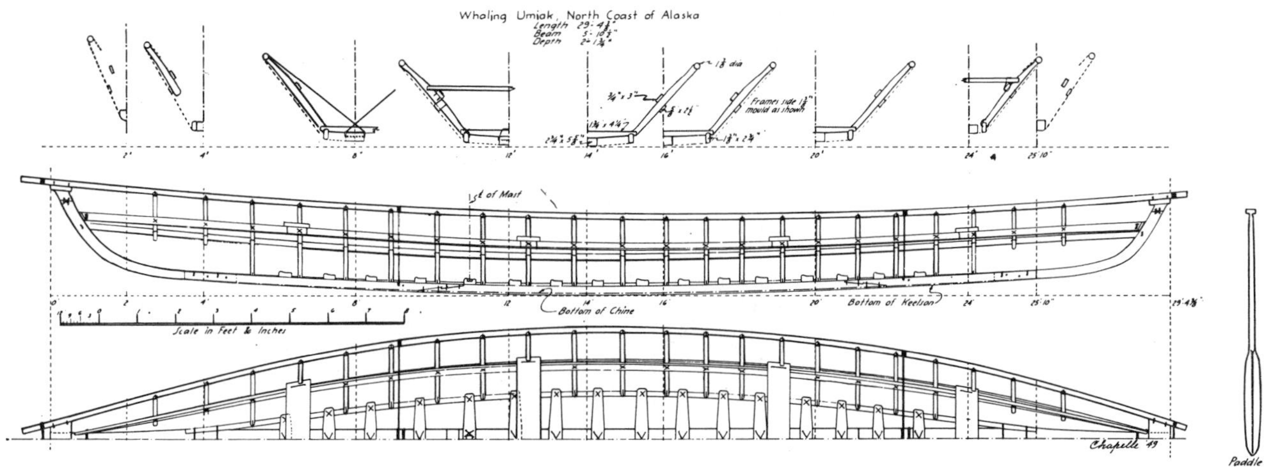


Figure 171

NORTH ALASKAN WHALING UMIAK of about 1890. Drawn from damaged frame, formerly in a private collection, now destroyed.

the Western Alaskan coast and among the islands.

The drawing on page 186 represents a large Alaskan umiak from King Island. Two boats of this model, but with modern metal fastenings, are in the Mariners' Museum, Newport News, Virginia, but the drawing shows the methods of fastenings used in 1886. The plan is of a burdensome model, such as is used for travel or other heavy cargo work. The boat is 34 feet 2½ inches over the gunwales, 8 feet ½ inch extreme beam, 2 feet 3⅜ inches deep and 2 feet 10 inches beam on the bottom over the chines. The construction follows the general plan of the small umiak just described, except that another method of fitting the floor timbers to the chines is employed. Due to the size and use of the umiak, two side battens are employed with a single riser. The thwarts are not notched over the frames, but instead fall between them. As diagonal thong braces from gunwale to keelson would be ineffective in this situation, two sets of wooden braces that resist not only tension but also compression are used to take the thrust off the thwart lashings. These brace-frames are staggered slightly to allow room to fit them at the keelson. The drawing, which requires no additional explanation, shows the plan of construction and the important lashings, and the method of fitting oars with thong thole loops.

Boats such as these carried a square sail lashed to a yard, the mast being stepped in a block on the keelson. No mast thwart is used; instead stays and shrouds of hide rope supported the mast, a method that made it easy to step or unstep the mast in a seaway. Early umiaks in this area are said to have had mat sails; later ones used sails of skin and drill. Modern umiaks of this class often have rudders hung on iron pintles and gudgeons and the floors fastened to the keelson with iron bolts or screws. The scarphs are also bolted, but the remaining fastenings are lashings in the old style, to obtain flexibility in the frame.

A North Alaskan whaling umiak, supposed to have been built about 1890, is represented in the drawing of figure 171. The remains of the boat were sufficient to permit reconstruction of the frame. This umiak is about the size of, and in profile greatly resembles, a New Bedford whaleboat. However, the model is that of the umiak, rather sharp-ended and strongly sheered. The boat is 29 feet 4⅜ inches over the headboards, 5 feet 10½ inches extreme beam, and 2 feet 1¾ inches deep. Umiaks of this model were used at Point Barrow and vicinity in offshore whaling, and were also used for travel and cargo carrying. Paddles were used in whaling, but in more recent times sail, oars, and outboard engines have been employed. The boats of this class appear to have been marked by a very graceful profile and strongly raking ends. Despite the resemblances of this type of umiak to the whaleboat, it is highly doubtful that its model was influenced by the white man's boat. In fact, it might just as well be claimed that since the whaleboat appears to have been first employed in the early Greenland whale fishery, the latter had been influenced by the umiaks found in that area. However, one might also point to the fact that the model of the early European whaleboat is much like that of a Viking boat, from which will be seen the danger in

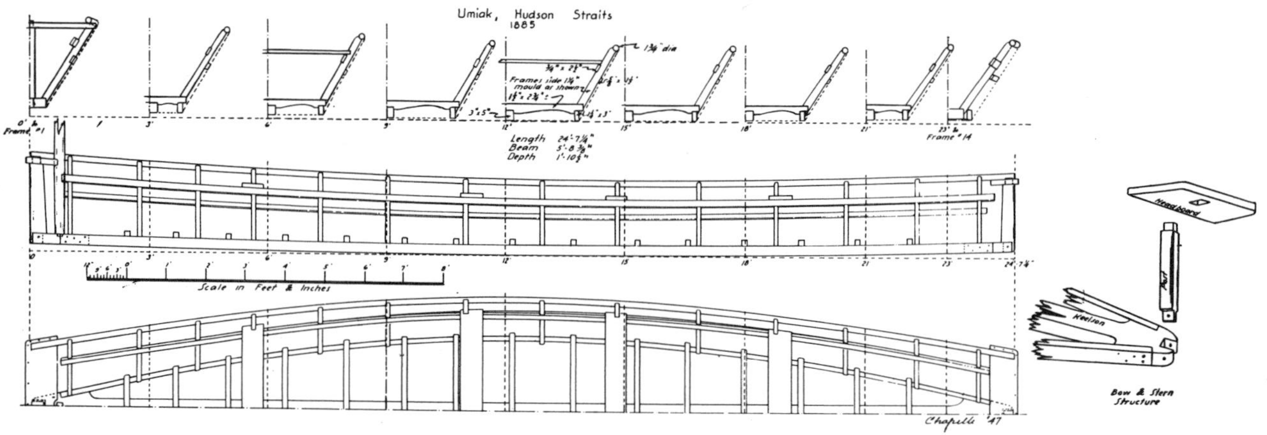


Figure 172

BAFFIN ISLAND UMIAK. Drawn from model and detailed measurements of a single boat.

accepting chance similarities in form or detail as evidence of relationship, particularly when it is not impossible that similarities in use and other requirements have produced similar boat types, the users never having come into contact.

The whaling umiak has been much used in the western Arctic by explorers and Arctic travellers, who regarded highly its lightness and strength, and its ability to be easily driven. It is much wider than the Chukchi umiak and has far more flare. From a study of models and numerous photographs it can be said that the amount of fore-and-aft camber in the bottom varies greatly between individual umiaks, some of which are almost straight on the bottom. The light framework and elastic construction often cause these umiaks to camber a good deal when heavily loaded; when sledged, they are sometimes fitted amidships with a support for a line from bow to stern, that forms a "hogging-brace," to prevent the boat from losing its camber. It is also apparent that there is no standard practice in fitting floors to the chines; Murdock* shows a rough sketch that indicates the floor ends are often tenoned into the chines, as in the small umiak. Treenailing of the floors and chines, and the keelson, is common, and sometimes both treenails and lashings are used in scarphs. In some umiaks both the single side batten and the riser are at the same height, but only the riser has its ends secured to the posts, the side battens being cut short and their ends lashed to the riser a few inches inside the posts.

*See bibliography.

The skin cover of the north Alaskan whaling umiak is made of bearded seal or of walrus hide, which has to be split, because of its weight. Occasionally polar-bear skins are used. Lashings of the frame are of whalebone, sinew, and hide. The skins are treated with seal oil and caribou fat, and when the whaling umiak is taken ashore it is usually stored on a stage to keep dogs from destroying the skin cover. In travelling, however, it is sometimes propped upside down on one edge and used as a shelter. In winter the skin is removed and stored; when it is necessary to be replaced on the frame, the skin cover is soaked in sea water for three to five days, after which it is laced on in the usual manner, dried, and then thoroughly oiled. Low, rather wide sledges are sometimes built to carry the umiak overland, or on the ice, but often the regular sledge is used. The boats cannot be sledged against a strong gale because of their windage.

The north Alaskan umiak is usually propelled by paddles, like the Chukchi umiak. These paddles range in length from about 50 to 76 inches, and as a rule have a rather long narrow blade, though a short and wide blade is occasionally found, particularly at Kotzebue Sound and Point Hope. Oars for the Alaskan umiaks range in length from 6 feet 3 inches to 8 feet 6 inches, and also have rather long narrow blades, 3 to 4 inches wide.

The three examples of Alaskan umiaks serve to show the features that are most common in the area. However, models in the U.S. National Museum suggest that there was a greater variety of form and appearance in the past. One model shows the

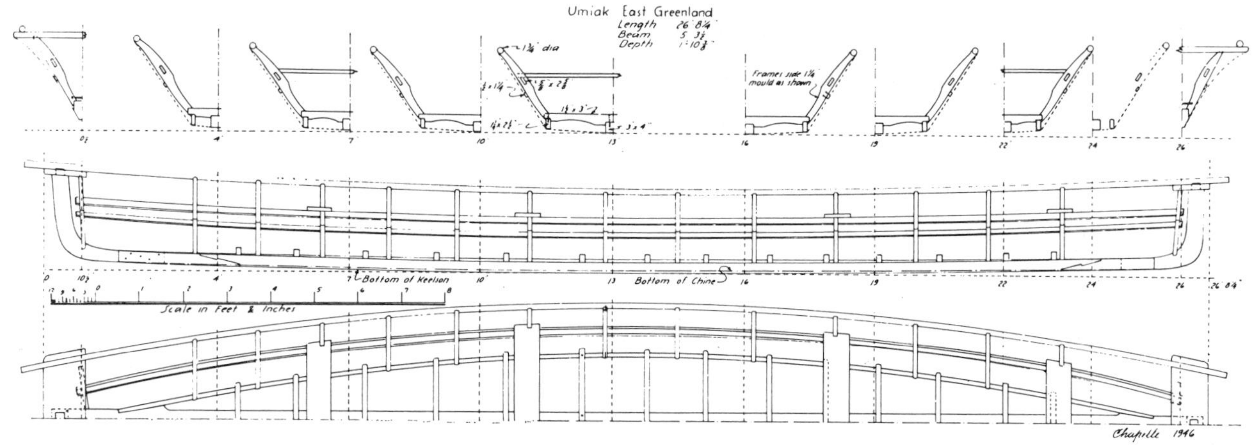


Figure 173

East Greenland Umiak, drawn from measurements taken off by a U.S. Army officer in 1945.

gunwale ends lengthened by pieces shaped very much like the projecting gunwales of the Malay prah. Some show extreme rake at the bow like that of the Koryak umiak but without the rounded gunwale ends. It is impossible to estimate how far the western Alaska umiak model has been affected by the early Russian traders in this area, but it is quite certain that the use of oars can be traced to this influence. The full-sized umiaks, and models and photographs, from the Bering Strait area give no real clues to the possible parentage or direction of spread of the Alaskan umiak types. Occasional details in fittings or construction, such as the gunwale extensions mentioned, seem to duplicate details in primitive Asiatic craft, but the evidence is too scanty to allow a hypothesis based on design and construction alone.

No models or photographs have been found of the extinct types of umiaks once used in the northern part of Hudson Bay and the sketches of early explorers are too crude to allow useful discussion. From such slight evidence it is impossible to say whether the umiaks in this area were of the western or eastern type.

The drawing of a Baffin Island umiak on page 188 is based on measured dimensions of a single boat and upon a small model in the U.S. National Museum. This model conforms in most respects with the drawings and sketches made by Boas.* The umiak is a small one, 24 feet 7¼ inches long, 5 feet 8⅜ inches extreme beam, 3 feet 10 inches wide over the chines, and 1 foot 10½ inches deep. These measurements show that the bottom of this type of umiak is wider than that of western types. The bottom is flat, and sheer and camber are both slight. The stem and stern are practically upright and are not formed of knees; rather, they are made by fitting the post into the keelson with an open tenon. Instead of the carved block headboards seen in the Alaskan umiaks, the Baffin Island boat has very wide headboards, and these are tenoned over the posts as in the Asiatic Koryak umiaks. The details of the rest of the framing are not dissimilar from those of the Alaskan umiaks, except that the Baffin Island umiak does not employ any short frames in the end of the hull. The framework is rather heavy and the square-ended appearance of this class of umiak makes it appear more clumsy than is actually the case. The side battens and risers stop short of the posts, and the risers used in this umiak are notched into the side frames, whereas in the Alaskan umiak only the lashings of the riser are let into the frames. The Baffin Island umiaks carry a square sail lashed to a yard, and the mast is placed right up in the eyes of the boat. Boas shows that some of these umiaks have rudders hung on metal pintles and gudgeons, a result of the influence of the white traders, whalers, and sealers who had operated in these waters long before Boas made his investigations. The umiak is rowed in the usual manner, using thong loops as tholes, and is usually steered with an oar or long paddle.

The ends of the gunwales of the Baffin Island umiak are cut off a little inside the forward edges of

* See bibliography.

the headboards, making this the only American type that does not have projecting gunwales at bow and stern. The projection of the gunwales undoubtedly serve a practical purpose in lifting the boat out of water, but obviously this is of minor importance. Probably the real reason for these projections is that they originally made building easier by providing space for a retaining lashing when the gunwales were being bent. As the headboards became wider and the spring of the gunwales, in plan view, became less acute, less strain was put on the lashings of the gunwales at the headboards, but by then the projecting gunwales and their retaining lashings were being utilized in lashing on the skin covering at bow and stern. Thus, beginning as a practical solution of a building problem, the projecting gunwales may have eventually become a traditional tribal feature of the umiak in many localities.

The drawing of an eastern Greenland umiak on page 189 was made from measurements taken off during World War II and checked against dimensions, photos, and descriptions of boats from the same territory. In general design and in construction this umiak differs little from umiaks of the southwest coast of the same island. The eastern Greenland boats are, on the average, much smaller than those on the southwest coast due to the more severe ice conditions met in the east. Some of the Greenland umiaks have flat bottoms like the Baffin Island boats, but the V-bottom appears to be more common. The chief characteristics of the Greenland umiaks are the slight rake in the bow and stern, the moderate sheer and camber, and the conservative flare of the sides. The drawing shows the important structural details seen in most of the Greenland umiaks. The floor timbers are on edge instead of on the flat as in Alaskan boats and this seems to be characteristic of all eastern umiak construction, as is the arching of the underside of the floors. Another common structural detail is the passing of the risers through the side frames; in some, however, the risers lie in deep notches fashioned in the inside of the frames. The eastern Greenland umiaks generally have rather wide headboards and somewhat more projection to the gunwales. Like the Baffin Island umiaks, the side battens and risers of the Greenland boats are cut short of the posts, but the ends of these members are commonly supported by frames placed very far fore and aft, and often these frames form brace-supports to the headboard, as in the drawing. The headboards of these umiaks are always tenoned over the top of the posts. Some of the Greenland umiaks have curved side frames which cause the side battens to form knuckles in the skin cover. The eastern Greenland umiaks rarely if ever carry sail, but this is common on the western and southwestern coasts, where a squaresail on a yard is popular, with the mast usually well forward. Hans Egede in 1729* found Greenland umiaks fitted with sails of seal intestines and also saw boats about 10 fathoms (60 feet) long; another early writer, Crantz* states that umiaks were commonly 36, 48, and even 54 feet long. In the larger umiaks two side battens were employed. The thongs and brace-frames seen in many Alaskan umiaks do not seem to have been used in eastern waters, the use of bracing-frames from stem or stern post to the gunwales probably serving the purpose, but it is noticeable that pictures of Greenland umiaks preserved in some European museums show that the hulls have a tendency to twist not seen in Alaskan boats. The old Greenland umiaks were built with lashed joints combined with pegging, or treenailing. In recent times the use of pegging has increased and iron fastenings are now quite common. Rigid fastenings of the peg and metal types are used only in scarphs and in securing the chines and keelson to the floors timbers, as in the modern Alaskan umiaks.

## *The Kayak*

The Eskimo hunting boat, the kayak, is more widely employed in the Arctic than the umiak, and its variations in model, construction, and appearance are more distinct and numerous. The kayak is a long, usually narrow, decked canoe and is commonly very well finished. In Alaska a few undecked skin-covered canoes, used in river, are built on kayak proportions, but the model of these is quite different from that of the Alaskan sea-kayaks; the river canoes are V or flat bottomed, much like the Greenland kayaks. A similar kayak-type canoe, flat bottomed but birchbark covered, is used by the Yukon Indians. Undoubtedly a number of such types once existed but most of these became extinct before any attempt was made to preserve models or canoes in museums.

Few Eskimo tribes are without kayaks, only those living inland or where the sea is rarely open are unacquainted with these hunting craft. In very recent times some tribes have ceased to use kayaks,

*See bibliography.

Figure 174

FRAME OF KAYAK, Nunivak Island, Alaska, with young owner beneath. (*Photo by Henry B. Collins.*)

employing purchased canoes instead. The kayaks of the Asiatic Eskimos, and those from the Mackenzie to Hudson Bay, are now crudely built and of inferior design. Both the Greenland and the Alaskan kayaks are highly developed. The Greenland kayaks are undoubtedly given more intricate equipment in the way of weapons and accessories than the Alaskan craft, but it would be difficult to decide which is superior in construction and design.

The basic models used in Eskimo kayaks are the multi-chine, the V-bottom and the flat bottom. The multi-chine models, except for the river kayak-canoe just mentioned, which probably should be classed as a true open canoe rather than a kayak, are employed throughout Alaskan waters. The geographic boundaries of each basic hull form are rather ill-defined. The multi-chine kayak appears as far eastward as the northwest coast of Hudson's Bay. In this area, however, a V-bottom kayak, now extinct, seems to have been in use on Southampton Island. A flat-bottom kayak, with the chines snied off much like a Japanese sampan, is in use in Hudson Strait, along the shores of Baffin Island and Labrador; a flat-bottom kayak shaped like a sharpie is used on the northwest coast of northern Greenland; and a V-bottom hull is employed on the eastern, southwest, and south coasts of Greenland.

According to the Danish classification of the coasts of Greenland, "Polar" is north of Cape York, "Northern" is above Disko Island, "Central" is from Frederikshaab to north of Disko Bay, "Southern" is from Julianhaab to Cape Farvell, and "East" is Angmagsalik and vicinity.

There are variations in each of the basic models, of course, as the tribal designs used vary a good deal. On the whole, the kayak is very carefully built to meet the local conditions of hunting, sea, and land or ice portaging. As a result, some types are far more seaworthy than others and the weight of hull varies a great deal, even within a basic model. The appearance of all the kayaks models, by tribal classifications, show the influence of tradition and, in many cases display, in either shape or decoration, a tribal totem or mark.

The basic requirements in nearly all kayaks are the same; to paddle rapidly and easily, to work against strong wind and tide or heavy head sea, to be maneuverable, and to be light enough to be readily lifted from the water and carried. The low freeboard required makes decking a necessity. In general, the kayak is designed to carry one paddler, but in Alaska are kayaks that can carry two or three paddlers, each in a manhole or cockpit, or a paddler and one or two passengers. It is generally conceded that the kayak built to carry three in this fashion is the result of Russian influence. Nunivak Island kayaks had large manholes that carried two people back-to-back. Where it is desirable to portage the kayak over ice or land for a great distance the boat is very light and is capable of being carried like a large basket, by inserting one arm under the decking at the manhole or cockpit, but where such a requirement is not an important factor, the kayaks are often rather large and heavy. In the majority of types, the degree of seaworthiness obtained is very great. Some types are built very narrow and sharp-ended; these usually require a skillful paddler. Others are wide and more stable, requiring less skill to use. In areas where severe weather is commonly met, the kayaks are usually very strong and well-designed. Where ice or other conditions do not allow a heavy sea to make up, the kayaks are often light, narrow and very low sided—more like racing shells than working canoes. Most Alaskan kayaks come stern to the wind when paddling stops, but most of the eastern craft come head to the wind. Nearly every type has been developed by long periods of trial and error, to produce the greatest efficiency in meeting the conditions of use in a given locality. This has made the kayak a more complicated and more developed instrument of the chase than is to be found in any other form of hunting canoe, due in part, perhaps, to the great craftmanship of the Eskimo.

The construction of the kayak follows a basic plan. In all kayaks the gunwales are the main strength members, longitudinally. A few designs employ, in addition, a stiff keel member, but most have rather slender and light longitudinal batten systems having little longitudinal strength value, but which in combination with very light frames, give transverse support to the skin cover. Even in the flat-bottom models, the kayaks, unlike the umiaks, depend entirely upon the gunwales for longitudinal strength. The frames are bent and in one piece from gunwale to gunwale in all but a few flat-bottom kayaks, of the sampan cross section; these employ bent frames. The longitudinal batten systems show great variety. The eastern kayaks of the flat-bottom and V-bottom models have three longitudinal battens (including the keel or keelson) in addition to the heavy and often deep gunwale members; these are supported at bow and stern either by stem and stern post of shaped plank on edge as in the Greenland V-bottom kayaks, or by light extensions of the keelson and small end-blocks as in the northern Greenland, Baffin Island, and Labrador types. The multi-chine types of the western Arctic have from seven to eleven longitudinals (including the keelson) in addition to the gunwales. In some of these kayaks there are no stem and stern posts, the battens and keelson coming together at a blunt point in small head blocks; but many types have rather intricate stem-pieces, carved from blocks of wood, and plank-on-edge stern posts. The Asiatic kayaks, curiously enough, exhibit the construction of both eastern and western Arctic kayaks, the crude, small Koryak kayak having a 3-batten V-bottom, while the Chukchi kayak is built like the kayaks on the east side of the Bering Strait. The decking of kayaks is of very light construction; usually there are two heavy thwarts to support the manhole and from one to three light thwarts afore and abaft these. The Alaskan kayaks from Kotzebue Sound southward have ridged decks supported by fore-and-aft ridge-battens from the ends of the hull to the manhole. Elsewhere the deck of the kayak is flat athwartship except at the manhole, where there is some crown or ridging to increase the depth inside the boat, particularly forward of the manhole. In the majority of these kayaks short fore-and-aft battens are laid on the thwarts forward of the manhole to support the skin cover in its sweep upward to the manhole. The transverse frames do not come into contact with the skin cover, to avoid transverse ridges being formed in it; and the longitudinal battens which support the skin cover form longitudinal ridges, or chines, in it.

The timber used in the Eskimo kayak building is usually driftwood. Fir and pine, spruce or willow are available in much of the Arctic for longitudinals. Bent frames are commonly of willow. Scarphing in the framework of kayaks was far less common than in umiaks; the scarphs when found are only in the gunwales. All scarphs are of the hooked type and are usually quite short (the hooked scarph is the best one when the fastenings are lashings). Sinew is generally used in all lashings and for sewing material. The heads of frames are commonly tenoned into the underside of the gunwales and are then either lashed or pegged with treenails of wood or bone to hold them in place. In the joining of frames and longitudinals, the lashings are commonly individual, but in some types of kayak continous lashings (connections in series using one length of sinew) are

Figure 175

FRAME OF KAYAK AT NUNIVAK ISLAND, Alaska, 1927. (*Photo by Henry B. Collins.*)

occasionally found. Where possible, the lashings are turned in so that the turns cross right and left. In some parts of the framework two pieces of timber are "sewn" together; holes are bored along the edges to be joined and a lacing run in with continuous over-and-over turns. These laced joints are common in the stems of the Alaskan kayaks. Gunwales and battens are most commonly lashed through holes bored in them and in the bow and stern members. Care is taken that all lashings are flush on the outside, so that the skin cover is smooth and chafing will be avoided. Bone knobs at stem and stern heads are used in the Coronation Gulf kayaks in the west and in many Greenland models. Bone stem bands are more widely employed, however, being in use at Kodiak and Nunivak Islands, in the Aleutians, at Norton Sound in Alaska, and in Greenland and Baffin Island in the east. It is probable that these bands were once in wider use than thus indicated. Strips of bone are also used to prevent chafing at gunwale in paddling and for strengthening scarphs in the manhole rim.

It will be noted that all Eskimo skin boats have a complete framing system, which is first erected and then fitted with the skin cover. This is a method of construction very different from that of the birch-bark canoes of the Indians living to the southward of the American Eskimo. The birch-bark canoe is built by forcing a framing system into an assembled cover and holding it in place there by a rigid gunwale structure, to which the bark cover is lashed. This basic structure is used even in the Alaskan area, where there are birch-bark canoes that in hull form and proportions strongly resemble the flat-bottom kayak. The basic difference between the two craft is illustrated by the fact that whereas the removal of the skin cover of the kayak leaves the frame intact, the removal of the bark cover of the kayak-like birch-bark canoes would result in the collapse of the framework, except for the gunwale-thwart structure or, in a few, the chine-floor structure. Because of this basic difference the superficial resemblance of some Indian bark canoes to kayaks has no meaningful relationship to the possibility of the influence of the kayak on the bark canoe, or vice-versa. Some Indian tribes have in fact built skin-covered canoes, as will be seen in chapter 8, but the framework and structural system used is always that of the bark canoe, never that of the Eskimo skin boat. Nor is there evidence that the Eskimo ever used the bark canoe frame-structure in their kyaks or umiaks. Hence, in spite of contact between these peoples, the watercraft of each remains basically different in structural design.

The almost universal method of constructing the kayak is first to shape and fasten together the gunwales and thwarts, with stem and stern pieces fitted as required, then to fit and place a few transverse frames to control the shape of the craft. Next the longitudinals are fitted and, finally, the remaining transverse frames are put in place. In some types the manhole rim is now fitted but in others the manhole rim is put on after the skin cover is in place, as some kayaks (the Alaskan) have the skin cover placed over the manhole rim and others have it passed under. The skin cover is stretched and sewn over the frame and is rarely secured to it by lashings except at the manhole. Due to the shape of bow and stern, in some types, difficult and tedious sewing is required to stretch the skins over the ends of the hull. Much of the sewing is completed after the skins are stretched over the hull and held by temporary lacings. The blind seam is used but in many kayaks the lap is very short, about ⅜ inch being common.

The covering most widely used in Alaskan kayaks was the bearded seal skin and with the Aleuts the skin of the sea lion was the most popular. Throughout the eastern Arctic seal skin was the preferred covering though caribou skin was occasionally used by the caribou Eskimos in the central Arctic. The heavy, thick hides were first piled and "sweated," until the hair became loose then the skins were scraped until they were clean. They were thin and light and could be air dried and stored until ready for use. The skins had to be well soaked before being stretched over the frame of a kayak or umiak. When dried out on the boat frame they were oiled in the usual manner. It is claimed by the Eskimos that walrus skin, though strong, is not as good as the bearded seal or the sea-lion skin for boat covers, as the latter two held the oil longer and did not become water soaked as quickly as the walrus hide.

The paddler's seat in most kayaks consists of a portion of heavy skin with fur attached. Sometimes this is supported by a few short, thin battens laced loosely together. These, and the fur seat sometimes are as long as the paddler's legs. No back rest is known to be used. The seat, and any batten supports, are loosely fitted and are not part of the permanent kayak structure.

The kayak is usually entered by floating the boat near a stone or low bank and stepping into it with one foot, which has first been carefully wiped. With the body steadied by placing the paddle upright on the shore, or outside the kayak, the other foot is then wiped and placed in the boat. The paddler then slides downward and works his legs under the deck until he is seated with his hips jammed into the manhole rim. Getting out of a kayak is almost the reverse of this process. Great care is exercised to avoid getting dirt into a kayak, as it might chafe the hide cover. Hence the care in wiping the feet before entering. The practice of entering the boat ashore and throwing man and kayak into the water, undoubtedly very rare, is said to have been practiced not only at King Island but in some parts of Greenland. Both Alaskan and Greenland hunters often lashed two kayaks together, in order to rest in rough weather. Many kayakers using the narrow models laid the paddle athwartships across the deck to help steady the kayak when resting or throwing a weapon; this is basically the same as holding the sculls of a racing shell in the water, to steady the boat. Lashing two kayaks side by side, or parallel with spacing rods, was commonly done to enable the craft to ferry persons or cargo across streams. Some Alaskan Eskimo thus converted kayaks into catamarans and then fitted a mast and sail, but such an arrangement was never used in rough water.

The methods used by a paddler to right a capsized kayak, without aid and while he was still in the cockpit, have aroused the interest of many canoeists. It was used by the King Islanders, some of the Aleuts, and the Greenlanders, who at times, it is said, would deliberately capsize their kayak to avoid the blow of a heavy breaking sea, then right it when the sea had passed. The Eskimo are reported to be gradually losing this skill, but in late years European and American kayakers have learned this method, called the "kayak roll," of righting a decked canoe with paddler in place. It follows in general the Greenland method. In the Appendix (p. 223) is an illustrated description of the kayak roll, supplied by John Heath.

Traditionally, the weapons used by kayakers were darts and harpoons, the bow not being employed, since wetting would damage the weapon. Various forms were used, and many were thrown with the "throwing-stick" to increase the range and force. An inflated bladder or skin was often carried to buoy the harpoon line and tire the game. Bolas and knives were also carried. All eastern kayaks appear to have been propelled with the double-blade paddle, but folklore suggests that the single-blade kayak paddle

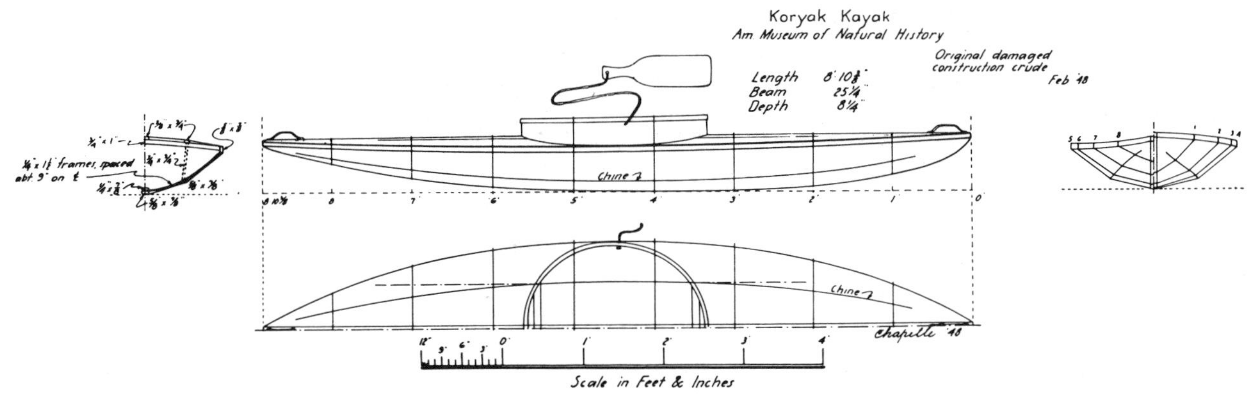


Figure 176

KORYAK KAYAK, drawn from damaged kayak in the American Museum of Natural History, 1948.

may have once been used. Greenland kayaks have been reported as carrying a small square sail, but this was actually a hunting screen, or camouflage, to hide the paddler and cause the seal to mistake the canoe for a cake of ice. It was a 19th-century addition, as was a fin attached to the kayak to counteract the effect of the screen in a beam wind. Any effect it had as a sail in a kayak was unintentional, of course; it was dismounted in strong winds or when not required for hunting.

Shown above is the plan of an Asiatic Koryak kayak. This type, used in the Sea of Okhotsk and on the Siberian coast of Bering Sea, is the only distinctive Asiatic type; the Chukchi of the Siberian side of Bering Strait uses a kayak that is on the same model as the one found at Norton Sound, in Alaska. The Chukchi kayak differs only in the ends, which are wholly functional and without the handgrips that distinguish the Alaskan type. There is also a crude Chukchi river kayak, covered with reindeer skin, but its design is not represented in an American museum.

The Koryak kayak is a hunting boat well designed for use in protected waters, but is rather weakly built. In general form it is much like the hunting and fowling skiffs formerly used in America. The plan idealizes the kayak somewhat, for the boat is crude in finish. The only example available for study, in the American Museum of Natural History, is in poor condition. The hull is short, wide and shallow, rather V in cross section, and there is a slight camber in the deck. The length of the Koryak kayak rarely exceeds 10 feet, the beam is from 24 to 26 inches, and the depth between 8 and 9½ inches. The manhole rim is of large diameter, high and without rake. The gunwales, although rather slight, are the strength members. The keelson, a thin, flat batten, forms the stem and stern posts; it is stiffened amidships by a short batten lashed inside the frames. The chine battens are also slight and do not reach the stem and stern. The frames are widely spaced and are wide and thin, in one piece from gunwale to gunwale. There are but two thwarts; these are strong and support the manhole rim, showing inside the cockpit. Two thin longitudinal battens afore and abaft the manhole, support the deck, in addition to a light centerline ridge-batten. On the kayak illustrated the outboard battens appear to have had additional support at one time from two pairs of stanchions standing on frames at the chines, with their heads secured to the deck battens; a pair being placed before and abaft the manhole. A small plank seat appears to have been used and the boat was propelled by two short one-hand paddles, secured to the manhole rim by lanyards made of thongs; these would be only efficient in smooth water. The cover is made from bearded seal skins and passes under the manhole rim being sewn to the rim on the inside at the top, by coarse sewing passed through holes bored in the manhole rim. There are two thong lifting handles or loops, one at bow and stern. This kayak is the most primitive of all types and the smallest as well. The Koryaks are not daring canoemen and do not venture into rough water. Nevertheless, this type of kayak is said to be fast and highly maneuverable.

Compared to the Koryak, the Alaskan kayak is tremendously advanced. The Aleuts are daring and accomplished kayakers, and their craft are among the finest in the Arctic. The Bristol Bay kayak

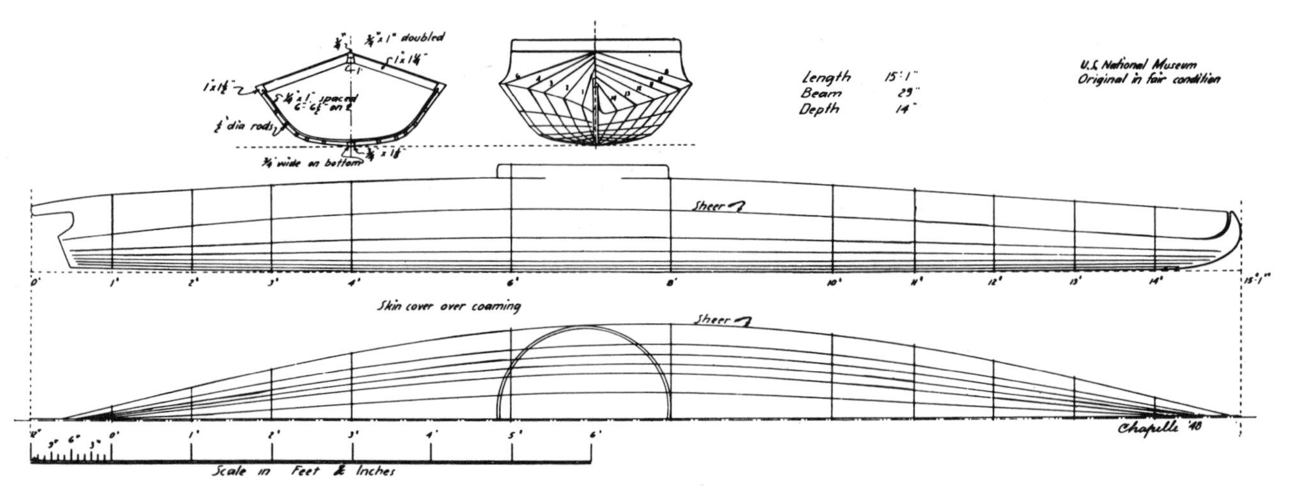


Figure 177

**Bristol Bay Kayak, 1885, in U.S. National Museum (USNM 76285). The identification of this kayak has been questioned by Henry B. Collins and John Heath, but it may represent an old form out of use in the twentieth century.**

of 1885, shown above, represents one type used in this area and that from Unalaska, shown below, the other. The Bristol Bay boat is rather short and wide, measuring 15 feet 1 inch in length, 29 inches beam and 14 inches depth to ridge batten of the deck just forward of the manhole. The boat has the humped sheer found in many Alaskan kayaks and is intended for use in stormy waters. Its large manhole, also a feature of the Nunivak Island kayak, permits two persons to be carried, one facing forward to paddle and the passenger facing aft, or the space can be used to carry cargo. The drawing shows the construction and requires no detailed explanation. Kayaks from the Aleutian Islands eastward to Kodiak use rod battens; only the gunwales and keelson are rectangular in section. The frames are thin flat strips bent in one piece from gunwale to gunwale. The ridge-batten of the deck is laminated, in two pieces. The deck beams and thwarts are notched into the ridge-batten and lashed. The bow piece is carved from a block, and the longitudinals are lashed to it, each in a carefully fitted notch. The sternpost is formed of a plank. The skin cover passes over the manhole rim and a line passed outside the rim holds the skin down enough to form a breakwater. The skin cover is sewn to the inside lower edge of the rim, thus covering it almost completely.

The Unalaska kayak of 1894 (below) is a better known type. This design is used throughout the Aleutians and on the adjacent mainland as far east as Prince William Sound. It was also employed in the Pribilof Islands and at St. Matthew, having been used by Aleuts engaged in sealing expeditions there. All kayaks of this type do not have the same bow and stern profiles as the example; some have the bifid

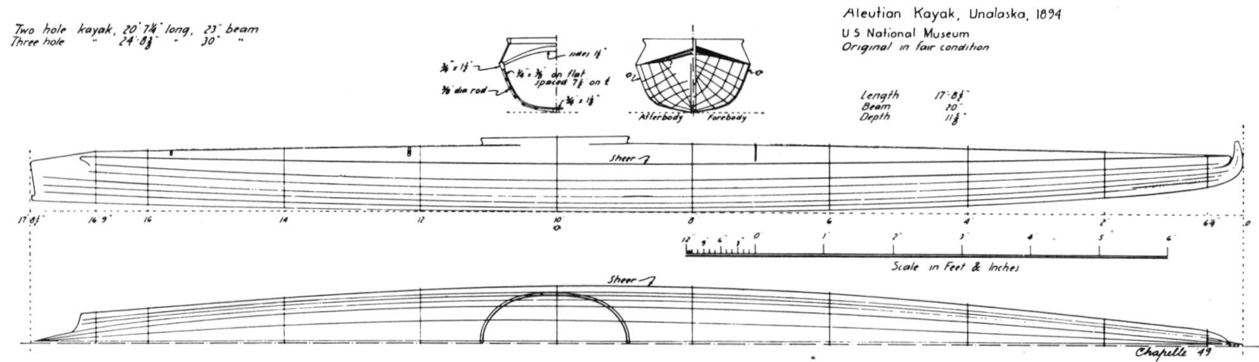


Figure 178

Aleutian Kayak, Unalaska, 1894, in U.S. National Museum (USNM 76282).

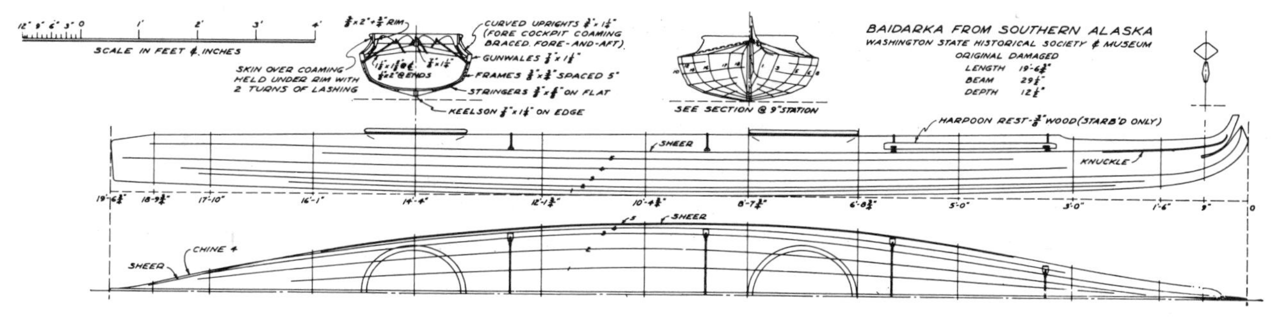


Figure 179

**Kayak From Kodiak Island, 2-hole Aleutian type, in Washington State Historical Society and Museum. Taken off by John Heath, 1962.**

bow built with the portion above the slit arched upward higher than the outer stem-piece and so more prominent; there are also minor variations in the stern. The shape of the hull, however, is consistently maintained throughout the area in which this type is used. Though the deck is ridged, it is relatively low compared to that of the Kodiak kayak, and the thwarts supporting the manhole are heavily arched and in one piece from gunwale to gunwale. The construction is like that of the Kodiak kayak, but the gunwales and upper longitudinal battens, instead of meeting the stern post, end on a crosspiece well inside the stern to give the effect of a transom stern. However, some Aleut kayaks have the normal sharp stern after the fashion of the Kodiak kayak, but without the projecting tail or handgrip, and nearly all have two thwarts between the after manhole thwart and the stern and three forward of the fore manhole thwart. The skin cover passes over the manhole rim as in the Kodiak type. The bow block is sometimes built up of two blocks sewn or laced together. Strengthening pieces of light plank are sometimes fitted from the bow block aft; these are laced to the top inside edge of the gunwales and pinned to the stem block to form long breast-hooks. In some kayaks with the square stern, only the gunwale is supported by the crosspiece on the stern, the two battens on each side being supported by the last frame only, about 6 inches inboard of the crosspiece.

This type of kayak is the only one known to have been built with more than one manhole. The two-hole kayak is an Aleut development used in whaling and sea-otter hunting, so far as is known; the paddler sits in the after manhole. Measurements of a two-hole kayak in the United States National Museum show it to be 20 feet 7¼ inches long, 23 inches beam, and 9½ inches deep to top of gunwale. The manholes are about 46 inches apart edge to edge and the foremost is about 8 feet from the bow.

The three-holer, commonly believed to have been introduced by the Russians, was used by Russian officers, inspectors, and traders in their explorations and travels on the Alaskan coast. One of these boats measures 24 feet 8⅜ inches long, 30 inches beam, and 10½ inches deep to top of gunwale. The center manhole is commonly larger in diameter than the other two and is used for either a passenger or cargo. The fore edge of the fore manhole is 8 feet to 8½ feet from the bow and the other manholes are from 4 to 4½ feet apart edge to edge. A large example of this class of kayak measures 28 feet 1½ inches long, 38½ inches beam and 12 inches deep to top of gunwale. Probably none exceed 30 feet in length. Both the single- and the double-blade paddle are used by the Aleuts, but the double blade is preferred in hunting. The paddle blades are rather narrow and leaf-shaped, with pointed tips.

The plan of a kayak from Nunivak Island (about due north of Unalaska and roughly half-way to St. Lawrence Island) is shown on page 198 (fig. 180). This type of kayak is obviously related to that of Kodiak Island, for it has approximately the same lines and proportions. Only the profiles of bow and stern exhibit marked differences. Perhaps the most striking feature of the Nunivak kayak is its bow, which might represent a seal's head; a hole through the whole bow structure forms the eyes and also serves functionally as a lifting handle. The stern profile is simpler than that used in the Kodiak kayaks. The example shows the mythological water monster Palriayuk, a painted totem that once distinguished the Nunivak kayaks; missionary influence has long since erased

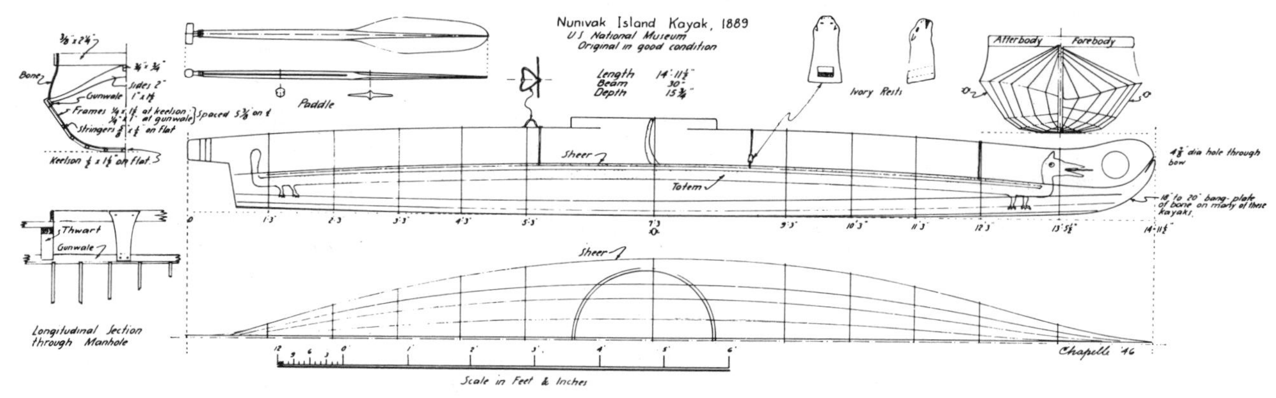


Figure 180

NUNIVAK ISLAND KAYAK, Alaska, 1889, in U.S. National Museum (USNM 160345), showing painted decoration of the mythological water monster Palriayuk.

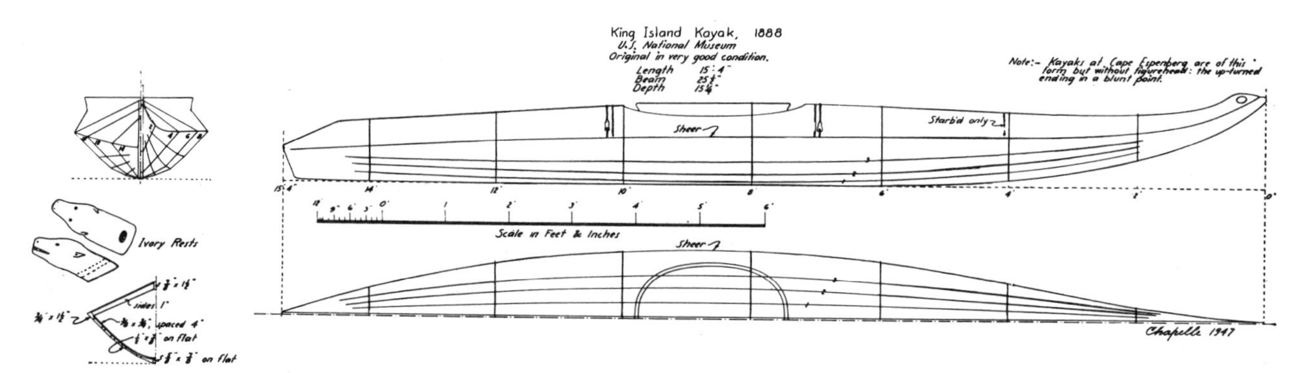


Figure 181

KING ISLAND KAYAK, Alaska, 1888, in U.S. National Museum (USNM 160326), collected by Capt. M. A. Healy, U.S. Revenue Steamer *Bear*.

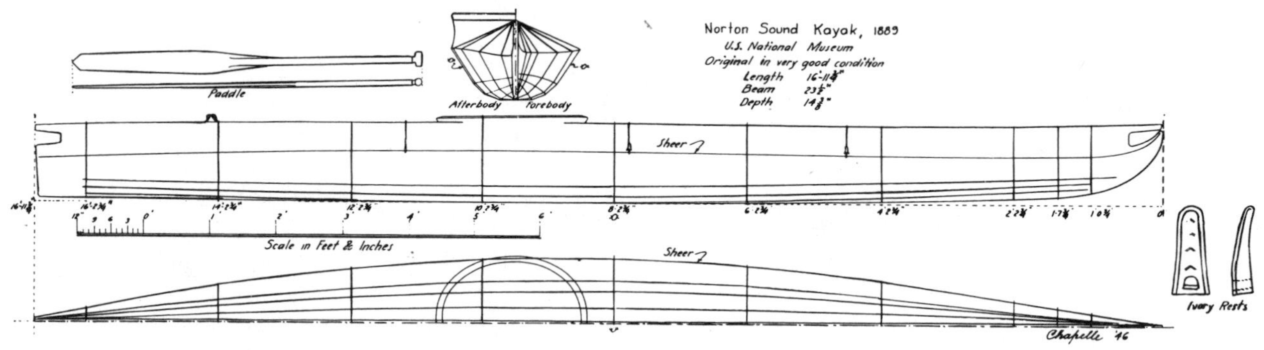


Figure 182

NORTON SOUND KAYAK, Alaska, 1889, U.S. National Museum (USNM 160175).

Figure 183

Nunivak Island Kayak with picture of mythological water monster Palriayuk painted along gunwale. (*Photo by Henry B. Collins.*)

such decorations from Alaskan kayaks. Whereas the Kodiak kayak has eleven battens (including the keelson) in its frame, the Nunivak kayak has nine, and all the longitudinals in it are rectangular in section. Differences in dimensions of Nunivak and Kodiak kayaks are remarkably slight; the greatest length reported for either type is about 15 feet 9 inches and the greatest beam is about 32 inches. Both types have a large manhole and carry a passenger back-to-back with the paddler. The single-bladed paddle is used. The kayak is sometimes transported over ice by means of a short sledge, by one man, but it is otherwise rather heavy to portage. Highly regarded by all who have had contact with it, this is generally considered one of the safest and most useful of the Alaskan kayaks.

King Island, at the entrance to Bering Strait, is the home of the kayak shown on page 198 (fig. 181). The King Islanders are noted as skillful kayakers and their kayak generally follows the Nunivak pattern, but is narrower and more V-shaped in cross section, and the stem and stern are also distinctly different. The King Island craft has a bold upturned stem ending in a small birdlike head, with a small hole through it to represent eyes and to serve for a lifting grip; the stern is low and without the projections seen in the Nunivak type. The fitting of the cockpit rim of the U.S. National Museum kayak is unusual; the rim is not supported by thwarts but rather is made part of the skin cover and therefore can be moved. This seemed to be intentional, for there is no evidence of broken or missing members, but John Heath considers this not typical. A watertight jacket with the skirt laced to the manhole rim is worn by the kayaker to prevent swamping. This practice was common among Eskimo working in stormy waters. A warm-weather alternate was a wide waist-band, with its top supported by straps

Figure 184

Nunivak Island Kayak in U.S. National Museum (USNM 76283) with cover partly removed to show framework. Collected by Ivan Petroff, March 30, 1894.

Figure 185

Western Alaskan Kayak, Cape Prince of Wales, 1936.
(*Photo by Henry B. Collins.*)

over the shoulders and the bottom laced to the manhole.

A somewhat similar but slightly smaller kayak was used at Cape Espenberg; in these the upturned bow ended in a simple point. The sterns were alike in both types. The Cape Espenberg kayak had a fixed cockpit rim however, as in the Nunivak type. Both types employed the single-bladed paddle.

A little to the South, in Norton Sound, the long narrow kayak shown on page 198 (fig. 182) is popular. These are somewhat like the Nunivak kayaks in cross section but with far less beam. They have a slight reverse, or humped, sheer and are very sharp ended. The peculiar handgrips at bow and stern are characteristic, though the shape and size of the grips vary among the villages; the style shown is that of St. Michaels. A single-bladed paddle is used. This type is very fast under paddle, but requires a skillful user in rough water. The Norton Sound kayaks are very well finished and strongly built.

From Kotzebue Sound, at Cape Krusenstern, along the north coast of Alaska to near the Mackenzie Delta, the kayaks are very low in the water, long, narrow, and spindle-shaped at the ends. They are distinguished by a very strong rake in the manhole rim, with an accompanying prominent swell in the deck forward of the manhole. They are built with seven longitudinal battens (including the keelson) in addition to the gunwales. In several examples seen, the latter are sometimes slightly channelled on the inside, but this may have been the result of shrinkage in the pith of the timber used and not intentional. These kayaks are very light and easily carried. Both single- and double-blade paddles are employed; the single blade is usually used in travelling.

On page 201 are shown a kayak from Cape Krusenstern (fig. 186) and one from Point Barrow (fig. 187). It is reported that these types have now gone out of use. In these boats no stem or stern posts exist, these usually being replaced by small end blocks. The only important difference in the two types shown is in the style of crowning the deck, which is ridged in the Cape Krusenstern kayak but more rounded in the Point Barrow kayak. In spite of their narrow beam and obviously unstable form, these kayaks are said to have been used by rather unskillful paddlers. In general, they were not employed in rough weather but were seaworthy in skillful hands.

Though the North Alaska type of kayak, as illustrated by the Point Barrow model (fig. 187), may be said to represent the structural design of kayaks to the eastward as far as Foxe Basin, the Mackenzie Delta kayaks are on an entirely different model. Due to migration of numerous groups of Eskimo to this area in the last seventy years, the design of kayaks here has undergone a great change. In figure 188 appears the plan of a modern Mackenzie Delta kayak.

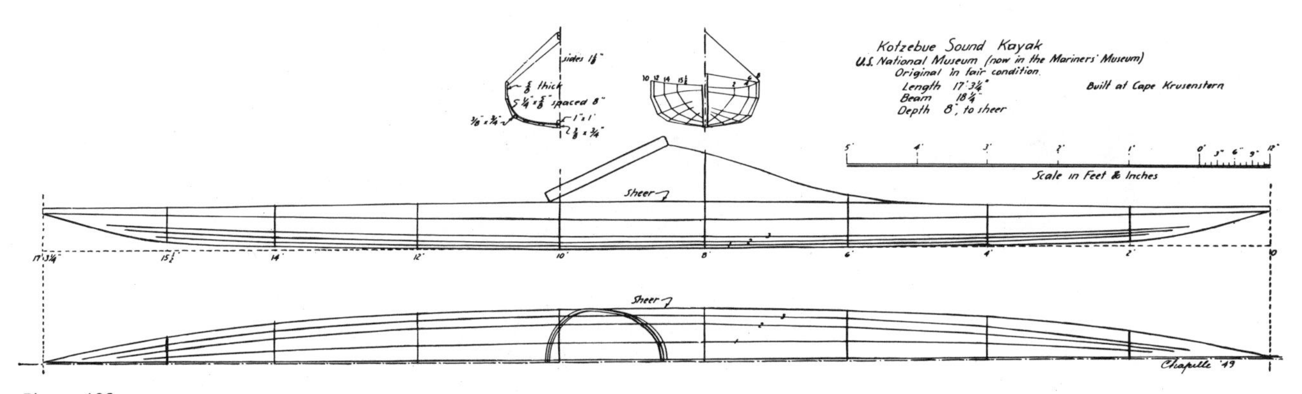


Figure 186

Kotzebue Sound Kayak (Cape Krusenstern), Alaska, formerly in U.S. National Museum, now in Mariner's Museum.

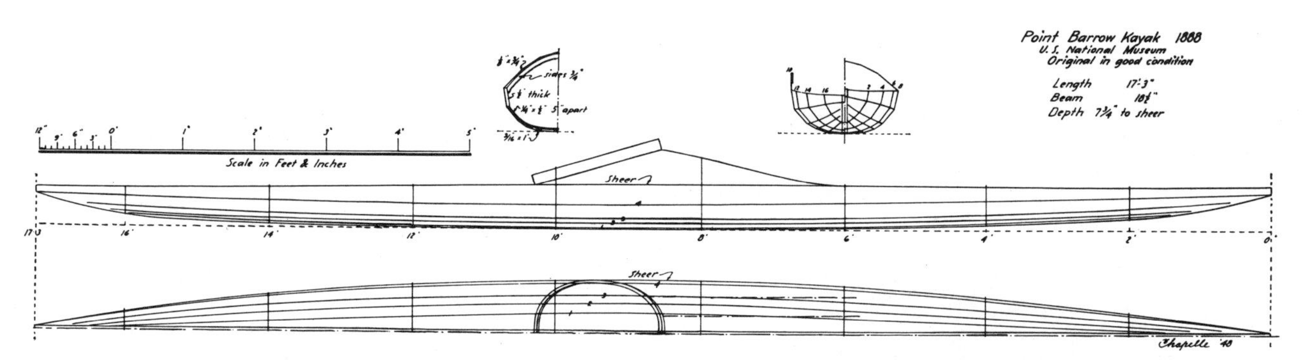


Figure 187

Point Barrow Kayak, Alaska, 1888, in U.S. National Museum (USNM 57773).

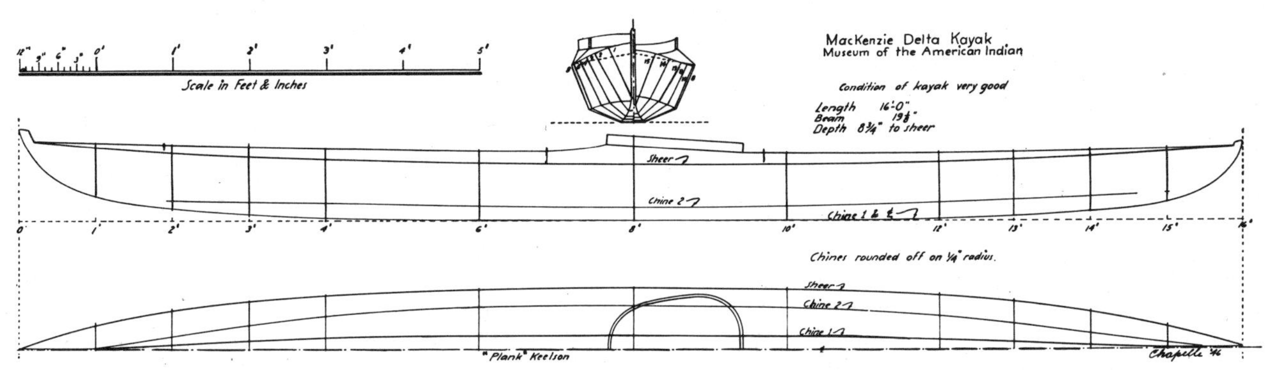


Figure 188

Mackenzie Delta Kayak, in Museum of the American Indian, Heye Foundation.

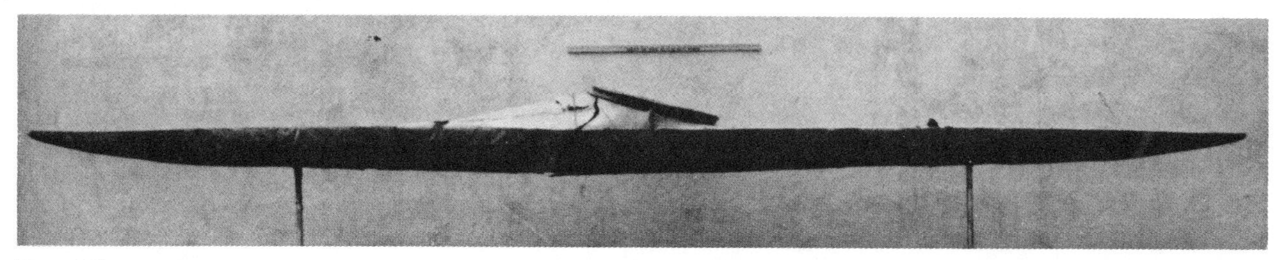

Figure 189

KAYAK FROM POINT BARROW, Alaska, in U.S. National Museum (USNM 57773). Collected by Capt. M. A. Healy, U.S. Revenue Steamer *Bear*, 1888. (*Smithsonian photo MNH-399-A.*)

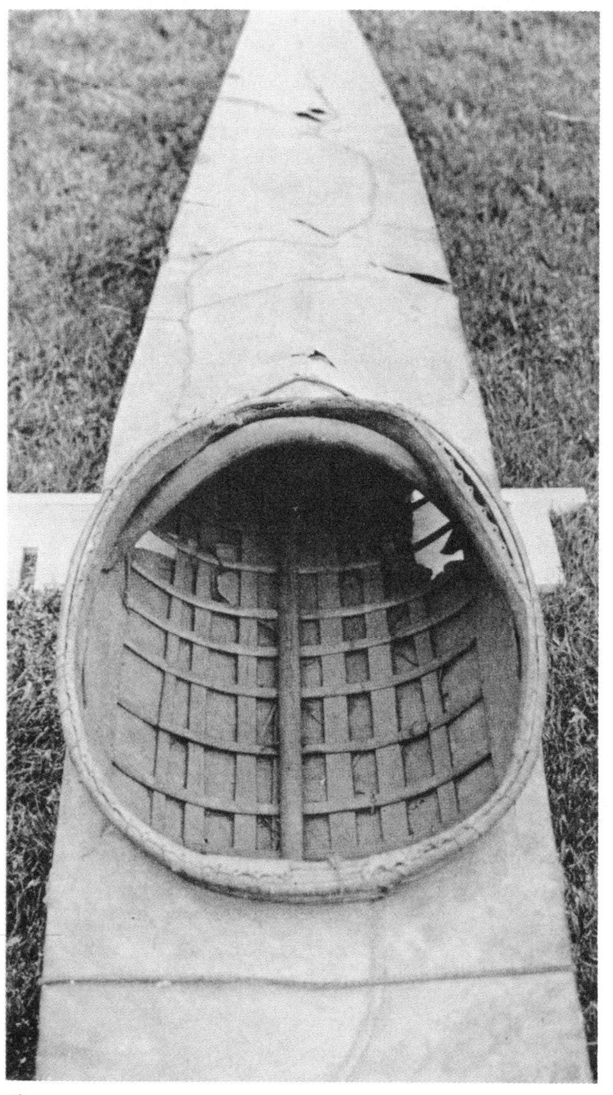

Figure 190

COCKPIT OF KAYAK from Point Barrow (USNM 57773), showing method of lashing skin cover to manhole. (*Smithsonian photo MNH-399.*)

The design is marked by a very narrow flat bottom or a wide keel combined with the V-bottom. These boats are well-built and are light and graceful. The wide keel is formed by a thick plank keelson which narrows at bow and stern and is bent up to form the stem and stern. The chine pieces run fore and aft and are lashed to the stem and stern thus formed. The gunwales are about ¾ by 1⅛ inches. The frames are about ¼ by ⅝ inch bent in a strongly U-shaped form, with their ends tenoned into the bottom of the gunwales. The keelson is only about ⅜ inch thick and the chines are rather wide thin battens; about 5/16 by 1¼ inch. Some kayaks have an additional batten in the sides above the chines. The deck is slightly ridged for nearly the length of the boat. The stem and stern are carried up above the sheer to form prominent posts; some builders carry them higher than shown. The construction is neat and light and the boat is very easily paddled. Its narrow beam makes it somewhat treacherous, however, in unskilled hands. A double-bladed paddle is generally used with this kayak. While the form appears to vary little among individuals of this class, the construction varies, particularly in the number and dimensions of the longitudinals. Frames are spaced rather consistently 5 to 6 inches apart.

The foregoing design differs greatly in every respect from the example in figure 191, collected by the U.S. Fish Commission in 1885 and identified as a Mackenzie River kayak. It is a large heavy boat compared to the one just described. The model of this old kayak, and the construction too, is on the eastern pattern, such as is used in Hudson Strait. The strongly upturned stern and less rising bow resembles the old Greenland kayaks. The V-bottom and 3-batten construction combined with heavy deep gunwales is not to be found in any of the known Alaskan kayaks. There is unfortunately no record

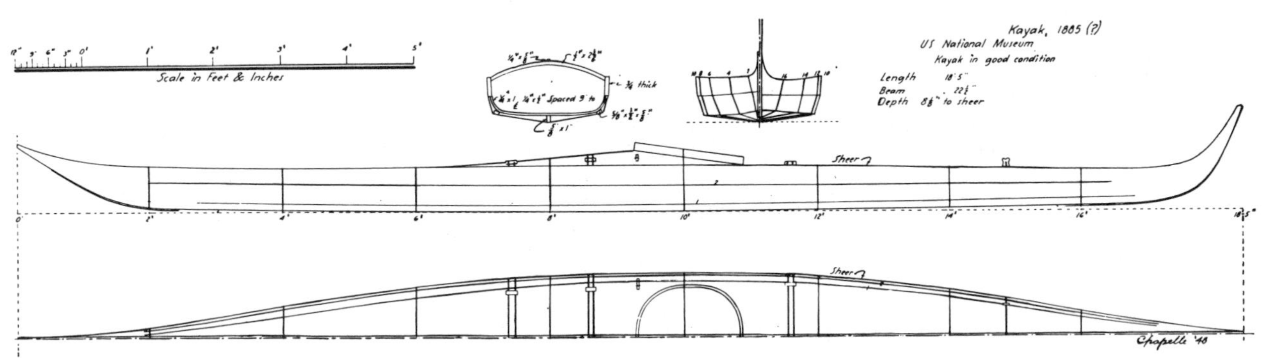


Figure 191

**Kayak in U.S. National Museum** (USNM 160325) **cataloged as from Mackenzie River area, 1885, but apparently an eastern kayak from Upernavik, West Greenland.**

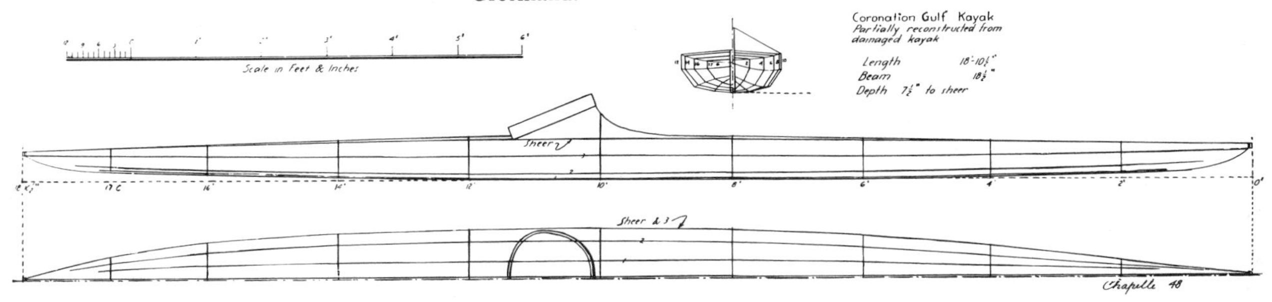


Figure 192

Coronation Gulf Kayak, Canada, partially reconstructed from a damaged privately owned kayak (now destroyed).

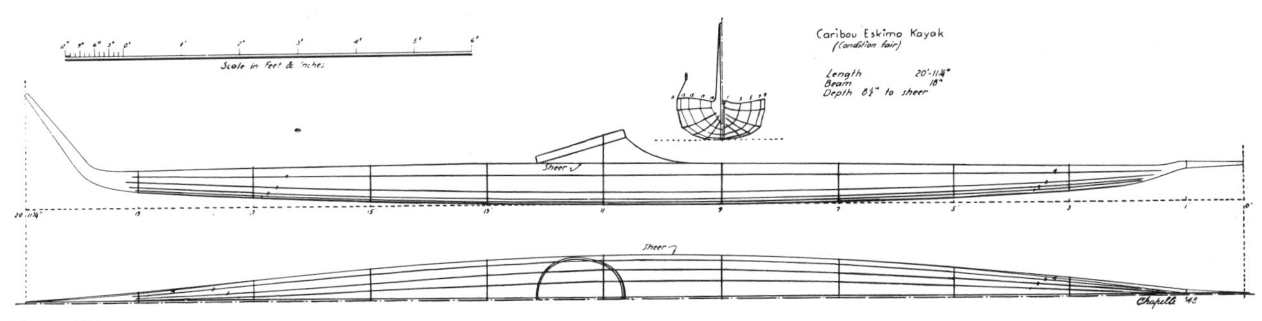


Figure 193

Caribou Eskimo Kayak, Canada, in American Museum of Natural History.

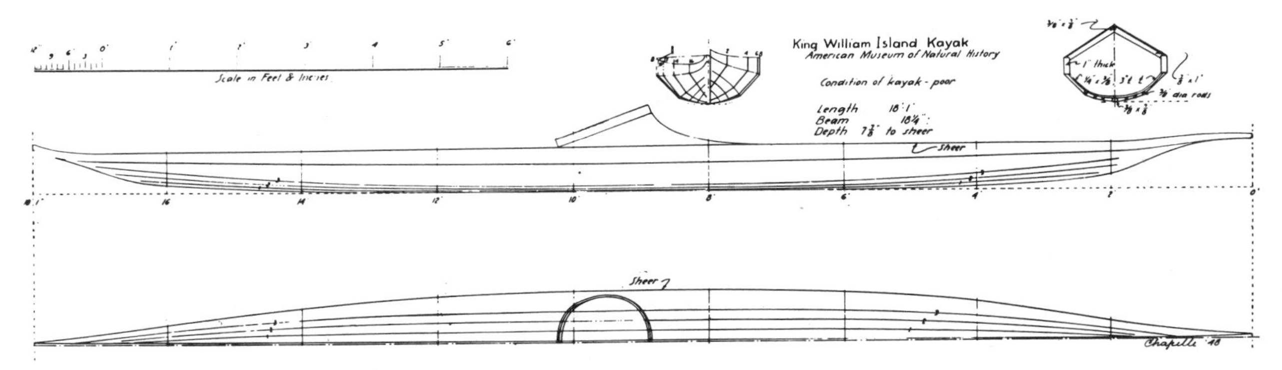


Figure 194

Netsilik Eskimo Kayak, King William Island, Canada, in the American Museum of Natural History.

of the exact location where this kayak was found, nor any information on the builders; if it is from the Mackenzie, the type now appears to be wholly extinct and there has been nothing in recent times in the vicinity faintly resembling it. The kayak is a well-built, safe, strong boat; the high stern would aid it in coming head to sea and wind when paddling stopped; and it resembles, more than most, the early explorers' drawings of Arctic kayaks. The very high ends indicate that it was not used where high winds are common, despite the otherwise seaworthy design and construction, and regardless of the documentation, it now seems certain that this kayak came from somewhere in the eastern Arctic.

To the eastward of the Mackenzie, the kayaks are narrow, spindle-shaped and very low sided, in the manner of the northern Alaskan boats. The drawing of figure 192 was made from the remains of a kayak from Coronation Gulf and to insure accuracy was compared with photographs and measurements of some Copper Eskimo kayaks. This kayak is characterized by a rather marked reverse sheer and a strongly raked manhole rim. The deck forward of the manhole sweeps up very sharply, but with a different profile than is seen on the north coast of Alaska; the deck of these eastern kayaks sweeps up in a very short hollow curve instead of the long convex sweep popular in Alaska. The ends of the hull finish in small bone buttons; the skin cover passes under the manhole rim, as in the Cape Krusenstern and Point Barrow types. A two-bladed paddle is commonly used. The hull design is more stable than that at Point Barrow and the ends are somewhat fuller, giving the boat a rather parallel sided appearance; it has longitudinal battens from the bottom of the hull, one the keelson; the gunwales are channelled on the inside and are very light and neatly made. The frames are split willows, round on the inside.

The Caribou Eskimo kayak preserved in the American Museum of Natural History is the best example of the type found. The drawing of figure 193 shows the features of this particular type; the construction is about the same as that of the Point Barrow kayak but is much lighter and weaker. The peculiar projecting stem is formed of a stem block, scarphed to the gunwales; to it the beak piece is attached with a lashing. The sharply turned-up stern is formed in a similar manner by two pieces joined together at the tip and lashed to the stern block; this stern construction is similar to that of the eastern Arctic kayak shown in figure 192. Both caribou hides and seal skins are used to cover the Caribou Eskimo kayak. The seams are rubbed with fish oil and ochre, a method also used extensively along the north coast of Alaska to paint the framework of both kayaks and umiaks.

The Netsilik Eskimo kayak is related to the Caribou, but is less stable and has different bow and stern profiles. The example shown in the drawing of figure 194 requires little discussion; the cover is of seal skin. These kayaks are used only in hunting caribou at stream crossings and are not employed in sealing. The very narrow bottom and narrow beam make this the most dangerous of all kayaks in the hands of a paddler unaccustomed to such craft. Neither the Caribou nor the Netsilik kayaks are very seaworthy and their construction is inferior. They are characterized by rather heavy gunwales but the other members of their structures are very slight.

No examples remain of the old kayaks once used on the Gulf of Boothia, at Fury and Hecla Strait, and on the west side of Foxe Basin. Early explorers in this area found kayaks, but the types used have been long extinct. One kayak, supposed to have been built at Southampton Island, had been preserved by a private collector, but when measured was in a damaged state. Shown in figure 195, it does not conform with the old description of kayaks from the Melville Peninsula but does agree reasonably well with the Boas model of a kayak from Repulse Bay in the U.S. National Museum (USNM 68126). On this basis it would appear that in Boas' time this form of kayak was also used on the east side of the Melville Peninsula. The design resembles to some extent the kayaks from the southwest coast of Greenland, but the stern is like that used in some Labrador craft. This old kayak was very light and sharp, rather slightly built, but very graceful in model so far as could be determined from the remains of the craft. The foredeck camber is ridged and carried rather far forward. If the identification of this kayak should be correct, it is apparent that the eastern model of the kayak once extended as far west as the west side of Foxe Basin.

The kayak of lower Baffin Island, in figure 196, is flat-bottomed, long, and rather heavy. The gunwale members are very deep and the keelson and chine battens are quite heavy. This type has a slight side-batten between chine and gunwale—in all, five longitudinal members besides the gunwales—hence this example is the sole exception to the 3-batten construction that may be said to mark the

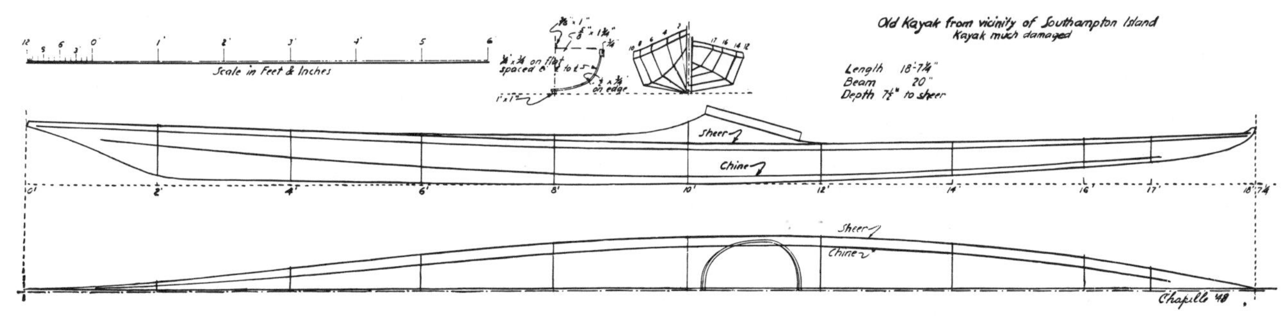


Figure 195

Old Kayak From Vicinity of Southampton Island, Canada. Plan made from a much damaged kayak, now destroyed, once privately owned.

eastern kayaks. The Baffin Island kayak is rather roughly built and the two examples found had many frames cracked at the chines. However, this kayak has many excellent features, being easily paddled, very stable, and seaworthy. The double-blade paddle used is like that of the Labrador kayak, very long with narrow blades. When the paddler is seated, these kayaks, like many of their eastern sisters, draw more water forward than the illustration would indicate (it should be remembered that the trim of the kayaks in the water is not indicated by the base lines used in the plans). The deeper draft at the bow, which allows the kayak to hold her course into the wind and to come head to the wind when at rest, gives a long easy run in the bottom toward the stern. The slight rocker in the bottom shown in the drawing is thus misleading. The stem is formed by the extension of the keelson, producing the "clipper-bow" seen in many eastern boats. The stern is shaped by a stern block of simple form into which the gunwales, keelson and chines are notched. The batten between chine and gunwale stops a little short of both bow and stern

A somewhat similar kayak is used on the Labrador side of Hudson Strait but, as shown in figure 197 on page 207, the appearance of the craft is distinctive. The kayak is flat-bottomed, with the snied-off chines seen in the Baffin Island boat, giving a cross section form like that of many Japanese sampans. The 3-batten system is used in construction, and the gunwales are very heavy and deep, standing vertical in the sides of the boat. The sheer is slightly reversed and there is little rocker in the bottom. One of the most obvious features of the Labrador kayak is the long "grab" bow, which is formed by a batten attached to the end of the keelson. The stern is formed with a very small block inside the gunwales, and to this the keelson is laced or pegged. It will be noticed that the rake of the manhole is very moderate. These kayaks are heavy and strong, paddle well, particularly so against wind and sea. Shown in the drawing is the type of long- and narrow-bladed paddle used.

This example illustrates better than the Baffin Island kayak the combination of deep forefoot and the greatest beam well abaft the midlength that marks many eastern models. When paddled, the craft

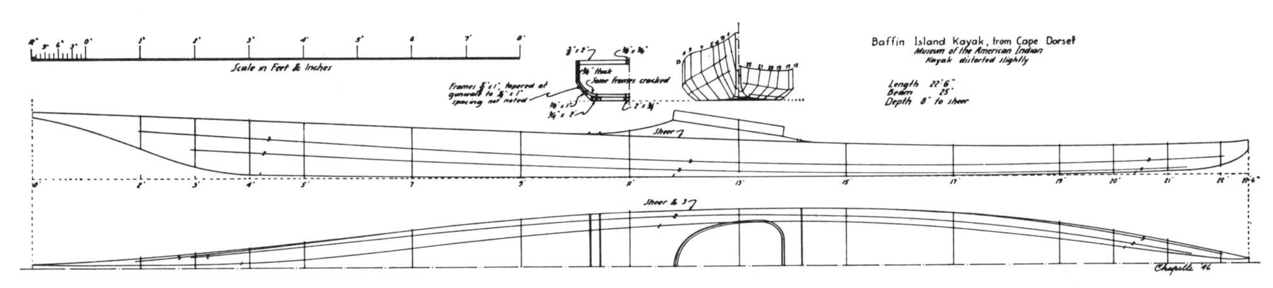


Figure 196

Baffin Island Kayak, from Cape Dorset, Canada, in the Museum of the American Indian, Heye Foundation.

always trims so that the kayak draws most water at the fore end of the keelson and the bottom of the stern is usually just awash. This makes the bottom sweep up from the forefoot in a very slight gradual curve to the stern, when the boat is afloat. As a result, the kayak may be said to be of the "double-wedge" form that has been popular in fast low-powered motor boats, since having the beam far aft gives to the bow a wedge shape in plan, while the deep forefoot and shallow stern produce an opposite wedge in profile. It would appear that this form had been found by trial and error to produce a fast, easily paddled rough-water kayak in an otherwise heavy hull. The North Labrador kayaks are the largest in the Arctic for a single person; some are reported as long as 26 feet. The long- and narrow-bladed paddle may be explained by the fact that the Eskimo never produced a "feathered" double paddle, with blades set at right angles to one another. To paddle against strong winds, he developed a blade that was very long and very narrow for a double-paddle, and therefore offered less resistance to the wind, yet could be dipped deep so that little propulsion effect was lost.

The kayak used on the northeast coast of Labrador, shown in figure 198, differs slightly from that of Hudson Strait. The northeast-coast kayak has a very slight V-bottom and a strong concave sheer with relatively great rocker in the bottom. While the craft trims by the bow afloat, the rocker probably makes it more maneuverable than the Hudson Strait kayak, though less easily paddled against strong winds. The V-bottom is formed by using a keelson that is heavier and deeper than the chines. The latter are thin, wide battens, on the flat. The V-bottom appears to help the boat run straight under paddle and may be said to counteract, to some extent at least, the effect of the strongly rockered bottom.

The Polar coast of Greenland is the home of sharpie-model kayaks having flat bottoms and flaring sides; the kayaks in figures 199 and 200 are representative of those used in the extreme north. These have "clipper" bows, with sterns of varying depth and shape, concave sheer and varying degrees of rocker in the bottom. Most have their greatest beam well aft and draw more water forward, as do the Labrador and Baffin Island types. The chief characteristic of the construction of this type is that the transverse frames are in three parts, somewhat as in the umiak. However, these kayaks depart from umiak construction in having the frame heads rigidly tenoned into the gunwales. This is done to give the structure a measure of transverse rigidity which would otherwise be lacking, since light battens are used for the keelson, stem, and chines. Figure 199 shows the details of the construction used.

These kayaks are highly developed craft—stable, fast, and seaworthy—and the construction is light yet strong enough to withstand the severe abuse sometimes given them. The cap on the fore part of the manhole is a paddle holder, for resting the paddle across the deck. Some Eskimos used this as a thole, and when tired, "rowed" the kayak with the paddle, to maintain control. It will be noted that oval or circular manholes are seldom found in the eastern types of kayaks already described; U-shaped manholes, or bent-rim manholes approaching this form, appear in those very stable types which do not require to be righted at sea by the paddler and in which the watertight paddling jacket or waistband is not used.

Farther south, on the northern coast of Greenland, and apparently also on the opposite coast of Baffin Island, a modified design of kayak is used. This type ,illustrated in figure 205, shows relationship to both the flat-bottom kayak of northern Greenland and to the northeastern Labrador type. In this model the "clipper" bow is retained but the stern and cross section resemble those of the Labrador kayaks. The construction, however, is fundamentally that employed in northern Greenland. As in the Labrador type, the deadrise in the bottom is formed by using in the keelson members tath are deeper than those in the chine. The gunwales do not flare as in the Greenland model, but stand vertical in the side flaring slightly at bow and extreme stern. The frame heads are rather loosely tenoned and are commonly secured to the gunwales with lashings. Transverse stiffness is obtained in this model by employing a rather heavy, rigid keelson fixed to the stern block, and by a tripod arrangement forward consisting of the stem batten and a pair of transverse frames placed at the junction of stem and keelson with their heads firmly lashed and tenoned into the gunwales. The construction, though strong, is rather rough compared to that of other Greenland types. The manhole rim in this type is not bent, but is made up of short straight pieces, as shown in the drawing; and the double-bladed paddle shown resembles that used in Labrador. This is a rather heavy kayak of very good qualities but not as maneuverable as some of the flat-bottom kayaks found farther north.

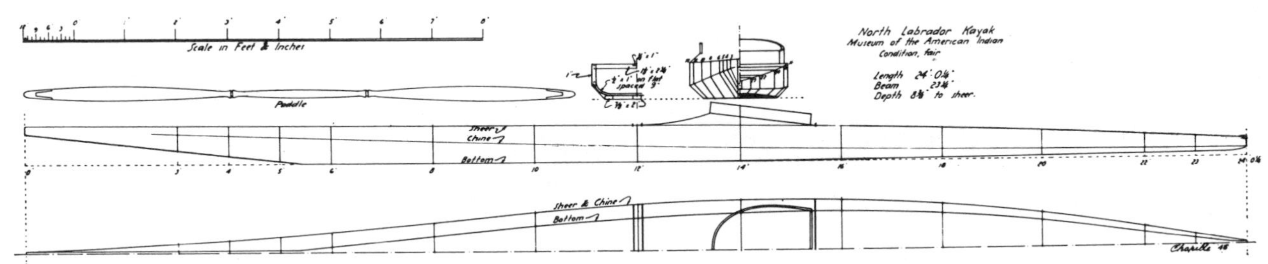


Figure 197

**Kayak from North Labrador**, Canada, in the Museum of the American Indian, Heye Foundation.

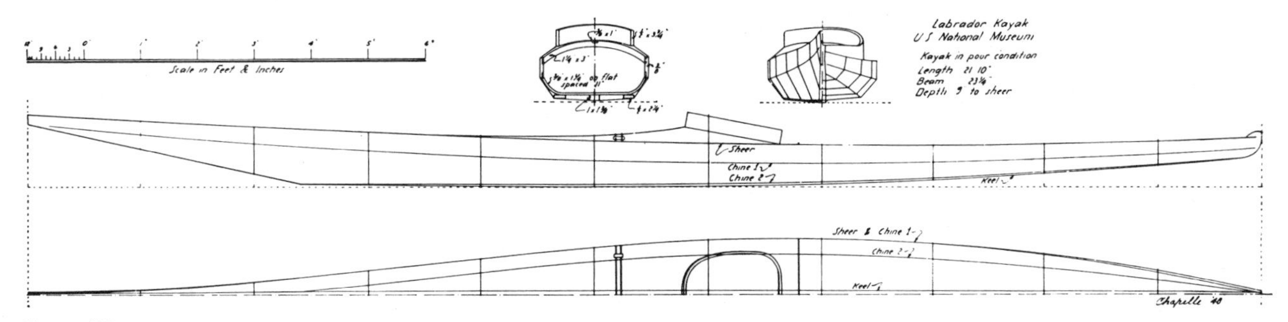


Figure 198

**Labrador Kayak**, Canada, in the U.S. National Museum (USNM 251693).

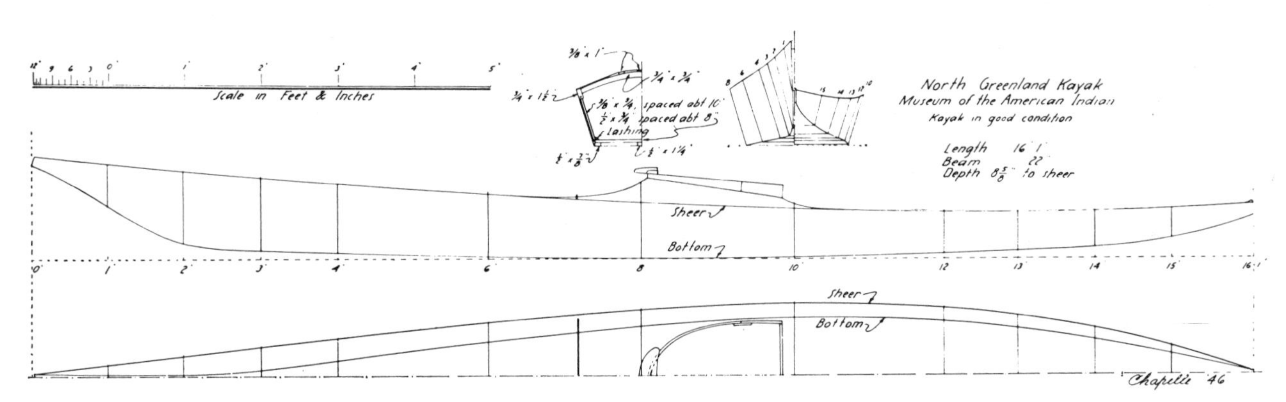


Figure 199

**North Greenland Kayak**, in the Museum of the American Indian, Heye Foundation.

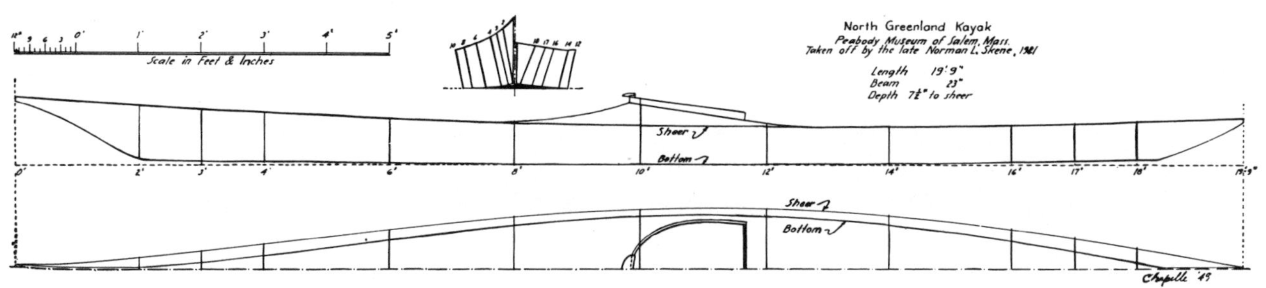


Figure 200

**North Greenland Kayak**, in the Peabody Museum, Salem, Mass. Taken off by the late Norman L. Skene, 1921.

Figure 201

PROFILE OF GREENLAND KAYAK from Disko Bay, in the National Museum (USNM 72564). Collected by Maj. Wm. M. Beebe, Jr., 1882. (*Smithsonian photo 15726–D.*)

Ross found that the Greenland Eskimos north of Cape York had ceased to use kayaks in 1818. Not until about 1860 was the kayak reintroduced here, by Eskimos from Pond Inlet, north Baffin Island, who walked over the sea ice. This fact probably accounts for the various sharpie and modified sharpie forms used along the northern and Polar coasts of Greenland.

The model of the kayak used on much of the central and southern coasts of Greenland has changed rather extensively since 1883, and this change has apparently affected the kayaks used on the east coast as well. In this part of the Arctic, the Eskimo are notable kayakers and the boat is not only well designed but also carries highly developed equipment and weapons for its work. The basic model used is a graceful V-bottom one, with raking ends and rather strong sheer. In the old boats represented by the drawings of figures 206 and 207, the sheer is strong at bow and stern, but this form has been gradually going out of favor. The kayaks are narrow but their shape gives them much stability. Pegged to the bow and stern are plates of bone to protect them from ice; in rare cases these bone stem bands, or bang plates, are lashed in place. The first drawing shows the construction used: light strong gunwales and a 3-batten longitudinal system with bent transverse frames. The keelson and chines—light, rectangular in section and placed on edge—are shaped slightly to fair the sealskin covering. The cover passes under the manhole rim. Bow and stern are made of plank on edge, shaped to the required profile. The gunwales are strongly tapered in depth fore and aft. Eight to twelve thwarts, or deck beams, are used in addition to the two heavy thwarts supporting the manhole; usually there is one more forward of the manhole than there is aft, and all are very light scantlings. The thwart forward of the manhole stands slightly inside the cockpit and is strongly arched; the after one is clear of the cockpit opening and has very little arch. Two light, short battens, or carlins, 24 to 36 inches long support the deck, where it sweeps up to the raked manhole, and usually there are two abaft the manhole as well. Lashings are used as fastenings except at the ends of the hull, where pegs secure the keelson to the stem and stern; at this point, on some kayaks examined, sinew lashings are also found. The whole

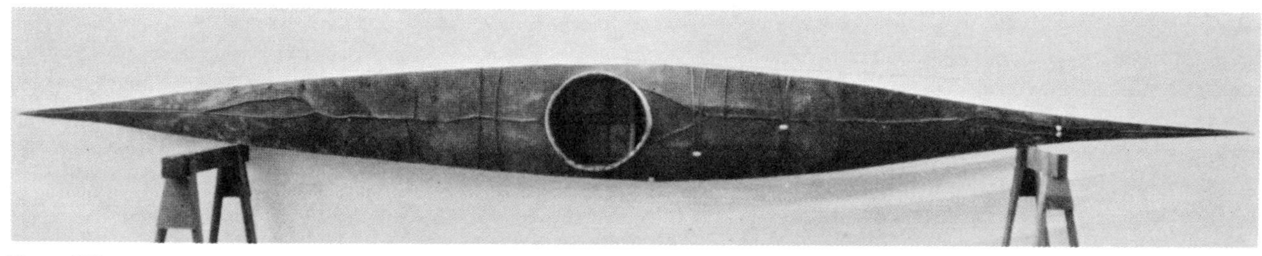

Figure 202

DECK OF GREENLAND KAYAK from Disko Bay (USNM 72564). (*Smithsonian photo 15726–C.*)

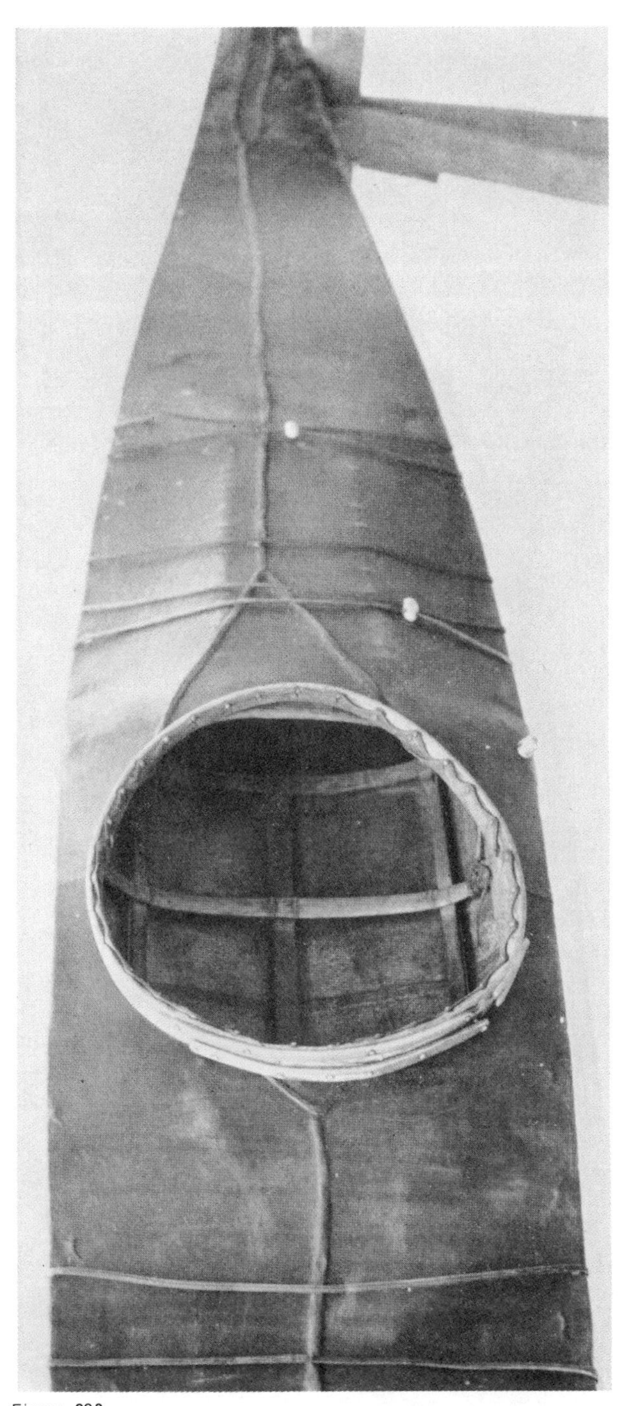

Figure 203

Cockpit of Greenland kayak from Disko Bay. (USNM 72564). (*Smithsonian photo 15726.*)

Figure 204

Bow View of Greenland kayak from Disko Bay (USNM 72564). (*Smithsonian photo 15726–A.*)

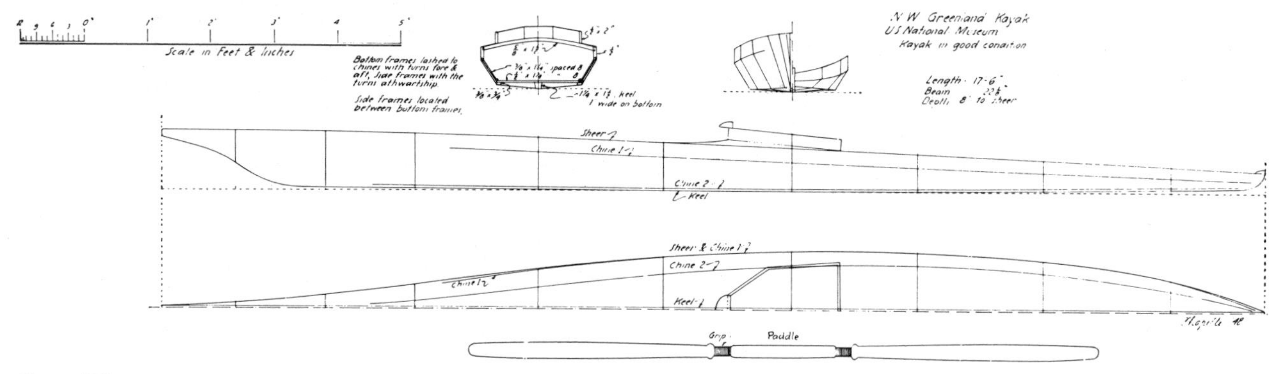


Figure 205

Northwestern Greenland Kayak, in the U.S. National Museum (USNM 160388).

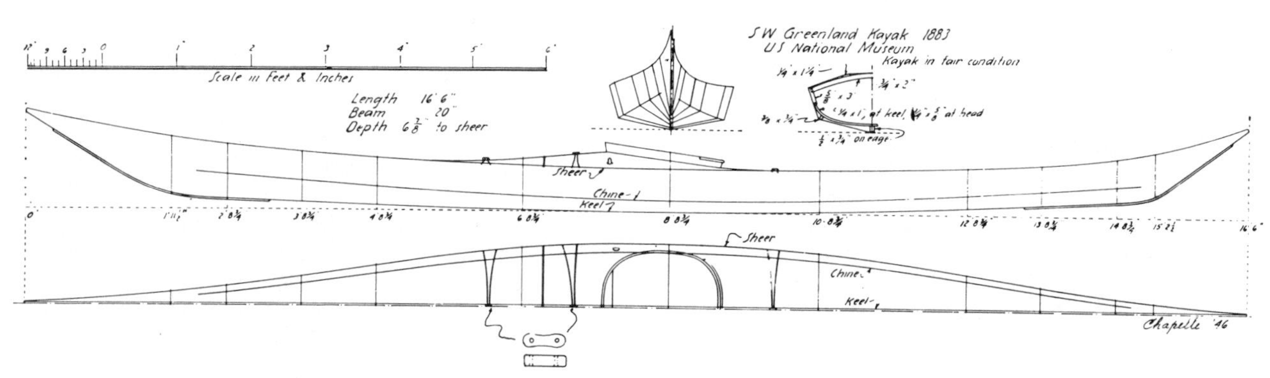


Figure 206

Southwestern Greenland Kayak, 1883, in the U.S. National Museum (USNM 160328).

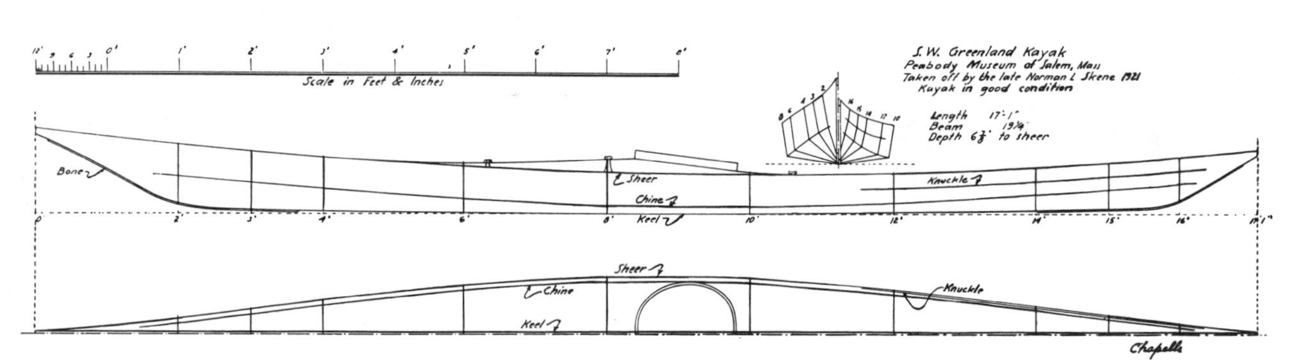


Figure 207

Southwestern Greenland Kayak, in the Peabody Museum, Salem, Mass. Taken off by the late Norman L. Skene, 1921.

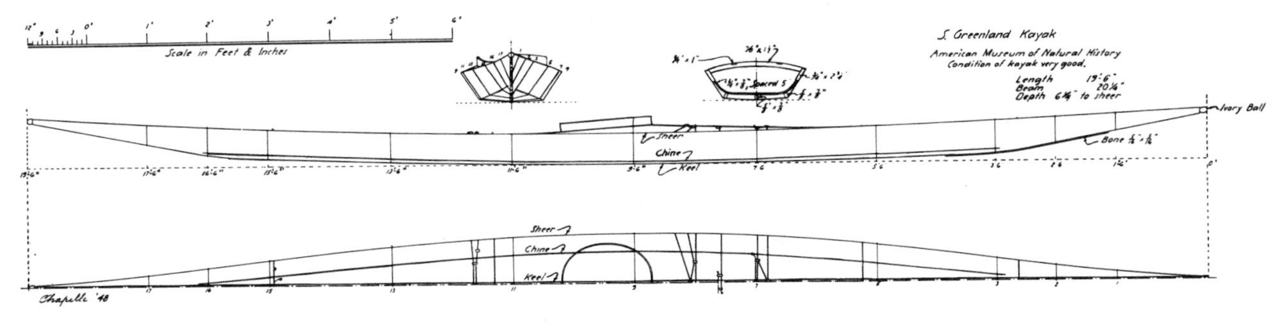


Figure 208

South Greenland Kayak, in the American Museum of Natural History.

framework is strong, light, and neatly made. In a few instances the gunwales do not flare with the sides the whole length and, thus, near the stern, a knuckle is formed in the skin cover, as in figure 207, opposite. The exact amount of flare and deadrise varies village to village. The old kayaks used in eastern Greenland had more rake in the bow than the examples illustrated, and also were marked by a sheer almost straight from the bow to within a foot or so of the stern, where it turned up sharply to a high stern, as in the drawing (fig. 191, p. 203.) These kayaks also had less flare and deadrise than most of the southwestern Greenland models. The amount of rocker in the keelson varies a good deal; that shown in figure 206, opposite, appears to have been about the maximum; a straight keelson does not seem ever to have been used. The manholes are fitted to allow use of the watertight paddling jacket; the projecting rim shown at the afterside of the manhole in the drawing is primarily to strengthen the manhole rim, but may also serve to prevent the drawstring holding the skirt of the jacket to the rim from slipping over the top. This old form of Greenland kayak, which has been widely described and much admired, was a fast and handy hunting boat; but it has become obsolete in most areas, and seems to have gone out of use more rapidly on the east coast than the west, where the type represented in the drawing was built as late as 1959 at Umanak Fjord.

Since the 1880's it has been gradually replaced by the type shown above. The modern version has the same construction as the old but, as can be seen, the model has undergone much alteration. The rake of the bow and stern have become much greater; the sheer is now almost straight. The flare of the sides has been increased and the deadrise in the bottom has been reduced. The new model is undoubtedly an improvement over the old type, being faster (particularly against a headwind) and quicker turning. However, it would probably be found to be somewhat harder than the old model to right when capsized. And although the new model is more stable than the old, it is not suited for unskilled users; a few American soldiers drowned during World War II through rashly venturing into rough water before becoming practiced in the use of these kayaks.

The intricate arrangement of deck lashings shown are required to hold weapons and accessories. Just ahead of the paddler a stand or tray on low legs holds the coiled harpoon line; and under the deck lashings are held such weapons as the lance, darts, and harpoons. Toggles of bone or ivory, often carved, are used to tighten and adjust these lines. The Greenland kayaks carry deck fittings and gear that are far better developed than those seen in any of the western types.

# *Chapter Eight*

# TEMPORARY CRAFT

USE OF TEMPORARY CRAFT SEEMS to have been confined to the Indians, who for the most part built them of bark, although some tribes used skins. However, very little in the way of information exists on the forms used by the individual tribes, for early travelers did not always have opportunities to see these emergency craft, and when they did they rarely took the trouble to record their construction and design.

## *Bark Canoes*

There is ample evidence to support the belief that a great many of the tribes building birch-bark canoes also used temporary canoes of other barks such as spruce and elm, as has been mentioned in earlier chapters. Invariably, the qualities of these other barks, particularly spruce, were such that their use was often somewhat more laborious and the results less satisfactory than with birch; but the necessities of travel and the availability of materials were controlling factors, and with care spruce bark could be used to build a canoe almost as good as one of birch bark. The forms of these canoes do not appear to have been as standardized as the tribal forms of the better-built bark canoes; rather, the model of the temporary canoe was entirely a matter to be decided by the individual builder on the basis of the importance of the temporary canoe to his needs, the limitation on time allowed for construction, and the material available.

The reasons for using substitute material are fairly obvious. In forest travel it was not always possible or practical to portage a canoe for a long distance simply to make a short water passage somewhere along the route. War parties and hunters, therefore, often found it necessary to build a temporary canoe, one that could be utilized for a limited water passage and then abandoned. Since such a limited use did not warrant expenditure of much time or labor on construction, the canoe was prepared quickly from readily available material and in order to meet these requirements many Indian tribes developed canoe forms and building techniques somewhat different from the more elaborate construction using birch or spruce bark.

It is obvious that much time and work could be avoided by use of a single large sheet of bark that was reasonably flexible and strong. But many of the barks meeting this specification had a coarse longitudinal grain that split easily, so forming a canoe by cutting gores was out of the question. This difficulty was avoided by folding, or "crimping," the bark cover along the gunwales at two or more places on each side of the canoe; this permitted the bottom to be flattened athwartships and the keel line to be rockered, both desirable in a canoe.

The problem of closing the ends also had to be solved. This was done by clamping the ends of the bark between two battens and, perhaps, a bark cord as well, and then lashing together the battens, bark ends, and cord with wrappings of root thongs. Cord made from the inner bark of the basswood and other trees could also be used for this purpose. The ends of the canoe could then be made watertight by a liberal application of gum or tallow, while grass, shavings, moss, or inner bark mixed with gum or even clay could be used to fill the larger openings that might appear in hurried construction.

Obviously, a simple wood structure was required by the specifications. Therefore, the gunwales were usually made of saplings with their butts roughly secured together or spliced. This allowed length to be obtained without the necessity of working down large poles to usable dimensions, a laborious and time-consuming undertaking with primitive tools.

The thwarts were commonly of saplings with the ends cut away so that the thin remainder could be wrapped around the main gunwales and lashed underneath the thwarts inboard. Ribs were usually of split saplings, but there is some evidence that in very hurriedly built canoes the whole small sapling was used. The kind of sheathing employed in these canoes during the pre-Columbian era is a mystery. It would be quite unlikely that time was taken to split splints such as were used in the late elm- and spruce-bark canoes, when steel tools were available. The writers believe that for small canoes it may have been the practice to use a second sheet of stiff bark inside the first and extending only through the middle two-thirds of the length, across the bottom and up above the bilge but short of the gunwales. This, with the ribs and a few poles lashed to each rib along the bottom, would have given sufficient longitudinal strength and a stiff enough bottom for practical use. However, in large canoes of the type reputedly employed by Iroquois warriors, a stronger construction seems necessary, and these canoes may have had a number of split or whole poles lashed to the ribs along the bottom.

With small variations in details, the general construction outlined above was employed by many North American Indians for building temporary canoes for emergency use. In at least one case, however, it was also used in canoes of somewhat more permanent status within the boundaries of the powerful Iroquois Confederation. On large bodies of water within their territory, the Iroquois used dugouts, but for navigating streams and for use in raiding their enemies they employed bark canoes. While some birch bark was available there, it was probably widely scattered; therefore these great warriors used elm or other bark for their canoe building.

Early French accounts show that the Iroquois built bark canoes of greater size than ordinary; Champlain wrote that their canoes were of oak bark and were large enough to carry up to 18 warriors; later French accounts, as we shall see, indicate that the Iroquois used even larger canoes than these. Champlain may have been in error about the Iroquois use of oak bark, as suggested earlier (p. 7), for experiments have shown that the inner bark of this tree is too thin and weak for the purpose; the canoes Champlain saw may have been built of white or red elm bark. The barks of the butternut, hickory, white pine, and chestnut might also have been employed, as they were usually suitable.

It was noted by the early French writers that the Iroquois built their bark canoes very rapidly when these craft were required by a war party in order to attack their enemies or to escape pursuit. In one case at least the canoes for a war party were apparently built in a single day. This was accomplished, it seems, by the excellent organization of their war parties, in which every man was assigned a duty, even in making canoes.

When it was deemed necessary to build a canoe, certain warriors were to search out and obtain the necessary materials in the order required for construction. To do this effectively, they had to know the materials in order of their suitability for a given purpose, for the most desirable material might not be available at the building site. Other warriors prepared the materials for construction, scraping the bark, making thongs, and rough-shaping the wood. Others built the canoe, cutting and sewing the bark, and shaping and lashing the woodwork. These duties, too, required intimate knowledge of the different materials that could be used in canoe construction. It would be natural, of course, to find that the methods used to construct a temporary craft for a war-party would also be employed at home by the hunter or fisherman, even when a rather more permanent canoe was desired. These were smaller craft and easily built. Only when a long-lasting watercraft was desired would the bark canoe be unsatisfactory; then the dugout could be built. The early French observers agree that though the Iroquois occasionally used birch-bark canoes. these were acquired from their neighbors by barter or capture and were not built by the tribesmen of the Confederation.

The details of the construction of elm canoes (and of other bark than birch) by the Iroquois are speculative, since no bark canoe of their construction has been preserved. This reconstruction of their methods is, therefore, based upon the incomplete accounts of early writers and upon what has been discovered about the construction of spruce- and elm-bark temporary canoes by other Eastern Indians.

In view of what has been reported, it must be kept in mind that the construction was hasty and that a minimum of labor and time was employed; hence, the appearance of the elm-bark canoe of an Iroquois war-party had none of the gracefulness that is supposed to mark the traditional war canoe of the Indians. The ends are known to have been "square," that is, straight in profile, and the freeboard low. The use

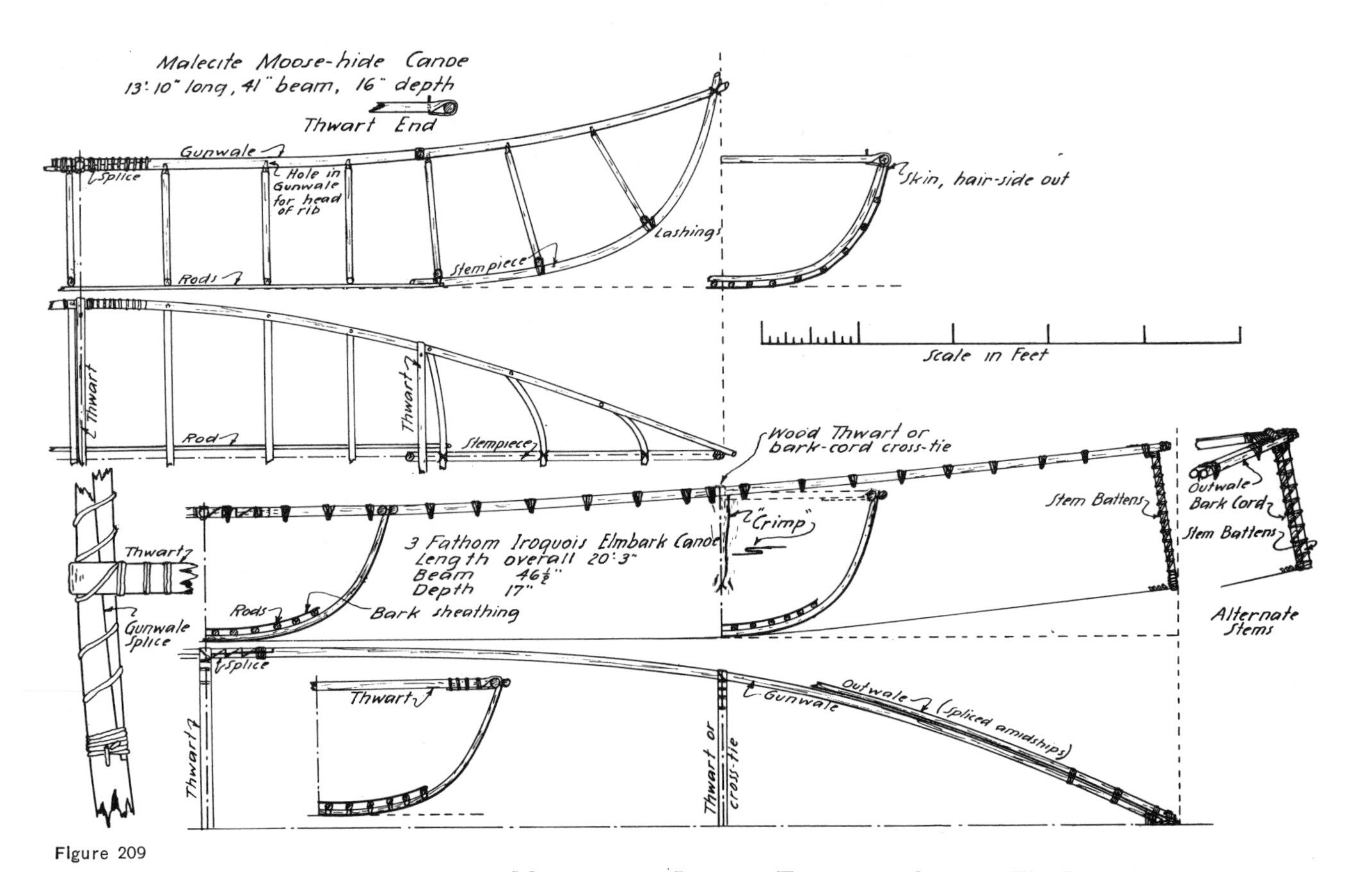


Figure 209

MALECITE AND IROQUOIS TEMPORARY CANOES. The Iroquois 3-fathom elm-bark canoe, below, is designed to carry ten to twelve warriors.

of saplings for the gunwales would cause an uneven sheer, and its amount must have been small; the high, graceful ends seen in some birch-bark canoes did not exist in the Iroquois model. The rocker of the bottom profile was not a fair curve, but was angular, made of straight lines breaking under the folds, or "crimps," in the bark cover at the gunwales. The amount of bark in each crimp and the location of the crimps fore-and-aft would determine the shape of the bottom profile and the amount of rocker, as well as the flatness of the bottom athwartships in the midbody. It appears that two crimps to the side were employed in most of these canoes, but perhaps more, say four to a side, might have been employed in a very large canoe. The tendency in forming these canoes must have been toward an almost semicircular midsection, a condition which would have produced an unstable craft if not checked.

The early French writers agree that the canoes of Iroquois war parties were sluggish under paddle. This was due to the fact that the hull form of these canoes was not good for speed, and also because the bulges at the bottom of the crimps caused them to be markedly unfair at and near the waterline. This handicap in their canoes may have been an inducement for the Iroquois to waylay their victims at portages when the travellers were usually spread out and easily cut down while burdened with goods. The Algonkin tribes countered by moving in very large numbers when within striking distance of Iroquois raiders. Hence there were very few recorded instances of battles in canoes; these took place only when sudden meetings occurred without preparation on either side, such as when war parties surprised canoemen in narrow waters. The shortcomings of their canoes did not seriously affect the deadliness of the Iroquois warriors, for their usual practice was to raid in winter, when they could travel rapidly on snowshoes and surprise their enemies in winter camps wholly unprepared for defense, a most pleasing prospect for the attacking warrior.

It would be a mistake, however, to assume that these factors made the Iroquois poor canoemen; the French repeatedly stated that they were capable in handling their craft and ran rapids with great daring and skill, showing that the apparently crude and weak elm-bark canoes were far better craft than they first appeared.

The theory that the Iroquois type of canoe was very like the emergency or temporary elm- and spruce-bark canoes of neighboring tribes is supported by some statements of the early French writers, as well as by a comparison of the rather incomplete descriptions of Iroquois canoes by later travellers with what is known about the spruce and other temporary bark canoes used in more recent times by the eastern Indians. M. Bacqueville de la Poterie, writing of the adventures of Nicholas Perrot in the years 1665 to 1670, tells of an instance in which Perrot's Potawatomi mistook the emergency canoes of some Outaouais (Ottawa) for Iroquois canoes.

LaHontan (1700) gives some general information as well as specific opinions on the speed and seaworthiness of Iroquois canoes, saying that—

the canoes with which the Iroquois provide themselves are so unwieldy and large that they do not approach the speed of those which are made of birch bark. They are made of elm bark, which is naturally heavy and the shape they give them is awkward; they are so long and so broad that thirty men can row in them, two-by-two, seated or standing, fifteen to each rank, but the freeboard is so low that when any little wind arises they are sensible enough not to navigate the lakes [in them].

LaFiteau, writing before 1724, stated definitely that the Iroquois did not build any birch-bark canoes, but obtained them from their neighbors, and that the Iroquois elm-bark canoes were very coarsely built of a single large sheet of bark, crimped along the gunwales, with the ends secured between battens of split saplings. He noticed that the gunwales, ribs, and thwarts were of "tree branches," implying that the bark was not removed from them. The most detailed description was by a Swedish traveller, Professor Pher Kalm, who gave extensive information on the construction of an elm-bark canoe in 1749; this account is particularly useful when interpreted in relation to the spruce- and elm-bark canoes of the eastern Indians. It is upon the basis of Kalm's account that the procedures used to build an Iroquois war canoe have been reconstructed.

The bark most favored by the Iroquois was that of the white elm. Next most favored was red elm, and then other barks—certain of the hickories and chestnut are mentioned in various early references. It was necessary to find a tree of sufficient girth and height to the first limbs to give a sound and fairly smooth bark sheet in the length and breadth required. If possible the bark was stripped from the standing tree; even after steel tools were available, felling was avoided for fear of harming the bark. Great care had to be taken in the operation, to avoid splitting or making holes in the bark, and often two or more trees had to be stripped before a good sheet of bark was obtained. In warm weather the bark could be removed without much difficulty, but in the spring and fall it might be necessary to apply heat; this was apparently done by means of torches or by the application of hot water to the tree trunk.

When the bark was removed from the tree, the rough outer bark was scraped away; if the builder was hurried this scraping was confined to the areas to be sewn or folded. The bark was then laid on a cleared piece of ground, the building bed, with the outside of the bark up, so that it would be inside the finished boat. The building bed does not appear to have required much preparation; apparently not raised at midlength, it was merely a plot of reasonably smooth ground, located in the shade of a large tree if building was to be done in summer.

It is not wholly clear from the descriptions whether the gunwales were shaped before or after being secured to the bark. However, extensive experiments in building model canoes show very plainly that it would be easiest to assemble the main gunwale frame and use it in building, after the fashion of eastern birch-bark canoe construction. With the main gunwales assembled, the stakes would be placed on the bed, the bark replaced, the frame laid on it and weighted, and the stakes then redriven in the usual way and their heads lashed together in pairs.

Each gunwale was formed either of two small saplings or of split poles, with the butts scarfed at the canoe's midlength. The canoe of an Iroquois war party would probably have gunwales of split saplings so that inwale and outwale for half the length of one side of the canoe would be from a single pole; this would allow the flat sides to be placed opposite one another, on each side of the edge of the bark, to form a firm gunwale structure. However, when a rather permanent craft was being built, the poles might be split twice, or quartered, to give pieces to make half of the gunwales of a canoe; these too might be worked nearly round before assembly.

That the gunwale joints were scarfed is reasonably certain. The elm-bark canoes of the St. Francis

Indians are known only from a model, as are the spruce-bark hunters' canoes of the Malecite, but the testimony of old St. Francis and Malecite builders support the evidence of the models; therefore it is probable that the use of scarfed gunwales was common in these canoes, and, hence, also in the canoes of the Iroquois, who dwelt nearby. The manner of scarfing is not certain. Probably the butts were snied off so that the lap would be flat face, as was usual in the Malecite spruce-bark canoes of this same class. The butts were secured together by lashings—apparently let into shallow grooves around the members. In a very hastily built canoe the butts might be merely lapped for a short distance, one butt above the other, and lashed; this, of course, would make a jog in the sheer, but do no harm, as the jog would occur in both inwale and outwale, and the bark would lay up between these and be trimmed to suit.

The thwarts were described in old accounts as very small saplings, or tree branches, with their ends sharply reduced in thickness so that they were thin and pliable enough to be bent around the gunwales and brought inboard under the thwart, as done by some Kutenai in the West (see p. 169). The thwart ends might be lashed or, as in some eastern spruce-bark canoes, brought up through a hole in the thwarts to the top where it could be jammed or lashed. In the Iroquois canoe it seems probable that the thwart ends passed around the main gunwales only and were secured under the thwarts for, as noted, the evidence strongly suggests that the main gunwale members were preassembled, a procedure that requires the thwarts to be in place. In the small hunters' canoes, however, some eastern builders apparently put in a temporary spreader in place of a single thwart until the canoe was completed to the point where the outwales were in place, then the thwarts were added, the ends passing over and around both inwale and outwale and through the bark cover below, to the underside of the thwart.

One requirement in building these canoes was to crimp the edges of the bark at the gunwales in such manner that the bottom of the canoe would be rockered and at the same time would be moulded athwartships. First steps in the process were to set into the building bed two heavy stakes on each side of the stems, a little inboard of the ends, and to tie the heads of each pair together with a heavy bark cord or a rawhide thong. Then a sling was made, the bight of which went under the bottom of the bark cover near its ends, and the ends of the sling were made fast to the heads of the stakes. By taking up on these slings, the ends of the bark cover were sharply lifted and then the folding of the bark along the gunwales could be easily accomplished, as they then formed naturally, without strain. The crimps were commonly located a fourth to a fifth the length of the canoe inboard of the ends, about where the end thwarts would be located. In small hunters' canoes the end thwarts were often replaced by twisted cords across the gunwales, but in the large Iroquois canoes there were probably five or seven or perhaps as many as nine thwarts according to length.

The ends of the gunwales were simply lashed together with cords or thongs in shallow grooves to prevent slipping. They were raised by a small inside post, its heel placed on the bark near the stem and its head brought under the gunwales, so that it served the purpose of a headboard in sheering the gunwales.

The procedure in building to this point, then, appeared to follow the general plan used in birch-bark construction. Next, the stakes were redriven in the bed around the gunwale frame, which was weighted on the bark with stones, and the sides of the bark cover were brought upright. Apparently only a few stakes were considered necessary—three or four to a side and two pairs of end stakes to raise the stems. The gunwale frame was then lifted to the required height of side and lashed temporarily to the side stakes, the ends of the bark cover were creased to form bow and stern, and the headboard posts were inserted to support the ends of the inwales and to sheer the canoe. Before this, of course, the ends of the bark cover had been raised by means of the slings to the end stakes.

The outwales of split saplings were now put into place, with the edges of the bark cover lashed between the flat surfaces of the inwale and outwale, the gunwales having been assembled with the flat face of the longitudinal members outboard. The lashings were in small groups spaced 5 to 7 inches apart so as not to split the bark, and these not only secured the bark in place but also held the inwales and outwales tightly together, to clamp the edges of the bark cover. At the thwarts, the outwales were notched on their inboard face to allow them to come up against the bark pressed against the face of the inwales (in some eastern canoes the bark cover was notched at the thwart ends to lay up smoothly there, and this may have also been done in the Iroquois canoes). In placing the outwales, the crimps were carefully formed and held by the clamping action of the inwale and

Figure 210

Hickory-Bark Canoe Under Construction, showing the sling with which the ends are elevated and the crimp which takes up the slack in the sides of the bark. Excess bark above the gunwales to be trimmed off. Completed model in The Mariners' Museum, Newport News, Va.

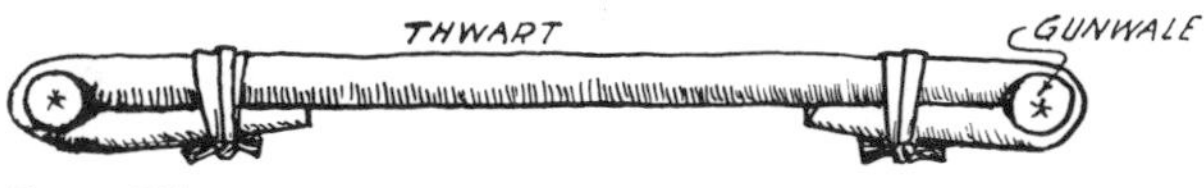


Figure 211

Detail of Thwart used in Malecite temporary spruce-bark canoe.

outwale, and reinforced by a lashing through the crimp or by two lashings close to the sides of the fold. The fold of the bark forced the outwale away from the inwale, and although this was counteracted to some extent by the lashings, the gunwales were unfair at these points. The crimps were formed so that the maximum fold in the bark took place at the gunwales; below this the fold tapered away to nothing, ending low in the side with an irregular bulge in the bark. Such a bulge could only be avoided by goring, which is impractical with elm, pine, chestnut, or hickory barks.

The ends of the canoe were closed, as has been mentioned, by use of split-sapling battens on the outside of the bark. The Iroquois and some other builders also employed at the stems a thong or a twisted cord made of the inner bark of some such tree as the basswood; this was wrapped around the ends of the bark cover abreast the headboard posts inside the canoe, so that the lashing stood vertically. Then the split battens were placed on each side of the bark cover, just outboard of the cord, and the whole was secured by a coarse spiral lashing of root or rawhide, which passed inboard of the cord lashing and the headboard post, as well as around them and the split battens outside of the bark cover. Some builders apparently added a split-root batten over the edges of the bark cover, as a sort of stem-band; this was secured by the turns of the stem closure lashing, which passed around them as well as the edges of the bark and the split side battens. It can be seen that this closure formed a strong stem structure. Watertightness was insured by merely forcing clay into the stems from the inside, or by forcing in a wad of the pounded inner bark of a dead red elm which would swell when damp.

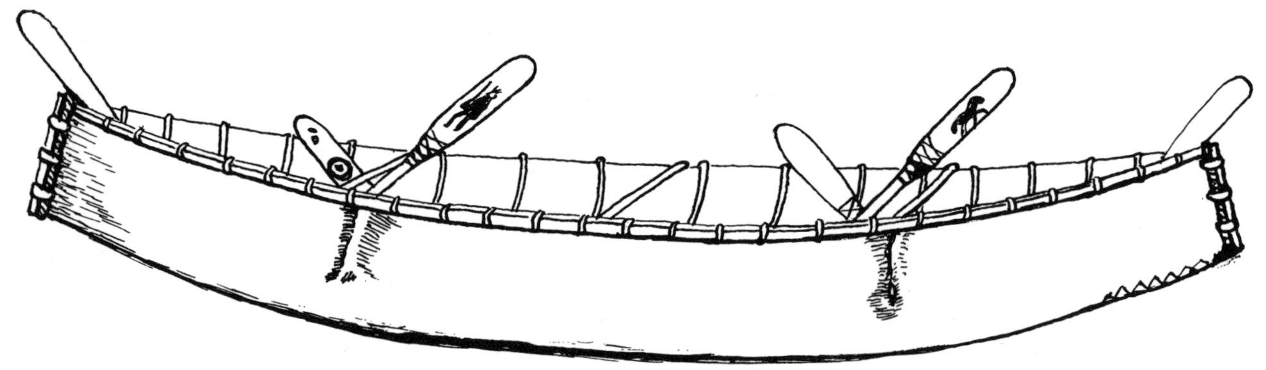

Figure 212

Iroquois Elm-Bark Canoe, after a drawing of 1849, equipped with paddles for a crew of six, with owners' personal marks on blades. Length of canoe 25 feet, with capacity for a war party of a dozen or more men. Note supporting piece of cord tied in with the end battens. Far gunwales are improperly sketched.

Still other methods included the use of grass or moss impregnated with warm tallow from the cooking pot. If available, the stems would be liberally smeared with spruce or other gum, of course.

While the ribs were customarily tree branches or small saplings, in some canoes the saplings were split and bent so their flat face was against the bark. In the East, hunters' canoes were often given the lath-like ribs of the birch-bark canoes, for when steel tools became available such ribs were easily made during the winter for use in the spring, when the temporary canoe would be needed.

According to the early reports, the ribs were placed some 6 to 10 inches apart in the bark cover, with the heads forced under the inwales against the bark, and were supported there by the outwales as well. No mention is made of any sheathing; Kalm refers to a piece of bark and some saplings or tree branches laid over the ribs to protect the bottom inboard. In the large Iroquois canoes it would have been possible and practical to employ a piece of bark inside the main bark cover, as noted on page 213; this inside piece needed to be only long enough to reach to the end thwarts, or abreast the crimps, and wide enough to cover the bottom and bilges up to 3 or 4 inches short of the inwales. With the ribs over this inner sheet, a stiff bottom would result. In a long canoe, split poles could be laid lengthwise inside the bottom of the canoe and fastened there by lashing them to a few ribs; these would serve to protect the bottom in loading and to stiffen the bark cover. However, in a small canoe the stiffness of elm bark when the rough outside layer was not fully scraped off would make sheathing of any kind unnecessary, and the bark mat inside the ribs, mentioned by Kalm, would be sufficient.

The difficulty in reconstructing the building methods of the large Iroquois canoes on the same basis is that Kalm's description is of a rather small canoe; the information on the temporary canoes of the eastern Indians also deals with short craft. It is evident, however, that poles were not usually placed between the bark and the ribs, as in temporary skin canoes built by Indians. It is also apparent that splints were not used by the Iroquois for sheathing large canoes.

The ends of the outwales in the Iroquois canoes seem to have been secured by snying them off on the outside face and holding these thin ends by the cord around the ends, as well as by the closure battens of the stems. In some eastern canoes, notably the elm-bark canoes of the St. Francis, the outwale ends projected slightly outboard of the stems and were lashed across them by a simple athwartship lashing which passed through the bark cover and under and over the lashing at the inwale ends.

In a drawing of an Iroquois canoe made about 1849, the cord around the stems is shown together with the outside stem battens and lashing; the ends of the outwales are apparently under the cord and perhaps under the stem battens. The stem batten is in one piece sharply bent under the stems in U-form. The end lashing shown seems to be in groups and the bottom, for a little distance inboard of the stems, is also shown as lashed. Three thwarts are shown.

It may be that this drawing was made not from a full-size canoe but from a model, for the proportions are obviously incorrect. This possibility casts some doubt on the picture as evidence of the building practices, for in Indian-built models simplified construction details not used in actual canoe building are often found.

According to early accounts and the statements of eastern Indians, these emergency canoes were often heavy and unsuitable for portaging. By 1750, at least, the Iroquois were using blanket square-sails in their elm-bark canoes.

## *Skin Boats*

Among the other forms of temporary or emergency canoes used by North American Indians, the most widespread was some form of skin boat. These would not require description here were it not for the fact that the Indian skin boats were usually built by bark-canoe methods of construction rather than by methods such as used by the Eskimo. To build their skin boats—kayaks and umiaks—the Eskimo first constructed a complete framework, and this was then covered with skins sewn to fit. This process of building required a rigid framework capable of not only standing without a skin covering but also of giving both longitudinal and transverse strength sufficient to withstand loading, without the slightest support from the skin covering. Hence, the framework of the Eskimo craft was made with the members rigidly lashed and pegged together. The majority of Indian skin canoes, however, required the covering to hold the framework together, as in a birch-bark canoe. An example is the Malecite skin-covered hunters' canoe. According to available information, the Malecite hunter would leave two or three moose skins on stretchers for use in building a skin canoe in the early spring. Sometimes the hair was removed from the hides and sometimes it was not. Spare time during the winter hunt might be spent in preparing the wooden framework, but if this were not done the delay would not be very great.

The gunwale frame was first made of four small sapling poles roughly scarfed at the butts. From a small sapling a middle thwart was made in the manner of the elm-bark canoe thwarts, the ends tapered enough to allow them to be wrapped around the gunwales and secured under the thwart by lashings. The ends of the gunwales were merely crossed and lashed. Where end thwarts would be placed, it was usual to use a cross tie made of twisted rawhide or cords of bark fiber. Holes were then drilled at intervals in the underside of the gunwale to take the heads of the ribs. Stem-pieces about 3 feet long were prepared of short saplings and bent to the desired profile; one builder used a full-length keel-piece, instead of the short stem-pieces. The ribs were usually of small saplings that could be bent green without the use of hot water. For sheathing a number of small saplings were also gathered, and from them were made poles in lengths about equal to three-quarters, or a little more, of the intended length of the canoe, which would be determined by the size of the skins available. The average canoe was about 12½ feet long, roughly 40 inches beam, and 14 to 19 inches in depth.

The skins were sewn together lengthwise, lapped about 6 inches or a little less, and secured by a double row of stitching. If the hair had not been removed, it had to be scraped away along the sewn edges. In such a case the hair would usually be on the outside of the finished canoe. Also, before work was started on assembling a canoe, the skins were worked pliable, and tallow and gum were accumulated.

When an emergency canoe was ready to be assembled a smooth place was prepared; either an open bit of ground or the floor of the hunter's hut, if large enough, might be used. The outlines of the gunwales were fixed by a few stakes temporarily driven around it and then pulled up. The skins were then laid on the bed and the gunwale frame placed on them and weighted with stones. Then the skins were left to dry for awhile until they became somewhat stiff; the proper condition was indicated by the curling of the edges.

When the skin was sufficiently stiff, the gunwale frame was lifted and temporarily secured to the stakes redriven in the bed, the sides of the skin were turned up, the skin was gored, and sometimes the ends of the gunwales were sheered up slightly at the end stakes; this latter was not always done, for in some canoes the sheer was quite flat.

The skins were now trimmed to the sheer of the gunwales and the edges lashed to these members with rawhide, the gores also having been sewn. Next the stem-pieces were put into place and the stem heads lashed inside the apex formed by the ends of the gunwales. Some ribs were then bent and forced down on the stiff skin cover, the rib ends being worked into the holes prepared for them on the underside of the gunwales. These ribs usually stood approximately square to the curve, or rocker, of the bottom. Now the skin could be trimmed to the stem profiles and sewn. The stitching was usually done so as to be outside the stem-pieces, with an occasional turn going around inside them to help hold the structure in place. Some builders first put in the stems temporarily and then trimmed the skins to match; after this was done the stem-pieces were removed to allow easy sewing. When they were replaced and secured permanently, a few more stitches were added along the stems to secure the woodwork.

The next step was to sheath the canoe inside with the small poles; these were placed a few inches apart transversely and their ends worked under the most inboard of the ribs on the stem-pieces, then held in place, while the necessary adjustments were made, by a few temporary ribs. Then the ribs were forced into place, one by one, each prebent to the desired section, just as in birch-bark canoe construction. In this final shaping, the skin cover might have to be wetted again to soften the material and to allow stretching. The seams were then payed with gum or tallow, and the canoe was ready for launching.

The description is for canoes of minimum finish; builders often used split and shaped gunwales, split ribs, and splint sheathing if these could be prepared during the winter. The construction of a skin canoe was not a specialized process in which a hunter consistently built this one type; the selection was determined by natural conditions. If he were to come out of the woods too early in the spring to make the construction of a spruce-bark canoe easy, then he would resort to skin construction; the statements of old Malecite hunters leads to the conclusion that as emergency craft they used spruce-bark canoes most often.

Perhaps the most primitive of the skin boats built by the North American Indian was the so-called bull-boat of the Plains Indians. These were not canoes but coracles—bowl-shaped and suitable only for use on streams, where ferrying would be the main requirement. The boats were covered with buffalo-hides and their framework was usually made of the willow shoots found along the streams. The framework followed, to some extent at least, the basketwork principle, a circular gunwale or rim being used. The ribs were set in two groups, half at right angles to the other half in very irregular fashion. This construction formed a sort of rough grating in the bottom. The ribs were lashed together with rawhide and apparently the craft was built up on the skin as were the Malecite skin canoes. Battens in circular form were used on the sides to fair the cover. The form of the bull-boat varied somewhat among individual builders; sometimes it assumed almost a dish shape with shallow flaring sides, but more commonly the sides were nearly upright; the bottom was always flat, or nearly so. These bull-boats appear always to have been small. Judging by the examples preserved, a bull-boat 5 feet over the rim or gunwale, or made of more than one skin, was extremely rare, and most examples are nearer 4 feet and built on a single skin. Many were too small to carry a person; these were intended to be loaded with cargo to be kept dry and towed by a swimmer. When they were large enough to be paddled, the paddler worked over the "bow," as in a coracle. Probably all the Plains Indians living near streams once used the bull-boat, but existing records show only the Mandan, Omaha, Kansas, Hidatsa, and Assiniboin to have used it. The Blackfoot (Siksika) and Dakota are said to have used some kind of a skin boat in which their tepee poles were employed as a temporary frame, but nothing is recorded of their form.

The use of spruce bark as a building material in the Northwest and throughout the extreme northern range of the birch-bark canoe has been discussed in earlier chapters (pp. 155 to 158). In these areas, the emergency canoe was usually built of caribou skin. On the Alaskan coast seal skin may also have been used, but generally it was used for the permanent kayak-type canoe and not for a hastily built temporary craft. The caribou-skin canoe was also built as a permanent type, in either kayak form or somewhat on the model of the spruce- or birch-bark canoe of the area. However, although references to temporary craft covered with caribou skin exist in early accounts of the fur trade, there is no record of their form or details of their construction. Early in the present century some of the Indians of the Mackenzie River country built skin canoes much like the modern canvas-covered freight canoes. Also, some of these skin canoes were built so that they resembled York boats or the whaleboats of the white man. No observer has described the methods used to construct the emergency canoe of the Northwest; we do not know whether they resemble those used in the Indian bark canoe or in the Eskimo skin boat.

# *Retrospect*

In view of the inclusion of skin boats in this discussion of bark canoes, it may be well to emphasize again the fact that the North American Indian's method of constructing bark canoes and of temporary skin canoes was on an entirely different principle than that used by the Eskimo in building their skin boats. This is even true of the kayak-form bark canoes of the Northwest, despite their superficial similarity in design and proportions to the Eskimo skin kayak.

As has been stated, the Eskimo construction required a rigid frame, with all members fastened together with lashings and pegs, the skin cover being merely the watertight envelope and not a strength member. This system of construction marks primitive skin-boat design in most parts of the world. The Indian bark construction, on the other hand, did not have a rigid frame, and all but a few of the structural members were held in place by pressure alone: the sheathing was held against the bark cover by pressure of the ribs; the stem-pieces, in most cases, were held in place by pressure of the ribs, gunwale sheering, or headboards. In fact without the bark cover in place, the greater part of the wooden structure of the bark canoe would collapse. Not only was the bark cover the fundamental basis of construction, it was to a great extent a strength member, though by clever design the loading of the bark was minimized.

This fundamental difference in construction must be recognized in comparisons of Eskimo and North American Indian watercraft. Here, too, it might be observed that one should view with skepticism any claim that widespread similarity of certain structural practices is evidence of some ancient connection between types of canoes. In most cases these similarities were imposed by the working characteristics of the materials employed. Similarly, limitations in materials available for construction have their effect upon building techniques.

The practice of employing pressure members in bark-canoe construction, particularly where birch bark was employed, was the result of the need to stretch this material by gentle and widespread pressure, whereas the skin cover could be stretched by the concentrated pull of stitching alone, or by force applied in a small area. Bark canoes built in areas where skin-kayak construction is carried on nearby show a greater rigidity of structure. Thus, in the lower Yukon Valley in Alaska the bottom frame of the canoes built there was a rigidly constructed unit, even though the side longitudinals were held in place by rib pressure alone. And it is reasonable to theorize that the Malecite, who through habit still employed bark-canoe construction practices in building their skin craft, would have eventually come to the Eskimo method of construction had conditions required them to use skins exclusively.

Figure 213

Large Moosehide Canoe of upper Gravel River, Mackenzie valley. (*Photo, George M. Douglas.*)

# *Appendix*

# The Kayak Roll

*John D. Heath*

THE MOST EXTRAORDINARY feat of kayak handling is the ability to right the craft after a capsize. This maneuver, called "rolling," is usually practiced by capsizing on one side and recovering on the other. Under emergency conditions, a kayaker will recover on whichever side is more convenient. When rolling, a kayaker wears a waterproof jacket having long sleeves and a hood. The waist, face, and wrist openings are fitted with drawstrings, so that when the waist opening is fitted over the cockpit rim, the kayak and kayaker become a waterproof unit. Thus equipped, the kayak is the most seaworthy craft of its size, this quality being limited only by the skill and stamina of the kayaker.

The art of kayak rolling was highly developed in Alaska and Greenland. Eskimos in both of these regions depended upon seal hunting by kayak as a major part of their economy, hence the ability to roll was an important means of survival. Very little detailed information exists regarding Alaskan kayakers, but the Greenlanders have been the object of intensive study by ethnographers and explorers. The earliest detailed record of rolling was that of David Crantz, a European missionary, who in 1767 enumerated ten methods of rolling in his *History of Greenland.** His description follows.

1. The Greenlander lays himself first on one side, then on the other, with his body flat upon the water, (to imitate the case of one who is nearly, but not quite overset) and keeps the ballance with his *pautik* or oar, so that he raises himself again.
2. He overturns himself quite, so that his head hangs perpendicular underwater; in this dreadful posture he gives himself a swing with a stroke of his paddle, and raises himself aloft again on which side he will.

These are the most common cases of misfortune, which frequently occur in storms and high waves; but they still suppose that the Greenlander retains the advantage of his *pautik* in his hand, and is disentangled from the seal-leather strap. But it may easily happen in the seal-fishery, that the man becomes entangled with the string, so that he either cannot rightly use the *pautik*, or that he loses it entirely. Therefore they must be prepared for this casualty. With this view

3. They run one end of the *pautik* under one of the cross-strings of the kajak, (to imitate its being entangled) over-set, and scrabble up again by means of the artful motion of the other end of the *pautik*.
4. They hold one end of it in their mouth, and yet move the other end with their hand, so as to rear themselves upright again.
5. They lay the *pautik* behind their neck, and hold it there with both hands, or,
6. Hold it fast behind their back; so overturn, and by stirring it with both their hands behind them, without bringing it before, rise and recover.
7. They lay it across one shoulder, take hold of it with one hand before, and the other behind their back, and thus emerge from the deep.

These exercises are of service in cases where the *pautik* is entangled with the string; but because they may also quite lose it, in which the greatest danger lies, therefore,

8. Another exercise is, to run the *pautik* through the water under the kajak, hold it fast on both sides with their face lying on the kajak, in this position overturn, and rise again by moving the oar *secundum artem* on the top of the water from beneath. This is of service when they lose the oar during the oversetting, and yet see it swimming over them, to learn to manage it with both hands from below.
9. They let the oar go, turn themselves head down, reach their hand after it, and from the surface pull it down to them, and so rebound up.
10. But if they can't possibly reach it, they take either the hand-board off from the harpoon, or a knife, and try by the force of these, or even splashing the water with the palm of their hand, to swing themselves above water; but this seldom succeeds.

* See bibliography.

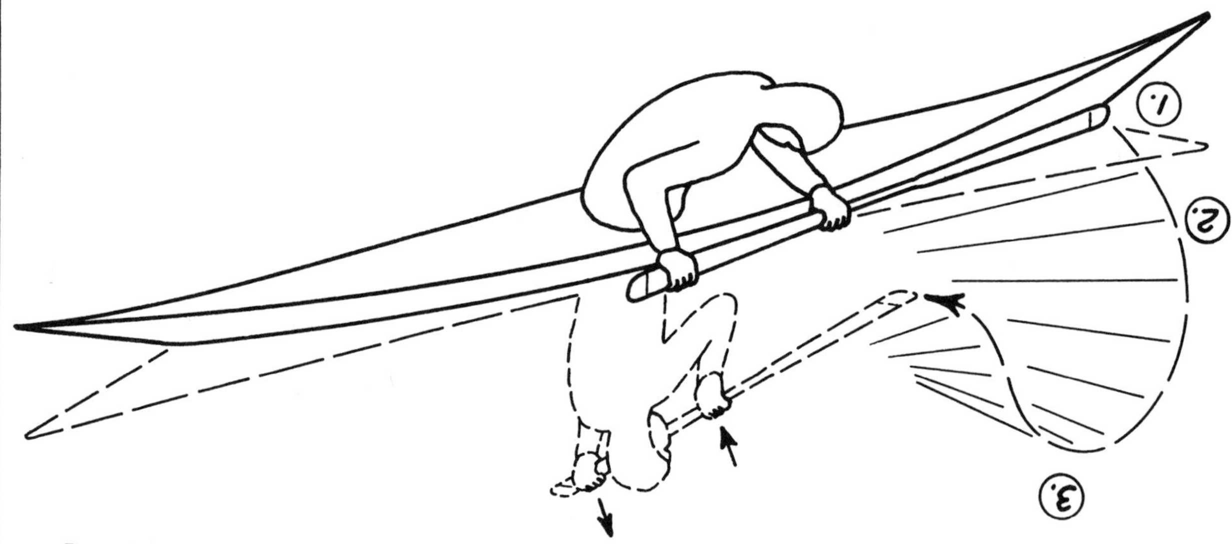


Figure 214

## THE STANDARD GREENLAND ROLL

The solid lines represent the starting position for a clockwise roll (disregard the phantom lines until later). The paddle is held blade-on-edge along the starboard gunwale, with one end near the right hip, and the other end toward the bow. The kayaker leans forward and faces slightly to starboard. His left forearm is against, or near, the foredeck, and his left hand reaches across the starboard gunwale to grasp the paddle near, but short of, the middle. The right hand holds the paddle near the end, about even with the hip. The palms of both hands pass over the paddle, so that the knuckles are outboard. The kayaker takes a deep breath, leans to starboard and capsizes.

(Now turn the page upside down)

The same lines which represented the starting position now represent a fish-eye view of the fully capsized position. The phantom lines represent the upright position, or goal.

To right himself, the kayaker—

(1) Flicks his wrists to swing his knuckles toward his face, thus causing the outboard edge of the paddle to assume a slight planing angle (not shown) with the water surface. The remaining steps constitute one continuous movement, to be done as quickly as possible.

(2) With his hips and right hand serving as pivot points, he sweeps his forward paddle blade, and his torso, outward in a 90-degree planing arc on the water surface, as shown from position (1) to (3), while pulling down on his left hand and pushing up on his right, thus lifting himself to the surface.

(3) Completes the roll by flicking his wrists to flatten the blade angle, then sharply increasing his opposing hand pressures, thus raising himself in a chinning attitude as the paddle blade sinks and is drawn inward. The roll is now completed.

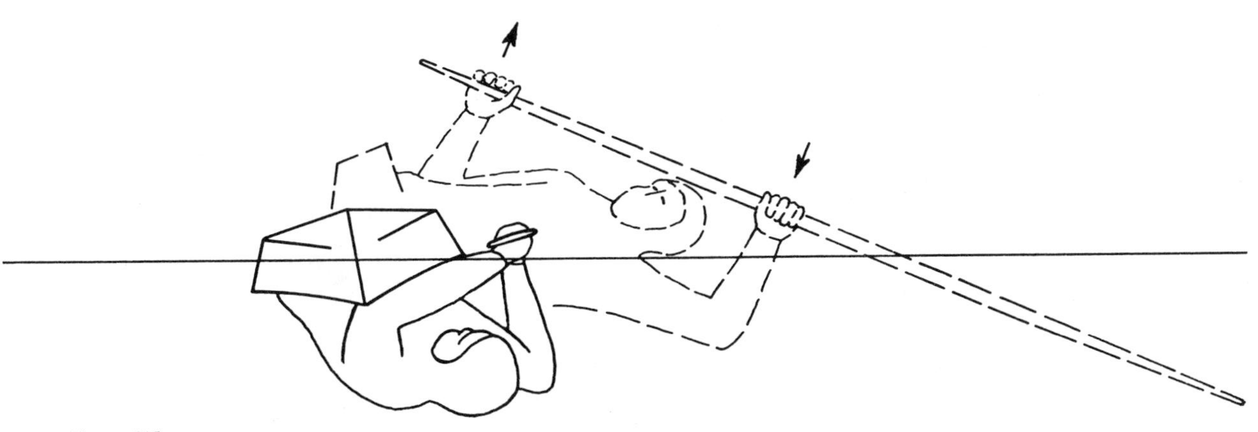

Figure 215

THE CRITICAL STAGE OF A CAPSIZE RECOVERY

The start (solid lines) and finish (phantom lines) of a planing sweep are shown head-on Success is almost certain if the kayaker has surfaced by the time he has completed the 90-degree sweep. Some minor refinements of rolling are apparent. The left forearm is shown right against the foredeck (a convenient means of orientation), the leading shoulder is nearer the surface (to gain lift when the torso is swung outward), and the hips right the kayak as far as possible while the torso is still partly submerged (to avoid having to lift torso and kayak at the same time).

Since Crantz's time, various authors have described kayak rolling. At least 30 methods of rolling have been known in Greenland. There are possibly many more, because the variations and combinations are numerous.

Although kayaking as a sport first became popular in the 1860's, it was not until the 1920's that the value of learning to roll began to be fully realized by the recreational kayaker. Interest has grown steadily since that time, and rolling instruction has been included as a regular part of many club training courses. A preliminary step in mastering the roll consists of using the paddle to prevent a capsize, by turning the blade parallel to the water surface and pressing down sharply on the side toward which the kayak is capsizing, while exerting an upward pressure with the other hand. This produces a rotary movement which restores the kayak to an even keel. Recreational canoeists call this maneuver a "paddle brace."

Most kayak rolls are based upon one or more of three basic movements. These are the paddle brace, the "sculling" stroke, from which lift is obtained by moving the paddle back and forth through a small arc with the leading edge of the blade at a slight planing angle, and the "sweep," from which lift is obtained by sweeping the blade through a large arc at a slight planing angle. The method of rolling shown in the sketches is the standard Greenland roll, so called because it is the most common roll encountered in Greenland. A slightly modified version of this roll is called by recreational canoeists the Pawlata roll in honor of the European who introduced it to them. Many skillful kayakers could not roll, and sometimes a highly skilled roller would fail to recover. Such men could be rescued by their companions by either of two common methods. One method was executed by placing the bow of the rescue craft within reach of the capsized paddler's hand, so that he could pull himself up by a one-handed chinning motion. The other method was executed by bringing the rescue kayak alongside the capsized kayak so that the two craft were parallel and about two feet apart. The rescuer then laid

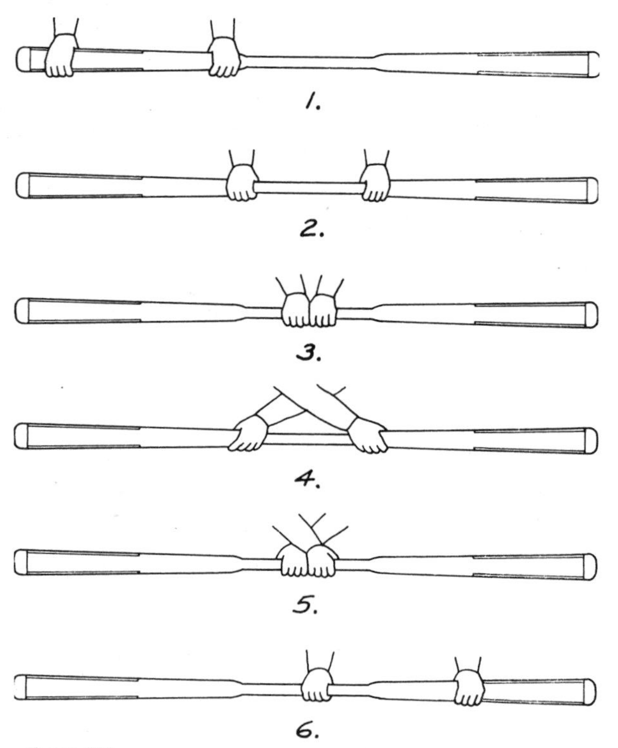


Figure 216

Hand positions used with the standard roll:

(1) The extended paddle position is the common method, and it gives maximum leverage. It is similar to the "Pawlata Roll" position used by recreational kayakers.
(2) The normal paddling position is more convenient, but gives less leverage. This is called the "Screw Stroke" position.
(3–6) Difficult trick positions demonstrated by Enoch Nielsen of Igdlorssuit, West Greenland, to Kenneth Taylor, a Scottish canoeist, in 1959.

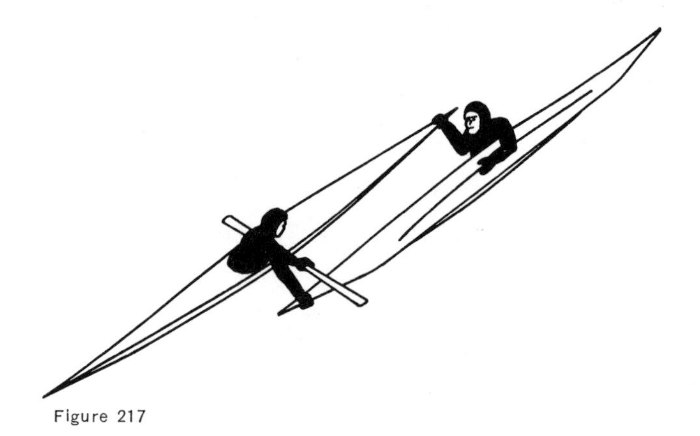

Figure 217

Kayak rescue, bow-grab method

Figure 218

Kayak rescue, paddle-grab method

his paddle across both craft and holding it with one hand, reached down and grabbed the capsized paddler's arm. He then pulled him up between the two kayaks. This method enabled an enfeebled or unconscious kayaker to be rescued.

Both of the above methods of rescue were completed with the capsized victim still in his craft. This prevented his kayak from swamping and also protected him from exposure, since his waterproof kayak jacket remained tied to the cockpit hoop. Little detailed information has been recorded on the methods of rolling known outside of Greenland, but there are many photographs of Bering Strait kayakers rolling with the single bladed paddle. A study of Alaskan rolling methods is now in progress, and it is hoped that much information can be recovered and preserved.

Figure 219

Preparing for Demonstration. Jonas Malakiasen puts on his tuvilik (a waterproof kayak jacket, pronounced in English "tooey-leek"). When it is fastened tightly about his face, wrists, and the cockpit hoop, he can capsize without getting water in the kayak. Igdlorssuit, West Greenland, summer 1959. (*Photo by Kenneth Taylor.*)

Figure 220

Getting Aboard. Enoch Nielsen, best kayak roller in the village of Igdlorssuit, West Greenland, wriggles into his kayak on the beach before embarking on a kayak rolling exhibition. Note that he is leaving the harpoon line stand and gun bag in place. (*Photo by Kenneth Taylor.*)

Figure 221

Pausing on Surface. Kayaker supports himself on the surface of the water by a sculling stroke before starting the roll. Note that Enoch Nielsen's body is twisted so that his shoulders are parallel with the surface, thus submerging as much of the body as possible in order to gain buoyancy. (*Photo by Kenneth Taylor.*)

Fully Capsized, view from forward quarter, looking aft. Enoch Nielsen prepares to roll up by the standard method. Note the planing angle of his paddle blade as he prepares for the next step, the planing sweep of the blade across the surface. (*Photo by Kenneth Taylor.*)

Figure 222

Emerging From Roll, view from forward quarter, looking aft. From the position of Enoch Nielsen's hands, this appears to be the standard roll. He has just completed the planing sweep and is halfway up. The inboard hand is a pivot point for the sweep and a fulcrum for the lift. (*Photo by Kenneth Taylor.*)

Figure 223

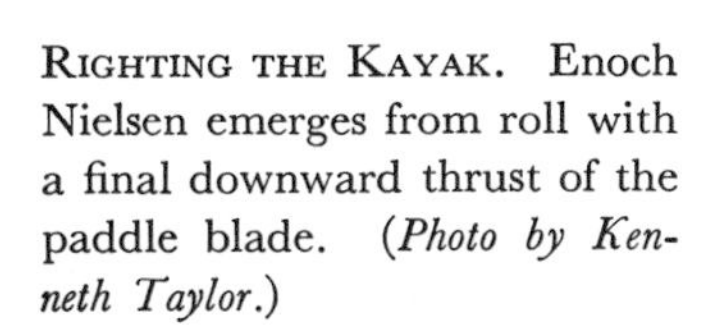

Righting the Kayak. Enoch Nielsen emerges from roll with a final downward thrust of the paddle blade. (*Photo by Kenneth Taylor.*)

Figure 224

# *Bibliography*

ADNEY, EDWIN TAPPAN. *Klondike stampede.* New York: Harper & Bros., 1900.

———. How an Indian birch-bark canoe is made. *Harper's Young People* (July 29, 1890). Supplement.

BEARD, DANIEL CARTER. *Boatbuilding and boating.* New York: C. Scribner's Sons, 1911. (Chapter 6, pages 48–61, is a revision of Adney's articles on canoe building.)

BEAUCHAMP, WILLIAM M. Aboriginal use of wood in New York. (New York State Museum Bulletin 89.) Albany, 1905, pp. 139–149.

BIRKET-SMITH, KAI. *Ethnography of the Egedesminde district.* (Vol. 66 of Meddelelser om Grønland, 1879–1931.) Copenhagen: C. A. Reitzel, 19?

BOAS, FRANZ. *The Central Eskimo.* (Pp. 409–669 in U.S. Bureau of American Ethnology, Sixth Annual Report, 1884–85.) Washington: Smithsonian Institution, 1888.

BOGORAS, VLADIMIR. *The Chukchi.* New York: G. E. Stechert, 1904–09. [Also as Memoir of the American Museum of Natural History, vol. 11; and as publications of the Jesup North Pacific Expedition, vol. 7, issued in three parts.]

CARTIER, JACQUES. *The voyages of Jacques Cartier.* (Canadian Archivist Publications No. 11.) Transl. H. P. Biggar. 1914. (Another edition: Ottawa: F. A. Acland, 1924.)

CARTWRIGHT, Captain GEORGE. *Captain Cartwright and his Labrador journal.* Edit. Charles W. Townsend. Boston: D. Estes & Co., 1911.

———. *A journal of transaction and events, during a residence of nearly sixteen years on the coast of Labrador . . . .* 3 vols. Newark, England: Allin and Ridge, 1792.

CHAMPLAIN, SAMUEL DE. *Les Voyages de la nouvelle France.* Paris: C. Collet, 1632.

———. Oeuvres de Champlain. 5 vol. in 6. Quebec: G. E. Desbarats, 1870.

CRANTZ, DAVID. *The history of Greenland.* Ed. and transl. [of part] John Gambold. 2 vols. London: Moravian Brethren Society, 1767.

DAWSON, GEORGE MERCER. Notes on the Shuswap people of British Columbia. *Proceedings and Transactions of the Royal Society of Canada for the year 1891* (Montreal, 1892), vol. 9, sec. 2, pp. 3–44.

DENYS, NICOLAS. *The description and natural history of the coasts of North America.* Transl. and edit. W. F. Ganong. Toronto: The Champlain Society, 1908.

DUNBAR, SEYMOUR. *A history of travel in America.* 2 vols. New York: Tudor Publishing Co., 1937.

DURHAM, BILL. *Canoes and kayaks of Western America.* Seattle: Copper Canoe Press, 1960.

*Early narratives of the Northwest, 1634–1699.* Edit. L. P. Kellogg. New York: C. Scribner's Sons, 1917.

ECKSTROM, [Mrs.] FANNIE HARDY. *The handicrafts of the modern Indians of Maine.* (Lafayette National Park Museum Bulletin.) Bar Harbor, 1932.

EGEDE, HANS. *A description of Greenland.* Transl. from the Danish. London: C. Hitch, 1745.

ELLIOTT, HENRY WOOD. *The Seal Islands of Alaska.* Washington (Government Printing Office), 1881.

FRANQUET, Col. Voyages et mémoires sur le Canada, par Franquet. *Institut canadien de Québec, Annuaire* (1889), pp. 29–129.

GODSELL, PHILLIP H. The Ojibwa Indian. *Canadian Geographical Journal* (January 1932).

HADLOCK, WENDELL S., and DODGE, ERNEST S.. A canoe from the Penobscot River. *American Neptune* (October 1948), vol. 8, no. 4, pp. 289–301. (Detailed description of early birch-bark canoe, with lines and numerous drawings of construction details.)

HEARNE, SAMUEL. *A journey from Prince of Wales fort in Hudson's Bay to the Northern Ocean, 1769, 1770, 1771, 1772.* Dublin: P. Byrne and J. Rice, 1796.

HENRY, ALEXANDER, Jr. *New light on the early history of the greater Northwest. The manuscript journals of Alexander Henry . . . and of David Thompson . . . 1799–1814.* Ed. Elliott Coues. New York: F. P. Harper, 1897.

———. *Travels and adventures in Canada and the Indian territories.* Ed. James Bain. Boston: Little, Brown and Company, 1901.

HOFFMAN, WALTER JAMES. *The Menomini Indians.* Part 1, (pp. 3–328 of U.S. Bureau of American Ethnology: 14th Annual Report, pt. 1, 1892–93.) Washington: Smithsonian Institution, 1896.

HOLM, GUSTAV FREDERIK. *The Ammassalik Eskimo, contributions to the ethnology of the East Greenland natives.* Vol. 1. Ed. William Thalbitzer. Copenhagen: B. Luno, 1914. [Also as vols. 39–40, Meddelelser om Grønland.]

HORNELL, JAMES. *British Coracles and Irish Curraghs.* London: Society for Nautical Research, 1938.

———. *Water transport, origins & early evolution.* Cambridge: The Cambridge University Press, 1946.

HOWLEY, JAMES PATRICK. *The Beothucks, or red Indians.* Cambridge: Cambridge University Press, 1915. (A very complete study.)

JENNESS, DIAMOND. *The Indians of Canada.* (Bulletin 65, Anthropological Series No. 15, National Museum of Canada.) 5th ed. 1960.

*The Jesuit relations and allied documents, 1610–1791.* Ed. R. G. Thwaites. Cleveland: Burrows Bros. Co., 1896–1901.

JOCHELSON, WALDEMAR [VLADIMIR]. *The Koryak.* New York: G. E. Stechert, 1908. [Also as Memoir of the American Museum of Natural History, vol. 9; and as Publications of the Jesup North Pacific Expedition, vol. 6, published in two parts, 1905–08.]

KALM, PEHR, *Travels into North America.* Transl. John R. Forster. 2 vols. London, 1770–71.

KROEBER, ALFRED LOUIS. The Eskimo of Smith Sound. *Bulletin of the American Museum of Natural History* (Feb. 19, 1900), vol. 12, art. 21, pp. 265–327.

LAFITAU, JOSEPH FRANÇOIS, *Moeurs des sauvages Ameriquains.* Paris: Saugrain, 1724.

LAHONTAN, LOUIS ARMAND, BARON DE, *Nouveaux voyages de M. le baron de LaHontan, dans l'Amerique septentrionale.* La Haye: Chez les Frères l'Honore, 1703.

LeClerq, Father Chrétien, *Nouvelle relation de la Gaspesie.* Paris, 1691.

Lyon, George Francis. *The private journal of Captain Lyon of N. M. S. "Hecla."* Boston: Wells and Lilly, 1824.

Mackenzie, Sir Alexander. *Voyages from Montreal, . . . to the frozen and Pacific Oceans; . . . 1789 and 1793.* 2 vols. From York: New Amsterdam Book Co., 1903.

Mason, Otis T., and Hill, Meriden S., *Pointed bark canoes of the Kutenai and Amur.* (Pp. 523–537 of Report of U.S. National Museum for 1899.) Washington: Smithsonian Institution, 1901.

Mitman, Carl Weaver. *Catalogue of the watercraft collection in the United States National Museum.* (U.S. National Museum Bulletin 127.) Washington: Smithsonian Institution, 1923.

Morgan, Lewis Henry. *League of the Ho-De-No-Sau-Nee, or Iroquois.* New York: M. H. Newman & Co., 1851.

Murdoch, John. *Ethnological results of the Point Barrow expedition.* (Pp. 3–441 of U.S. Bureau of American Ethnology, 9th Annual Report, 1887–88.) Washington: Smithsonian Institution, 1892.

Murray, Alexander Hunter. *Journal of the Yukon, 1847–48.* Ottawa (Government Printing Bureau), 1910.

Nansen, Friotjof. *The first crossing of Greenland.* 2 vols. Transl. Nubert M. Gepp. London: Longmans, Green and Co., 1870.

———. *The Norwegian north polar expedition, 1893–1896.* 6 vols. London: Longmans, Green and Co., 1900–06.

———. *In northern mists.* Transl. Arthur G. Chater. 2 vols. New York: Frederick A. Stokes Co., 1911.

———. *Farthest north.* 2 vols. New York: Harper Brothers, 1897.

Nelson, Edward William. *The Eskimo about Behring Strait.* (Pp. 3–518 of U.S. Bureau of American Ethnology, 18th Annual Report, 1896–97.) Washington: Smithsonian Institution, 1899.

Paris, Edmond. *Essai sur la construction navale des peuples extra-européens.* Paris: A. Bertrand, [n.d.].

Parry, Sir William Edward. *Journal of a second voyage for the discovery of a northwest passage.* London: J. Murray, 1824.

Patterson, Rev. George. The Beothiks or red Indians of Newfoundland. *Proceedings and transactions of the Royal Society of Canada for the year 1891* (Montreal, 1892), vol. 9, p. 137.

Poterie, Claude Charles Le Roy Bacqueville de la. *Historie de l'Amerique septentrionale . . .* 2 vols. Paris, 1722.

Richardson, Sir John. *Arctic searching expedition: A journal of a boat voyage . . . in search of discovery ships . . . of Sir John Franklin.* London: Longman, Brown, Green, and Longmans, 1851.

Ritzenhalter, Robert Eugene. The building of a Chippeway Indian birch-bark canoe. *Bulletin of the Public Museum of the City of Milwaukee* (November 1950), vol. 19, no. 2, pp. 53–90. (The modern method of building.)

Rosier, James. *A true relation of the most prosperous voyage made this present year, 1605, by Captain George Waymouth in the land of Virginia.* Londini; 1605.

Ross, Alexander. *The fur hunters of the Far West.* London: Smith, Elder and Co., 1855.

Schoolcraft, Henry Rowe. *Information respecting the history conditions and prospects of the Indian tribes of the United States.* 6 vols. Philadelphia: Lippincott, Grambo, 1852–57.

SCHOOLCRAFT, HENRY ROWE. *The Indian tribes of the United States.* 2 vols. Philadelphia: J. B. Lippincott & Co., 1884.

SKINNER, ALANSON BUCK. Notes on the eastern Cree and northern Saulteaux. (*Anthropological Papers of the American Museum of Natural History*, vol. 9, pt. 1). New York, 1911.

SNELL, GEORGE F, Jr. Pine country Hiawatha. *Sports Afield* (August 1945), vol. 120, no. 2. (Describes modern Ojibway canoe building.)

STEFANSSON, VILHJALMUR. *My life with the Eskimo.* New York: The Macmillan Company, 1913.

———. *Ultima Thule.* New York: Macmillan Company, 1940.

TURNER, LUCIEN MCSHAW. *Ethnology of the Ungava District: Hudson Bay Territory.* Edit. John Murdoch. (Pp. 159–350 of U.S. Bureau of American Ethnology, 11th Annual Report.) Washington: Smithsonian Institution, 1894.

WARREN, WILLIAM WHIPPLE. *History of the Ojibways.* St. Paul: Minnesota Historical Society Collections, 1885. Vol. 5, pp. 21–394.

WHITBOURNE, Sir RICHARD. *Westward hoe for Avalon in the new-found-land.* Edit. and illus. T. Whitburn. London: S. Low and Marston, 1870.

WILLOUGHBY, CHARLES CLARK. *Antiquities of the New England Indians.* Cambridge, Mass.: Peabody Museum of American Archaeology, Harvard Univ., 1935.

WISSLER, CLARK. *Indians of the United States.* New York: Doubleday, Doran & Co., 1940.

*Wood: A manual for its use in wooden ships.* Washington: U.S. Department of Agriculture, 1945.

*Wood handbook.* Washington: U.S. Department of Agriculture, 1955. (Basic information on wood as a material for construction.)

# *Index*

☆ U.S. GOVERNMENT PRINTING OFFICE : 1978—O-259-428